Grassland Restoration and Management

David Blakesley and Peter Buckley

Pelagic Publishing | www.pelagicpublishing.com

Published by Pelagic Publishing
www.pelagicpublishing.com
PO Box 725, Exeter EX1 9QU, UK

Grassland Restoration and Management
ISBN 978-1-78427-078-0 (Pbk)
ISBN 978-1-907807-80-0 (Hbk)
ISBN 978-1-78427-071-1 (ePub)
ISBN 978-1-78427-072-8 (Mobi)
ISBN 978-1-78427-073-5 (PDF)

British Library Cataloguing in Publication Data
A catalogue record for this book is available from the British Library.

Cover images: Main picture: South Downs National Park.
Pyramidal Orchid; Duke of Burgundy; wildflower seedlings in a nursery

Foreword

Since the mid-twentieth century, species-rich habitats in the UK have been destroyed or degraded at an alarming rate. In the case of grasslands the losses have been catastrophic – over 97% of semi-natural grasslands were lost in England and Wales between 1930 and 1984.

The key cause of these losses has been the conversion of unimproved to 'improved grassland' or arable land, which supports higher levels of food production. However, it is now well recognised that while semi-natural grasslands may be less 'productive', there is a positive correlation between the number of plant species and the provision of ecosystem services, such as pollination, pest control and carbon storage. Of course, these habitats also have a high cultural value and positive benefits for human health and well-being.

It is pleasing, therefore, that in the last decade losses have declined substantially and a range of initiatives have been launched which have led to improved protection for remaining sites, a range of restoration and habitat creation projects, and a recognition that conservation management is vital to maintain the quality of this important habitat.

The UK Native Seed Hub was launched in 2011 to mobilise the resources and expertise of the Royal Botanic Gardens, Kew and the Millennium Seed Bank in support of conservation and habitat restoration in the UK. To date that project has focused on grassland restoration, particularly the chalk and Weald grasslands that are found close to our home at Wakehurst Place in West Sussex. More information about our work to increase the quality, quantity and diversity of native plants and seeds available for conservation and habitat restoration can be found in Chapter 6 (Box 6.2).

Given our passion for facilitating and improving the number, quality and success of grassland restoration projects, we are thrilled to be involved in this book. The text provides a timely and very welcome contribution to the renaissance of the grassland habitat in the UK. It gives the reader an authoritative review of the current scientific literature related to grassland restoration, but at the same time provides a wealth of practical advice as to how to actually go about restoring and managing dry grassland. This very informative book should prove useful to anyone with an interest in the restoration and management of unimproved grassland, and will hopefully inspire more people to get involved with UK grassland restoration.

Clare Trivedi
UK Conservation Partnerships Co-ordinator
Millennium Seed Bank Partnership, Royal Botanic Gardens, Kew

Acknowledgements

We must start by expressing our sincere thanks to the Eden Project, the Royal Botanic Gardens Kew, the Royal Society of Wildlife Trusts and Jonathan Swire for their generous sponsorship of this book. Without the backing and support of these organisations and individuals, this book would not have been possible. We would especially like to thank Kate Hardwick, Tony Kendle, Nigel Symes and Paul Wilkinson who helped us to develop the idea for the book.

We are particularly grateful to the following individuals for supplying information or drafting text for the featured case studies that bring grassland restoration to life: John Adams, Ted Chapman and Kate Hardwick (Royal Botanic Gardens, Kew), Daniel Bashford (Historic England), Dawn Brickwood (Weald Meadows Initiative), Patrick Cashman (Royal Society for the Protection of Birds), Alan Cathersides (Historic England), Andy Coulson-Phillips (Berkshire, Buckinghamshire and Oxfordshire Wildlife Trust), Keith Datchler (Beech Estate), Don Gamble (Yorkshire Dales Millennium Trust), Sally Marsh (High Weald Area of Outstanding Natural Beauty) and Mark Schofield (Lincolnshire Wildlife Trust). We would also like to thank Clare Trivedi (Royal Botanic Gardens, Kew) for her thoughtful Foreword.

Other organisations who have kindly allowed the reproduction of information include: Agricultural Research and Education Centre Raumberg-Gumpenstein, Austria (Bernhard Krautzer); Butterfly Conservation; Defra; Joint Nature Conservation Committee; Natural England; Rare Breeds Survival Trust, Grazing Animals Project; Royal Society for the Protection of Birds; and United Nations Environment Programme-World Conservation Monitoring Centre.

We would like to thank the many individuals who have kindly read sections of the book or provided valuable comments, including Keith Alexander, John Altringham, Nicola Bannister, Tim Blackstock, Nigel Bourn, Lee Brady, Ruth Dalton, Rob Fuller, Gareth Griffiths, Richard Jefferson, Hilary Macmillan, Colin Morris, Simon Mortimer, Andrew Powling, Henry Schofield, Roger Smith and Nigel Symes.

While the majority of photographs were taken by author David Blakesley, others who have kindly contributed images for the book include: Malcolm Ausden, Tone Blakesley, Peter Buckley, Patrick Cashman, Keith Datchler, Nicola Evans, Jason Fridley, Don Gamble, Robert Goodison, Andy Hay (http://www.rspb-images.com/), Michal Hejcman, Lincolnshire Wildlife Trust, Ross Newham, Pippa Rayner, Mathew Roberts, Juliet Rogers, Rothamsted Research, Roger Smith, Trustees of the Royal Botanic Gardens, Kew and Dave Vandrome. We are also especially grateful to Tharada Blakesley for her illustrations.

Finally, we are indebted to Nigel Massen and Thea Watson (Pelagic Publishing) for the design and layout of the book. And we are both grateful for the support and understanding of our families during the writing of this book.

David Blakesley and Peter Buckley

Contents

Foreword   iii
Acknowledgements   iv

1. Grassland character and communities   1

1.1   Introduction   1
1.2   The origins of semi-natural dry grassland communities   3
1.3   Dry grassland succession   7
    1.3.1   Primary succession   7
    1.3.2   Secondary succession   9
1.4   Semi-natural dry grassland types   9
    1.4.1   Lowland calcareous grassland   10
    1.4.2   Lowland dry acid grassland   12
    1.4.3   Lowland meadows (mesotrophic grasslands)   17
    1.4.4   Upland calcareous grassland   20
    1.4.5   Upland hay meadows   23
1.5   Specialised semi-natural dry grassland types   26
    1.5.1   Calaminarian grasslands   26
    1.5.2   Machair   27
    1.5.3   Limestone pavements   28
1.6   Dry grassland scrub communities   30
    1.6.1   Lowland communities   30
    1.6.2   Upland communities   31
1.7   Semi-improved dry pastures and meadows   32

2. Grassland wildlife   34

2.1   Invertebrates   34
    2.1.1   Butterflies   37
    2.1.2   Implications for management   43
2.2   Birds   46
    2.2.1   Lowland birds   46
    2.2.2   Upland birds   50
    2.2.3   Conservation   50
    2.2.4   Implications for management   52
2.3   Reptiles and amphibians   54
    2.3.1   Reptiles   54
    2.3.2   Amphibians   55
2.4   Mammals   57
    2.4.1   Bats   58
2.5   Fungi   62

2.5.1   Macrofungi   63
2.6   Assessing the conservation value of a site   64
2.6.1   Phase 1 habitat survey   65
2.6.2   Wildlife surveys   65

3. Semi-natural dry grassland management   69

3.1   Grazing   69
3.1.1   Livestock species and breeds   70
3.1.2   Cattle   72
3.1.3   Sheep   73
3.1.4   Horses and ponies   75
3.1.5   Stocking density   77
3.1.6   Timing and duration of grazing   79
3.1.7   Animal management   81
3.1.8   Grazing by Rabbits   83
3.2   Cutting   89
3.2.1   Cutting methods   90
3.2.2   Timing   90
3.2.3   Cutting and aftermath grazing   91
3.2.4   Cutting to replace grazing   93
3.2.5   Combined cutting and grazing regimes   94
3.3   Weeds and herbicides   94
3.3.1   Non-chemical control   95
3.3.2   Chemical control   96
3.3.3   Common Ragwort control   97
3.3.4   Bracken control   100
3.3.5   Control of other species   102
3.4   Scrub management   104
3.4.1   Assessment   104
3.4.2   Management techniques   105
3.5   Fertiliser application   109
3.5.1   Farmyard manure   109
3.5.2   Lime   110

4. Grassland restoration: threats and challenges   115

4.1   Threats to semi-natural dry grassland communities   115
4.1.1   Agricultural improvement and land conversion   115
4.1.2   Lowering soil fertility   115
4.1.3   Habitat deterioration   122
4.2   Climate change and dry grassland   123
4.2.1   Climate change in Britain   123
4.2.2   Impacts on grassland habitats and wildlife   124
4.3   Challenges in dry grassland restoration   130
4.4   Limits to natural colonisation   136
4.4.1   Impoverished seed banks   138
4.4.2   Limited dispersal opportunities   140
4.4.3   Missing trophic levels   142
4.5   Assessing hydrological and topographic constraints   144

5. Opportunities in grassland restoration

147

5.1   Conservation of semi-natural dry grassland habitats   147
5.2   Opportunities for dry grassland restoration   148
5.3   Reinstating traditional management   150
    5.3.1   Grazing effects   150
    5.3.2   Cutting effects   152
    5.3.3   Influences of restoration management on sward diversity   155
    5.3.4   Impacts of restoration management on invertebrates   157
5.4   Site limitations and solutions   163
    5.4.1   Fallowing   163
    5.4.2   Plant and animal offtake   164
    5.4.3   Chemical manipulation   167
    5.4.4   Immobilisation   169
    5.4.5   Reinstating soil communities   169

6. Plant material for dry grassland restoration

172

6.1   Surveying the restoration site   172
    6.1.1   Soil analysis   172
    6.1.2   Vegetation   173
6.2   Surveying reference sites   175
6.3   Selecting plant material   176
    6.3.1   Use of ecological traits   176
    6.3.2   Complex or simple mixtures?   178
    6.3.3   Hemiparasites: Yellow-rattle   180
6.4   Sourcing plant materials   182
    6.4.1   Nursery production   184
    6.4.2   Wild harvesting   185
    6.4.3   Adverse impacts of seed collection   191
6.5   Sowing and the role of soil disturbance   204
6.6   Sowing practice   208
    6.6.1   Sowing mechanics   208
    6.6.2   Complete and partial sowing   209
    6.6.3   Arable reversion and grassland enhancement protocols   210
    6.6.4   Transplants   213

7. Defining success in grassland restoration

215

7.1   Plant introductions   215
7.2   Preparation for sowing   218
7.3   Management techniques   218
7.4   Long-term vegetation development   218
7.5   Cost-effectiveness in restoration   220
7.6   Monitoring success   222

References   232
Species Index   254
Subject Index   262

Species-rich calcareous grassland at Old Winchester Hill National Nature Reserve, Hampshire (Tone Blakesley)

1. Grassland character and communities

1.1 Introduction

Grasslands are open habitats, naturally dominated by grasses and herbaceous plants, with sedges, rushes, bryophytes and, sometimes, occasional shrubs. Most swards have a continuous cover, in some cases reaching heights of up to 1 m. Grassland vegetation extends over a broad range of soil types, usually categorised as acidic (calcifugous), neutral (mesotrophic), calcareous (calcicolous) or marshy and wet.

A further major subdivision defines the extent to which the habitat has been modified by agricultural management practices. Collectively, some 37% of the land area in the UK is classified as grassland (Countryside Survey 2009), although the vast majority of this is either improved, or semi-improved. The most 'natural' grasslands are semi-natural (sometimes referred to as unimproved), representing just 2% of the UK grassland area, a total of about 1.7 Mha (Bullock *et al.* 2011). These communities have long histories of traditional, usually low-intensity agricultural management. They are often species-rich, supporting only naturally occurring species; many were designated as areas of conservation concern and included in the UK Government's Biodiversity Action Plan (BAP) priority habitats (Figure 1.1). The UK BAP was succeeded in 2012 by the UK Post-2010 Biodiversity Framework, which represents a broad enabling structure for conservation action across the four countries of the UK, at country level. In contrast, most improved grasslands have been ploughed, fertilised or reseeded, usually with varieties of Perennial Rye-grass and White Clover. Transitional between these two extremes are semi-improved grasslands that have been partially modified but retain some of their original floral composition.

Soil conditions are governed by the parent material, i.e. the immediately underlying geology or drift deposits, and also by topography, drainage and land use. One of the most critical factors appears to be pH, which governs basic fertility and the solubility and availability of a wide range of elements (Hopkins 2003).

Calcium-rich limestone and chalk support calcareous grassland and some of the most species-rich communities. Base-rich igneous rocks also occasionally support calcareous grassland. Calcareous soils with pH values above 6.5 are mostly humic grey and brown rendzinas. Soils associated with eroding limestone may have high levels of free calcium carbonate, for example, supporting Sheep's-fescue–Carline Thistle grasslands, classified as calcareous grassland (CG)1 by the National Vegetation Classification (NVC). In contrast acidic, siliceous, low-calcium soils overlying sandstones, igneous rocks such as

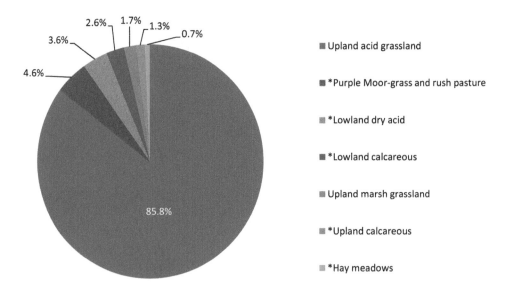

2.6% 1.7% 1.3%
3.6% 0.7%
4.6%
85.8%

■ Upland acid grassland

■ *Purple Moor-grass and rush pasture

■ *Lowland dry acid

■ *Lowland calcareous

■ Upland marsh grassland

■ *Upland calcareous

■ *Hay meadows

Figure 1.1 Estimated extent of semi-natural grassland habitats in the UK. The majority consists of upland acid grassland; hay meadows account for only 0.7% of the total area, mostly located in the lowlands. Data from Bullock *et al.* (2011; © United Nations Environment Programme's-World Conservation Monitoring Centre). * Former UK BAP Priority Habitats.

granite, or superficial deposits such as sands and gravels, support acid grassland. Dry acid grassland soils are usually free-draining, base-poor mineral soils within the pH range of 4–5.5. Acid grasslands contain tolerant species such as Mat-grass and Tormentil, known as 'calcifuges' (lime-avoiders), while the greater diversity of calcicoles (lime-lovers), such as Salad Burnet and Common Rock-rose, are specialists of species-rich calcareous grassland. Where the common factor is oligotrophic soils, a few species may occur in both acid and calcareous grasslands, for example, in chalk heaths. Soils that are neither strongly acid nor base-rich make up neutral grassland communities that can also be species-rich. Their productivity is much higher than most calcareous and strongly acidic soils, which is why so much neutral grassland has undergone agricultural improvement. These simple definitions based on soil pH imply sharp boundaries between communities on the three soil types; in reality the composition of plant species changes subtly over a continuum of soil pH. An example of this is the transition between calcareous and mesotrophic grassland vegetation, where the thin rendzina soils of a steep chalk escarpment give way to deeper soils of (upper) periglacial deposits or (lower) valley colluvium along the slope. Even at a single location, variations in different grassland communities may be found relating to natural differences in soil conditions.

Other properties of soil, such as depth and moisture levels, are also important in determining the vegetation type. Drainage has a key role in determining the species composition of British grasslands. Many communities, such as the lowland calcareous grasslands, are classified as 'dry grasslands' and are clearly associated with free-draining soils. In contrast, fen meadow and rush-pasture communities, such as the Soft Rush/Jointed Rush–Common Marsh Bedstraw (NVC M23) and Purple Moor-grass–Meadow Thistle (NVC M24) are classified as 'wet' grasslands. In other cases, the distinction between dry and wet communities may be less obvious. In lowland England, the gradient between the Crested Dog's-tail–Common Knapweed and Meadow Foxtail–Great Burnet hay meadow communities is largely determined by fluctuations in the water table. In this book, we focus on the restoration and management of drier grassland communities, as described in Sections 1.4–1.7.

1.2   The origins of semi-natural dry grassland communities

Several classifications have been used to describe dry grassland vegetation types in Britain, all of which have their merits. The NVC, widely adopted by ecological practitioners, provides systematic descriptions of the major grassland communities (Rodwell 1992). Within these, dry grassland types may comprise 28 semi-natural communities, plus three semi-improved communities.

The origins of shade-intolerant plants in grassland communities are still not well understood. Some authorities believe it unlikely that any higher plants or animals survived in Britain during the maximum of the last glaciation, between approximately 18,000 and 14,500 radiocarbon years before present (BP) (Buckland *et al.* 2005). Others have suggested that the tundra of southern Britain could have supported an arctic flora of sedges, dwarf willows and dwarf birches at this time (Marren 1999), while Peterken (2013) goes further, listing a number of herbs present during early interglacial periods, many of which survived in refugia during the last glaciation. The climate then began to warm, and plants and animals are likely to have migrated north from refugia in southern and south-eastern Europe. A 'brief' return to cooler, then glacial conditions over c. 13,500 to 10,000 radiocarbon years (Buckland *et al.* 2005), a time known as the 'Younger Dryas', would have temporarily slowed the process of colonisation.

Pollen evidence has been used to investigate the vegetation in Britain during the Younger Dryas and early Holocene periods. Some of today's open-ground species, notably Mountain Avens, after which the Dryas was named, were present in the steppe-tundra vegetation of the Younger Dryas. The climate warmed rapidly in the early Holocene (starting around 10,000 BP), probably due to changes in ocean currents in the North Atlantic, suddenly creating very favourable conditions for the continued northward migration of plants and animals. After the retreat of the ice, the tundra vegetation in southern Britain developed into grassland, then scrub and eventually woodland; descendants of ancient lowland 'alpine meadows' may still survive today in the Pennine Dales and in some Scottish glens (Marren 1999). Marren also highlighted the herb-rich meadows found on steep and precipitous carboniferous limestone slopes in northern England. These are dominated by Meadowsweet and False Oat-grass (MG2), in which Jacob's-ladder is a constant in one subcommunity, which Pigott (1958) identified as a very rare example of 'natural' grassland.

By 9,300 years ago, as the climate continued to warm and soils developed, forest extended over most of Britain, and a period of stability continued for about 2,800 years (Rackham 1990). Historical evidence suggests the so-called 'wildwood' formed a well-connected mosaic of forest cover, prior to any significant human disturbance. For many years, historians assumed that 'grassland communities' survived along river margins, on inland and sea cliffs, coastal dunes and above the treeline. Within the wildwood itself, they would have been confined largely to glades created by falling trees, maintained by large herbivores such as deer and Aurochs. Although relatively small in area when compared to the vast expanse of the wildwood, these open habitats may still have supported a large pool of plants, insects and other open-ground species, able to recolonise areas cleared by

Mountain Avens is a dwarf shrub that was present in the steppe-tundra vegetation of the Younger Dryas, and can still be found in Britain today on mountainous ledges and crevices, and in high-altitude grassland and grass/heath.

humans at the start of the Neolithic in Britain, roughly 4,000 BCE. At this time, the creation of pasturage, and the subsequent adoption of arable farming would have opened up the landscape, allowing shade-intolerant plants which had survived in woodland glades to flourish in the grazed landscapes, and some to become arable weeds.

The Great Orme, a carboniferous limestone headland in North Wales exemplifies the importance of coastal refuges. Cliff faces and ledges of every aspect support a range of locally rare grassland plants, together with endemic races of the Silver-studded Blue and Grayling butterflies, strongly implying that these communities may have existed since the early Holocene (Thomas 2009). Similar limestone crags across Britain would have collectively supported a diverse range of calcareous species during the peak of forest cover. Today, many rare grassland plants are confined to shallow soils on limestone, where they would presumably have avoided the heavier shading of the wildwood, even though the limestone pavement was probably sparsely wooded (Piggot and Piggot 1959). The closest present-day habitats to these prehistoric natural grasslands are those occupying harsh and inaccessible sites, such as rock ledges and outcrops where the soil is too thin and the site too exposed to have been colonised by shrubs and trees. Peterken (2013) identified just such a site on the Seven Sisters, a group of limestone towers in the Wye Valley. The grassland that occupies the tops, clefts and ledges, is species-rich and includes many rare plants. Similarly, many wetland species might have survived in the wet meadows, mires and bogs of river margins and valley bottoms.

It has been suggested that the wildwood of northwest Europe resembled a 'half-open' landscape, with large areas of grassland and heathland maintained by large numbers of large herbivores (Vera 2000). Others have concluded that while some parts of the British wildwood may have resembled modern wood pasture, the majority would probably have been closed high forest, with a mixture of temporary and more permanent glades (Kirby 2005). This was further supported by an extensive review of palaeoecological data from European primeval forests (Mitchell 2005). The smaller open areas, which undoubtedly

The Great Orme is believed to have represented an important coastal refuge for grassland plants and animals since the early Holocene.

An endemic race of Grayling found on the Great Orme may have evolved since the early Holocene.

existed, must have allowed the persistence of some plants that we would today consider as natural meadow species, and may have contained meadow-like grassland (Peterken 2009; 2013).

When people first started to fell trees in the wildwood to graze animals, new assemblages of plants must have developed under the new management systems. Species already present might have included plants which are widespread in woodland today, such as Yorkshire-fog, Common Bent, Creeping Buttercup and Tormentil. Similarly, species less common in lowland meadows, but also found in woods today, include Common Twayblade and Early-purple Orchid. Species we normally associate with 'woodland', such as Wood Anemone and Bluebell, are also meadow species in some parts of Britain. Shade-intolerant species present in woodland clearings will have been

Bluebells are usually thought of as a woodland species, but in some parts of Britain, such as the south-west, they are known as meadow species.

augmented by colonisation from grassland refuges outside the wildwood, such as cliffs, dunes and wetland river margins.

Evidence from pollen studies on the Isle of Skye (Birks 1973) showed that tall herb communities present on the island during the Late Glacial Maximum period contained both woodland species, such as Dog's Mercury and Herb-Paris, as well as others characteristic of modern hay meadows, such as Globeflower and Melancholy Thistle. Today, some of these species are more characteristic of woodland, others of hay meadows. Peterken (2009) identified a strong link between pre-Neolithic natural communities and historic Yorkshire Dales hay meadows, also known as upland hay meadows (predominately MG3). He describes these meadows, with their characteristic Wood Anemone, Pignut, Bluebell, Wood Cranesbill, Globeflower, Water Avens and

Leyburn Old Glebe Nature Reserve (Yorkshire Wildlife Trust), a species-rich traditional hay meadow that would have once been common throughout the Yorkshire Dales (© Roger Smith).

Melancholy Thistle, as 'woodlands lacking only the trees'. Although some species were lost in the conversion and others were gained from refuges outside, it is likely that these northern hay meadows are closely associated with the original wildwood.

The link between woodlands and meadows is clearly strong, despite the fact that they are usually considered as distinct ecological communities, and treated as such by ecologists and other professionals. It is important to remember that virtually all our natural grassland has been modified at some stage by man. The landscape in Britain today, and indeed the climate, no longer resemble those during the Late Glacial or early Holocene, or the time when the wildwood covered most of Britain (Thomas 2009).

1.3   Dry grassland succession

1.3.1   Primary succession

Succession describes a series of developments in ecosystem structure and composition over time known as seral stages, which in Britain's temperate climate usually result in a relatively stable 'mature' woodland ecosystem. Succession starting with bare substrates that have not previously supported an ecological community is termed primary succession. Sand dunes represent a good example of primary succession in Britain, as it is possible to see all stages in a single location. The first dune colonists are pioneers, such as Sand Couch and Lyme-grass, which may be outcompeted by Marram as it stabilises windblown sand dunes at the 'yellow dune' stage. Marram may be accompanied by species such as Sand Sedge, Sea Bindweed, Portland and Sea Spurges, and plants typical of waste ground, such as various hawkbits, hawkweeds, thistles and ragworts. During the subsequent 'grey dune stage', Marram dies out as new sand deposition ceases and the dunes are colonised by mosses, lichens and calcareous plants, such as Viper's-bugloss, Wild Thyme and Common Centaury. Older dunes may become increasingly acidic, and characterised by species such as Gorse, Tormentil and Heather. With the build-up of humus, the soil becomes capable of supporting woody plants, and low scrubby growth dominated by Creeping Willow will eventually develop, possibly followed by woodland. Each seral stage is accompanied by different communities of plants and animals, increasing biomass and usually increasing species richness.

Sand Couch embryo dune community (NVC SD4), with a Marram mobile dune community (NVC SD6) behind. Right: further along the same beach, Sea Rocket–Sea Sandwort strandline community (NVC SD2), again with a Marram mobile dune behind.

Back from the shoreline, Red Fescue–Lady's Bedstraw (NVC SD8a) community, characteristic of calcareous fixed sands and dunes.

Sea Holly with Lady's Bedstraw (left) and Sand Sedge (right), found in Marram mobile dune communities, and in transitional areas between NVC SD6 and SD8.

The dunes and associated dune slacks and coastal grassland at Sandwich Bay in Kent support a distinctive flora with species including Wild Onion, Viper's-bugloss, Sea Holly and Restharrow, while the nationally rare Lizard Orchid and Bedstraw Broomrape have their largest British colonies here.

1.3.2   Secondary succession

Succession on cleared land such as farmland, where a soil layer has already formed, is called secondary succession. Much of the grassland we find in the modern landscape will have gone through alternate phases of arable and pasture, as the fortunes of farmers changed over hundreds of years. In the past, arable land would have been allowed to 'tumbledown' from seed sources in the wider landscape, and through subsequent grazing would gradually develop into grassland. Farmers may also have strewn barn sweepings from other parts of the farm, and animals or agricultural machinery would have carried seeds from field to field. New grassland would develop a flora similar to that of adjacent fields, from seed of local origin. Despite these alternating cycles, grassland that has not recently been improved, with communities containing a high proportion of native grasses and herbs, may still be classed as semi-natural grassland.

With the exception of rare clifftop, ledge and outcrop communities, most semi-natural dry grasslands are plagioclimax communities which depend on continued management, such as cutting, grazing or burning to prevent succession to heath, scrub or woodland. Meadows are generally allowed to grow ungrazed from late spring and cut from late June to early August, after which the regrowth is usually grazed. Pastures are grazed by farm animals throughout the summer and not cut for hay. However, many farmers do not adhere to these strict definitions. Some 'hay meadows' are grazed early in the season but still cut later in the year. Farmers may also introduce animals into a meadow in midsummer, rather than take a hay cut.

1.4   Semi-natural dry grassland types

To manage or restore semi-natural dry grassland, it is important to have some understanding of species assemblages and community types. If restoration involves reintroducing species, then these should reflect the community that would naturally occur on the site, unless changes caused by agriculture or urban and industrial development have significantly altered soil conditions. Species selection may be guided by semi-natural grassland communities present locally, or so-called 'reference communities'. This section introduces the main dry grassland communities in Britain, most of which are UK priority habitats, i.e. habitats of principal importance for the conservation of biodiversity in Britain (formerly known as UK BAP priority habitats).

Dry grassland communities vary enormously in structure and composition, depending on a range of factors including geology, soils and climate. Almost every area of grassland is unique in its composition and abundance of plant species. Plant assemblages may also vary within a meadow or grassland, depending on local variations in moisture or fertility, for example. Grasslands may also occur in mosaics with other habitats, such as heathland or scrub. The widely used NVC (Rodwell 1992) describes 47 grassland communities, most of which have two or more associated subcommunities. These are grouped into three main categories, based on soil types: mesotrophic (moderately base-poor to neutral); calcicolous (or calcareous); and calcifugous (acid) and montane communities. Within each type, individual communities are separated based on altitude, soil characteristics, wetness, fertility and species richness. The NVC grassland types vary considerably in area across Britain: some are relatively localised, such as the Blue Moor-grass–Limestone Bedstraw grassland community (CG9); whereas Crested Dog's-tail–Common Knapweed (MG5) is found throughout Britain.

Although the practical guidance in this book focuses on the more widespread communities, localised inland dune communities and upland calcareous grassland communities (montane heath, dwarf-herb and ledge) are included in this section for completeness. More specialised calaminarian grasslands, machair and limestone pavement communities follow in Section 1.5.

1.4.1 Lowland calcareous grassland

Semi-natural lowland calcareous grassland communities are found on lime-rich soils derived from parent rock primarily composed of calcium carbonate. The soils are usually thin, free-draining rendzinas, poor in nutrients, with a pH ≥ 6.5. They are mainly found overlying Oolitic limestone, Carboniferous limestone and chalk, although some Breckland soils are also derived from chalky boulder clay. Grasses found in these communities include Sheep's-fescue, Meadow Oat-grass, Upright Brome, Downy Oat-grass, Crested Hair-grass and Quaking-grass, with forbs such as Salad Burnet, Common Rock-rose, Wild Thyme and Lady's Bedstraw. The calcareous grassland flora can be very rich, with up to 20–50 vascular plants per square metre in some communities, making this one of the most diverse habitats in Europe (Hutchings and Stewart 2002). These swards also contain many nationally rare or scarce plants: for example, Monkey Orchid

Table 1.1 Summary of the characteristics of lowland calcareous dry grassland types (Rodwell 1992; Rodwell *et al.* 2007)

NVC type	Description
CG1	Sheep's-fescue–Carline Thistle: short, open turf of warm, rocky limestone slopes, with shallow, well-draining soils; notable as refuges for rare plants during the early Holocene. Heavy grazing is often important in maintaining the characteristically open sward. Scattered sites around the coasts of the south and west where the climate is warm and mild.
CG2	Sheep's-fescue–Meadow Oat-grass: very rich diversity of herbs and grasses in a continuous, closed sward. Typical species include Meadow Oat-grass, Quaking-grass, Salad Burnet and Common Rock-rose. Grazing, particularly by sheep and Rabbits prevents successional changes. Widespread across warmer, drier lowlands, including South Downs chalk and Carboniferous limestone in the Mendip Hills, Derbyshire and Wales, where it is frequently associated with other calcareous communities, such as CG1 and CG6.
CG3	Upright Brome: Upright Brome ≥ 10% of cover with fewer light-demanding herbs. Salad Burnet, Common Bird's-foot-trefoil, Ribwort Plantain and Glaucous Sedge frequent among the bulky grasses.
CG4	Tor-grass: Tor-grass ≥ 10% of cover, Glaucous Sedge and Sheep's-fescue are the other constants. Quaking-grass, Harebell, Salad Burnet, Common Bird's-foot-trefoil and Fairy Flax are frequent. Impoverished counterpart of CG2, with a similar range, resulting from lack of grazing and shading by Tor-grass.
CG5	Upright Brome–Tor-grass: bulky grasses co-dominate this open or closed rank sward, with Sheep's-fescue, Quaking-grass and Glaucous Sedge constants. Herbs, a constant feature, can be locally abundant. Characteristic of Jurassic limestones, centred on the Oolite of the Cotswold scarp and Northamptonshire/Lincolnshire.

and Military Orchid are country-level priority species found on chalk; Pasqueflower is found on chalk and Jurassic limestone. A diverse invertebrate fauna includes endangered priority species such as Hazel Pot Beetle, Phantom Hoverfly and the Wart-biter Bush Cricket. As a consequence, lowland calcareous grassland has a high conservation value in addition to its considerable aesthetic appeal. Most of these grasslands are still grazed by sheep, cattle and, occasionally, horses.

Lowland calcareous grassland (formerly a BAP priority habitat) includes ten NVC types (CG1–10) (Table 1.1), the first eight of which are mainly distributed in the lowlands and confined to England and Wales. However, Blue Moor-grass–Limestone Bedstraw grassland (CG9) and Sheep's-fescue–Carline Thistle grassland (CG10) are strictly submontane or montane communities (Rodwell 1992), but were included in the lowland BAP as they are occasionally found below the upper limits of agricultural enclosure, almost to sea level in north-west Scotland. The conservation status of lowland calcareous grassland is also recognised in Annex I of the 1992 EU Habitats and Species Directive (listed as habitat 6210: Semi-natural dry grasslands and scrubland facies on calcareous

Table 1.1 continued

NVC type	Description
CG6	Downy Oat-grass: dominated by Red Fescue, with Downy Oat-grass and Meadow Oat-grass. Occurs on deeper, moister, mesotrophic soils, and reflects a transition to ranker meadows, with herb constants limited to rosette species. Sites may have been disturbed in the past. Scattered over southern, central England and Wales.
CG7	Sheep's-fescue–Mouse-ear-hawkweed–Wild Thyme: Sheep's-fescue is often abundant, in open or closed swards characterised by herbs. Characteristic of thin, free-draining oligotrophic soils of continental eastern areas with low rainfall, such as Breckland and southerly exposures of Carboniferous limestone. Several stands occur in Wales.
CG8	Blue Moor-grass–Small Scabious: closed sward, species-rich, often dominated by Blue Moor-grass, with Wild Thyme and Common Rock-rose dominant. Restricted to free-draining, calcareous soils over the Magnesian limestone of eastern Durham.
CG9	Blue Moor-grass–Limestone Bedstraw: submontane or montane community restricted to free-draining, lithomorphic soils on Carboniferous limestone in the North Pennines. The Hoary Rock-rose–Squinancywort subcommunity (CG9a), found in lowland areas has a characteristically open sward, often on rock ledges, dominated by Blue Moor-grass and Sheep's-fescue. Wild Thyme, Horseshoe Vetch and Common Rock-rose are characteristic.
CG10	Sheep's-fescue–Common Bent–Wild Thyme: essentially a submontane community of base-rich, often moist brown earths overlying a variety of calcareous bedrock, including base-rich igneous rocks, but descends to near sea level in north-west Scotland. It is also found in colder, wetter climatic conditions in Wales. Maintained by grazing. The White Clover–Field Wood-rush subcommunity (CG10a) found in lowland areas generally has a short, close-grazed sward, dominated by grasses such as Red Fescue and Sweet Vernal-grass. The community is also associated with species characteristic of unimproved neutral grasslands and Mat-grass grasslands such as Tormentil and Heath Bedstraw.

Lowland calcareous swards, such as this CG4 grassland in Kent, can be rich in orchids; species present include, clockwise from top left, Fragrant Orchid, Common Spotted-orchid, Common Twayblade, Monkey Orchid, Musk Orchid and Greater Butterfly Orchid. Monkey Orchid (Tone Blakesley).

substrates (*Festuco-Brometalia*)). Annex I habitat 6211 also includes important orchid sites of lowland calcareous grassland, particularly in CG2–5 communities.

There has been a sharp decline in the area of lowland calcareous grassland over last 50 years. This has led to habitat fragmentation throughout lowland Britain, with estimates for the 2013 UK Habitats Directive Report indicating 48,700 ha in England, 270 ha in Scotland and 740 ha in Wales. Most of this grassland is found on chalk, with major concentrations in Wiltshire, Dorset and the South Downs. Downland in Wiltshire, including Parsonage Down, Porton Down and Salisbury Plain covers some 13,000 ha, the largest expanse of calcareous grassland in Britain. Because of extensive agricultural improvement, such grassland is now restricted to steeper slopes on limestone or chalk, such as dry valleys or dales, and scarp slopes, although there are some areas of flatter topography, for example, on Salisbury Plain and in the Breckland.

A Joint Nature Conservation Committee (JNCC) report focusing on the special features for which Sites of Special Scientific Interest (SSSIs) were designated, found that only 29% of lowland calcareous grassland sites were in favourable condition, although 40% were 'unfavourable recovering' (Williams 2006). For non-statutory sites in England, just 28% were reported to be in favourable or unfavourable recovering condition (Hewins *et al.* 2005). On many lowland calcareous grassland sites, grazing with domestic animals is critical to maintaining habitat quality. Where changes are made, either by reducing or increasing grazing intensity, these fragile grassland ecosystems can quickly be threatened (see Section 3.1). Relaxing or abandoning grazing on CG2 grassland can allow bulky grasses such as Tor-grass to take hold, and push the community towards CG3, CG4 or CG5 status, which may be difficult to reverse. The loss of wild grazers, such as Rabbits, also poses a serious threat to the continuity of swards. At the extreme, an invasion by coarse grasses may be followed by shrubs, resulting in the loss of important shade-intolerant herbs if the scrub is left unchecked. Eventually, left to itself, the habitat would revert to woodland. Where there have been past attempts at agricultural improvement, this can be recognised by the increased frequency of neutral grassland species such as White Clover and Crested Dog's-tail. In future years, climate change may also become an increasingly important threat to lowland calcareous grassland. More severe droughts and milder winters could lead to the invasion by drought-tolerant ephemerals into vulnerable communities such as CG1, CG7 and CG9 (Rodwell *et al.* 2007).

1.4.2   Lowland dry acid grassland

Semi-natural lowland dry acid grassland communities are defined as enclosed or unenclosed communities below 300 m, normally managed as pasture (Maddock 2008). Communities occur on free-draining, base-poor mineral soils with a pH of 4–5.5, derived from acid rocks, such as sandstones, or superficial deposits of sands and gravels. The moisture status of the soils varies, giving rise to a gradient between the very dry Sheep's-fescue–Common Bent–Sheep's Sorrel grassland (U1) and the moister Sheep's-fescue–Common Bent–Heath Bedstraw grassland (U4b). Enclosed lowland dry acid grassland is normally managed as pasture, while unenclosed grassland often forms an integral component of lowland dwarf shrub heaths, contributing considerably to the diversity of these habitats. Characteristic grasses and sedges include Sheep's-fescue, Common Bent, Wavy Hair-grass, Sand Sedge and Bristle Bent, with herbs including Heath Bedstraw, Sheep's Sorrel, Heath Dog-violet, Harebell and Tormentil. Heathers may also occur, but at lower frequencies than in heathland. Lowland dry acid grassland

Fragments of CG3 (left) and CG2 grassland (right) support species such as Round-headed Rampion (inset) in the largely agricultural landscape of the South Downs.

is generally less species-rich than some calcareous and mesotrophic communities; Wavy Hair-grass communities are particularly species-poor, and often considered to represent degraded dwarf shrub heath. However, U1 grassland can support a wide range of grasses and herbs including a large number of scarce or rare priority species, such as Deptford Pink and Spanish Catchfly. Lowland dry acid grassland also supports many specialist invertebrates, such as *Gryllus campestris* or the Field-cricket, which is a priority species. Where dry acid grasslands form a mosaic with heaths, they also support rare reptiles and amphibians, such as the Smooth Snake and Sand Lizard.

Four lowland dry acid grassland communities (U1–4) were included in the former UK BAP priority habitat designation (Table 1.2). The BAP also included two inland dune communities, SD10b and SD11b, which are highly localised and confined to inland sites in Breckland and Lincolnshire; SD11 is the only lowland dry acid grassland vegetation type recognised in Annex I of the EU Habitats and Species Directive (habitat 2330: Inland dunes with open *Corynephorus* (Grey Hair-grass) and *Agrostis* (Bent) grasslands). All categories of lowland dry acid grassland are becoming increasingly rare, having undergone a considerable decline in the twentieth century due to intensive agriculture and forestry, although no precise figures are available for Britain as a whole. There are an estimated 20,142 ha in England, 4,357 ha in Scotland (UK BAPs Targets Review 2006) and 39,500 ha in Wales (Blackstock *et al.* 2010; Stevens *et al.* 2010). Estimates for the 2013 UK Habitats Directive Report indicate just 120 ha of SD11 in England.

A JNCC report found that only 38% of designated lowland acid grassland sites were in favourable condition, although 32% were 'unfavourable recovering' (Williams 2006). For non-statutory sites in England, just 21% were reported to be in favourable or unfavourable recovering condition (Hewins *et al.* 2005). Lowland acid grassland in poor condition is usually the result of undergrazing and abandonment, often leading to scrub and Bracken encroachment. Undergrazing in the past has resulted from agricultural economics and policies, exacerbated by animal diseases such as foot-and-mouth (Williams 2006). Semi-improved acid grasslands, particularly those under high grazing pressure, are recognised by high frequencies of neutral grassland species such as White Clover and Yorkshire-fog. Nitrogen (N) deposition from atmospheric pollution may also be a factor, particularly in the north and west. Climate change may also threaten lowland dry acid grassland. A drier climate, particularly in the east, could reduce species diversity, facilitate invasion of continental species and reduce frost heaving – a key feature of some Breckland grasslands (Rodwell *et al.* 2007).

Table 1.2 Summary of the characteristics of lowland dry acid grassland types (Rodwell 1992; Rodwell *et al.* 2007)

NVC type	Description
U1	Sheep's-fescue–Common Bent–Sheep's Sorrel: open sward, with patches of bare ground, sometimes with abundant herbs, on base-poor, oligotrophic, summer-parched soils. Open structure dependent on grazing by domestic stock and Rabbits. Widespread, but local and increasingly rare across the warmer, drier lowlands of southern Britain.
U2	Wavy Hair-grass: impoverished swards are dominated by tussocky Wavy Hair-grass on base-poor, free-draining but sometimes moist soils. Probably derived from woods, heaths and mires. Species-poor, with few common associates except heathers. Grazing important. Widespread but local across moderately oceanic lowlands, more common towards humid upland fringes, usually in association with heaths, woods and drying peatbogs.
U3	Bristle Bent: swards dominated by Bristle Bent, but Sheep's-fescue, Common Bent and Heath-grass can be frequent. Scattered shrubs may also be common. Found on moist, base-poor podzolised soils in response to burning and grazing. Confined to milder, oceanic parts of southern and south-west Britain, where it forms an integral part of heathland communities.
U4	Sheep's-fescue–Common Bent–Heath Bedstraw: submontane community of short, often close-cropped swards dominated by intimate mixtures of grasses and herbs; generally species-poor. The main lowland subcommunity found within the limit of enclosure is Yorkshire-fog–White Clover (U4b), a semi-improved type of more fertile soil, where Red Fescue often replaces Sheep's-fescue, while Yarrow and Common Mouse-ear are preferential. Occurs on better-drained, more base-poor mineral soils. The typical subcommunity (U4a) also occurs on the upland fringes of Wales, although it has been widely recorded in the lowlands, even though relatively small in extent. Dependent on grazing.
SD10b	Sand Sedge: pioneer vegetation of fresh deposits of sand among sheltered dunes. Open or closed impoverished swards dominated by Sand Sedge, accompanied by Sheep's-fescue (SD10b). Occurs inland in Breckland and Lincolnshire, particularly on degenerating heath (H1) or dune grasslands (SD11 and SD12).
SD11b	Sand Sedge–*Cornicularia aculeata*: lichen-rich swards, drought prone and impoverished, on fixed, acid sands in which Sand Sedge is the only constant plant, with frequent Sheep's-fescue, Common Bent and Field Wood-rush. Maintained by heavy Rabbit grazing, but lichen carpet vulnerable to larger animals and human traffic. Occurs inland on Breckland sands.

Semi-natural U4 grassland associated with rocky knolls and old anthills in a formerly grazed pasture in Cumbria.

Semi-natural but species-poor U1 acid grassland in unfavourable condition, West Sussex.

Harebells are often found in acid grassland communities.

1.4.3 Lowland meadows (mesotrophic grasslands)

Lowland meadows encompass semi-natural mesotrophic (or neutral) grasslands found on moderately fertile circumneutral soils, typically brown earths or alluvium, which are neither strongly acid nor basic. They may be dry, or periodically inundated with water. Most are managed as traditional hay meadows or pasture, with low-intensity grazing and occasional manuring. The dry grassland component of lowland meadows comprises only one NVC type: Crested Dog's-tail–Common Knapweed grassland (MG5). Two wet grassland types, (MG4 and MG8) (Table 1.3), are also included in the lowland meadow designation and all are probably much more variable than the communities defined by the NVC (Rodwell *et al.* 2007). MG4 flood meadow is the only vegetation type recognised in Annex I of the 1992 EU Habitats and Species Directive (habitat 6510: Lowland hay meadows).

MG5 meadows are the most widespread of the three types and are typical of low-input agriculture. Although they are mainly found on farmland, important relicts may also be found in woodland rides and glades, roadside verges, churchyards and parkland areas. They are variable in appearance, but species-rich, diverse swards typically support an intimate mix of wildflowers and fine-leaved grasses. Typical grasses include Common Bent, Crested Dog's-tail, Red Fescue and Yorkshire-fog. Most of the herbs – which can form up to 95% of the cover – are widespread: Common Bird's-foot-trefoil, Ribwort Plantain and White Clover are often abundant and are frequently accompanied by species such as Yellow-rattle, Rough Hawkbit, Oxeye Daisy, Cowslip, Red Clover, Common Knapweed, Yarrow, Cat's-ear, Autumn Hawkbit and Bulbous Buttercup. Different subcommunities are often associated with more or less acidic soils; the Heath-grass subcommunity (MG5c), for example, is found in more acidic conditions, whereas the Lady's Bedstraw subcommunity (MG5b) prefers more base-rich sites. Lowland dry meadows also support a number of nationally rare or scarce plants, including Fritillary, Greater Yellow-rattle, Sulphur Clover, Green-winged Orchid and Pepper Saxifrage.

The dramatic decline in the area of unimproved lowland meadows over the last 50 years may still be continuing (Hewins *et al.* 2005). The decline has been largely due to agricultural improvement through the application of fertilisers, ploughing and reseeding, and drainage of wetter pastures and meadows. There are an estimated 7,282 ha in England, 980 ha in Scotland (UK BAPs Targets Review 2006) and 1,322 ha in Wales (Blackstock *et al.* 2010). Areas may have diminished slightly since the 2006 review (www.ukbap.org.uk/BAPGroupPage.aspx?id=98). Estimates for the 2013 UK Habitats Directive Report indicate just 1,510 ha of SD11 in the UK.

Many designated unimproved lowland meadows in England and Scotland are in an unfavourable condition, with only 42% in favourable condition (Williams 2006). For non-statutory sites in England, just 16% were reported to be in favourable or unfavourable recovering condition (Hewins *et al.* 2005). Many sites classified as unfavourable have management plans in place to reverse this situation, but more restoration management is needed. The major reasons for poor condition are likely to be undergrazing, neglect of mowing and abandonment, leading to the invasion of coarser grasses such as False Oat-grass, a scarcity of herbs such as Rough Hawkbit and Common Bird's-foot-trefoil, and the eventual encroachment of scrub and sometimes Bracken. Some sites also suffer from overgrazing and nutrient enrichment (Hewins *et al.* 2005; Williams 2006). In addition, some unimproved meadows may still be undergoing agricultural improvement, particularly those on better-drained ground (Rodwell *et al.* 2007), and may be recognised by the presence of Perennial Rye-grass and few herbs.

Table 1.3 Summary of the characteristics of lowland meadow types (Rodwell 1992; Rodwell *et al.* 2007)

NVC type	Description
Dry grassland communities	
MG5	Crested Dog's-tail–Common Knapweed: herb-rich hay meadow of variable appearance depending on grazing intensity, from short swards to tall, lush growth. Circumneutral brown soils of loamy to clayey texture. Fine-leaved grasses such as Crested Dog's-tail, Sweet Vernal-grass, Common Bent and Red Fescue are common; herbs form a substantial component. Traditionally managed as hay meadows, with hay cutting followed by low-intensity grazing in autumn and winter, sometimes with light dressing of manure; also sometimes managed as pasture. Widespread but increasingly rare throughout lowlands of Britain.
Wet grassland communities	
MG4	Meadow Foxtail–Great Burnet: herb-rich floodmeadow, characteristic of seasonally flooded land with alluvial soils. Mixture of grasses, usually with no single dominant species; herbs dominate by midsummer, when grasses may be overtopped by abundant Great Burnet and Meadowsweet. Meadows with abundant Fritillaries now rare. Traditionally managed as hay meadows, with hay cutting followed by low-intensity grazing in autumn and winter, sometimes with light dressing of manure. Almost restricted to the Midlands and southern England, with a few sites in Wales, and becoming increasingly rare.
MG8	Crested Dog's-tail–Marsh Marigold: species-rich flood-pasture, characteristic of seasonally flooded land by rivers and streams, with gley soils of rather silty texture, above and sometimes with a humose topsoil. Crested Dog's-tail, Red Fescue, Yorkshire-fog, Rough Meadow-grass and Sweet Vernal-grass may all be abundant; herbs well represented, with Marsh Marigold prominent in the sward. Traditionally managed as pasture for cattle and horses, with an occasional hay crop. Widespread but local throughout lowlands of Britain as many water meadows have been drained or improved. Also occurs locally in high-altitude meadows in the Pennines and Cumbria.

Hoe Road Memorial Meadow, owned and managed by the Hampshire & Isle of Wight Wildlife Trust has been identified as a 'Coronation Meadow', representing an outstanding example of a semi-natural lowland meadow in Hampshire (http://coronationmeadows.org.uk/) (Tone Blakesley).

Semi-natural lowland meadow (MG5) in East Sussex, with Dyer's Greenweed and Betony growing along the margins.

Semi-natural hay meadow in Kent supporting scarce species, such as Green-winged Orchid (in vast numbers, left) and Adder's-tongue (right).

1.4.4 Upland calcareous grassland

Upland calcareous grasslands are found on shallow, lime-rich soils derived from a wide range of underlying lime-rich bedrock. The most widespread is Carboniferous limestone in Wales and the North Pennines, but other limestones support calcareous grassland in the uplands of northern England and Scotland. Various types of shale, sandstone, schist and basic igneous rock locally contain calcium carbonate, which also gives rise to calcareous soils. Soil pH is usually above 6, but can be as low as 5. Most upland calcareous grasslands occur above the level of agricultural enclosure, some 250–350 m above sea level in the submontane and montane zones of Britain, although Mountain Avens–Glaucous Sedge heath (CG13) is found almost at sea level in north-west Scotland. The upland calcareous grassland flora is exceptionally rich in rare plants that accompany the more widespread lime-loving plants.

Two of the upland types, CG9 and CG10 also occur in the lowlands, although the latter is most abundant in the unenclosed uplands. Four montane heath, dwarf-herb and ledge communities (Table 1.4) were also included in this category under the UK

Table 1.4 Summary of upland calcareous dry grassland communities, with montane heath, dwarf-herb and ledge communities (Rodwell 1992)

NVC type	Description
Dry grassland communities	
CG9	Blue Moor-grass–Limestone Bedstraw grassland: submontane or montane community restricted to free-draining, lithomorphic soils on Carboniferous limestone in the North Pennines. Species-rich closed or open swards, short when grazed or taller and tussocky. Crested Hair-grass and Sheep's-fescue are constants, together with Wild Thyme and Common Rock-rose. Limestone Bedstraw and Common Bird's-foot-trefoil are also frequent. Supports a number of nationally rare plants, such as the arctic-alpine Spring Sandwort.
CG10	Sheep's-fescue–Common Bent–Wild Thyme grassland: essentially a submontane community of base-rich, often moist brown earths overlying a variety of calcareous bedrocks. Occurs at a mean altitude of 335 m in areas with a cool, moist and cloudy climate. Generally closed swards, often heavily grazed. Sward dominated by Common Bent and Sheep's-fescue, with frequent Red Fescue. Wild Thyme is the most common herb, often abundant, with other constants including Harebell, Tormentil and Common Dog-violet.
Montane heath, dwarf-herb and ledge communities	
CG11	Sheep's-fescue–Common Bent–Alpine Lady's-mantle: a 'grass-heath' community where Alpine Lady's-mantle is often the most abundant species. Grasses such as Common Bent and Sheep's-fescue may be frequent or abundant in the sward between the Alpine Lady's-mantle plants. Confined to northern England and Scotland, it is usually a high-altitude community typical of free-draining, often moist brown earths, maintained by grazing in a montane climate.
CG12	Sheep's-fescue–Alpine Lady's-Mantle–Moss Campion: a 'dwarf-herb' community with a short, open sward and a rich diversity of grasses and herbs, including a wide variety of montane Arctic-alpines, some very restricted, such as Alpine Mouse-ear and Alpine Pearlwort. Confined to the high montane regions of Scotland above 700 m on moist, skeletal mull soils of only moderate base status, where climate is cold, wet and windy.
CG13	Mountain Avens–Glaucous Sedge: heath with a low patchy cover of shrubs over a submontane grassy sward. Mountain Avens is abundant, with Sheep's-fescue and Glaucous Sedge making up the bulk of the sward. Most species also common in other submontane calcareous grassland communities. It supports a relatively small complement of Arctic-alpines. Restricted to calcareous lithomorphic soils, mainly on Durness limestone, in the cool, oceanic lowlands of north-west Scotland, up to 500 m.
CG14	Mountain Avens–Moss Campion: a 'ledge' community with a rich diversity of dwarf shrubs, tall herbs, sedges and grasses among a carpet of cushion herbs and bryophytes. Mountain Avens is the most frequent component, often with one of a number of rare Arctic-alpine willows, such as Net-leaved Willow. Confined to ungrazed crags and ledges, often on lithomorphic, base-rich soils in the montane regions of Scotland.

BAP. Of these, CG10 and CG11 are included in Annex I habitat 6230 of the EU Habitats and Species Directive (Species-rich *Nardus* (Mat-grass) grassland, on siliceous substrates in mountain areas), together with more species-rich subcommunities of U4 Sheep's-fescue–Common Bent–Heath Bedstraw and U5 Mat-grass–Heath Bedstraw grassland. Calcareous dwarf-herb (CG12), heath (CG13) and ledge (CG14) communities are also recognised in Annex I habitat 6170 (Alpine and subalpine calcareous grasslands).

Data on the distribution and condition of upland calcareous grassland is scarce; however, there are an estimated 16,000 ha in England, 5,000 ha in Scotland and just 700 ha in Wales (UK BAPs Targets Review 2006). Estimates for the 2013 UK Habitats Directive Report indicate that included in these totals, are 681 ha of alpine and subalpine calcareous grasslands in Britain and 5,126 ha of species-rich *Nardus* grassland in Britain.

On many upland calcareous grassland sites, as with their lowland counterparts, grazing with domestic animals is a critical factor in maintaining the quality of the habitat. A JNCC report found that only 25% of upland calcareous grassland sites were in

Calcareous grassland growing on limestone scree in Lathkill Dale, Peak District National Park, Derbyshire (Peter Buckley).

Blue Moor-grass–Limestone Bedstraw grassland at Cow Green, Teesdale (© Robert Goodison).

Bird's-eye Primrose sward (CG9) at Great Close Allotment (© Robert Goodison).

favourable condition (Williams 2006). The unfavourable condition of the majority of sites at this time may have resulted from overgrazing, possibly due to overstocking or deer, which destroys grassland structure, allows coarse grasses to invade and can lead to the loss of both plants and insects. Undergrazing is also problematic in some areas, allowing Blue Moor-grass in particular to form rank, species-poor swards.

1.4.5   Upland hay meadows

Upland hay meadows in the submontane zone of northern Britain are typically represented by the distinctive Sweet Vernal-grass–Wood Crane's-bill community (MG3). This occurs in traditionally managed, species-rich upland hay meadows on the lower slopes and level ground, often on moderately fertile brown soils. Such meadows are now almost entirely restricted to northern England, typically between 180 and 400 m, usually as part of the 'in-by' land of Pennine and Lake District hill farms. Outlying fragments can still be found in Northumberland and Scotland. Species-rich grassland can also be found on unfarmed land, such as banks and roadside verges. These sites are important sanctuaries for upland hay meadow communities, especially in areas where the farmland meadows have been completely lost to agricultural improvement. The upland hay meadow priority habitat also includes upland forms of MG8; MG5 is not included in this definition, but upland forms may occur locally, while U4c may occur on banks within upland hay meadow sites, but is again not included in the definition.

In the North Pennines, several other species-rich grassland communities have been recognised, which include northern, montane species such as Mountain Pansy and Alpine Bistort. They comprise both recognised dry and wetter grassland communities (summarised in Table 1.5) and several other communities that are yet to have a published description (O'Reilly 2010).

The typical upland meadow communities contain mixtures of grasses, which comprise a small proportion of the vegetation, including Sweet Vernal-grass, Cock's-foot, Red Fescue, Common Bent, Rough Meadow-grass and Yorkshire-fog, usually with no single dominant species. In contrast, herbs are more abundant and species-rich; Wood Crane's-bill is dominant, with Great Burnet, Pignut, various Lady's-mantle

Table 1.5 Summary of the characteristics of upland hay meadows (Rodwell 1992; Rodwell *et al.* 2007; O'Reilly 2010)

NVC type	Description
Designated dry grassland priority habitat types	
MG3	Sweet Vernal-grass–Wood Crane's-bill: herb-rich, upland hay meadows on level to moderately sloping sites, often on brown soils. Mixtures of grasses comprise a small proportion of the vegetation, and include Sweet Vernal-grass, Cock's-foot, Red Fescue, Common Bent, Rough Meadow-grass and Yorkshire-fog, usually with no single dominant species. Herbs dominate the sward in variety and abundance; e.g. Wood Crane's-bill, with Great Burnet, Pignut, Smooth Lady's-mantle and Hogweed among the most frequent. Differences in NVC subcommunities mainly attributable to agricultural treatment.
Other species-rich upland grassland types	
MG5	Crested Dog's-tail–Common Knapweed: dry grassland community lacking northern montane and woodland species, but sharing many other species with MG3, such as Burnet-saxifrage, Oxeye Daisy and Common Bird's-foot-trefoil (see Table 1.3).
MG6	Perennial Rye-grass–Crested Dog's-tail: improved dry grassland usually assumed to be species-poor and dominated by grasses, but some MG6 meadows reported to be herb-rich in parts of the North Pennines, though lacking characteristic northern meadow species such as Wood Crane's-bill. Probably derived from MG3 hay meadows that have been fertilised (O'Reilly 2010) and thought to have the best potential for restoration to species-rich MG3 types.
U4c	Sheep's-fescue–Common Bent–Heath Bedstraw; Bitter-vetch–Betony subcommunity: rare upland meadow community of acid soils, often found on unmown banks. Fine-leaved grasses, such as Sheep's-fescue and Sweet Vernal-grass, are accompanied by a range of herbs, including northern montane species such as Mountain Pansy, Alpine Bistort and rarely, Small-white Orchid.
MG8	Crested Dog's-tail–Marsh Marigold: damp meadows of variable quality, reported to be the main vegetation type in 15% of North Pennine meadows (O'Reilly 2010). Characterised by wet meadow species such as Marsh Marigold and Meadowsweet, but richer meadows may contain species such as Globeflower and Ragged-Robin (see also Table 1.3).

species and Hogweed among the most frequent. Other characteristic species include Globeflower and Melancholy Thistle. The differences that may be found in the NVC subcommunities are mainly attributable to agricultural treatments (Rodwell *et al.* 2007).

The conservation status of upland hay meadows is also recognised in Annex I habitat 6520 (Mountain hay meadows) of the EU Habitats and Species Directive. Estimates for the 2013 UK Habitats Directive Report indicate only 1,000 ha in England and 50 ha in Scotland. Jefferson (2005) estimated that around 770 ha of upland hay meadow were protected in SSSIs in England, with only 5 ha being afforded similar protection in Scotland.

Extensive losses of upland hay meadows probably resulted from subsidies for agricultural improvement during the latter half of the twentieth century. But even in recent decades, some fields have been lost, while others have declined in quality, possibly due to more homogenised hay meadow management – less variability in cutting dates – that may lead to a build-up of annual 'weeds' (O'Reilly 2010). Traditionally, hay meadows were managed by cutting from late June onwards, with grazing from late summer through to early spring. Changes in the intensity or timing of grazing are almost certain to affect sward composition. Increasing the rate of nutrients added, rather than the form in which they are applied, may also be an important factor effecting changes in species richness in hay meadows. Many stands are becoming more eutrophic, impoverished grassland communities (Rodwell *et al.* 2007). In a survey carried out by the Hay Time project, significant declines were found in many SSSIs previously thought to be in favourable condition (O'Reilly 2010). Hewins *et al.* (2005) found only 7% of a random sample of non-statutory sites to be in favourable condition.

Northern hay meadow with abundant Globeflower (Pen-y-ghent, Yorkshire Dales, in the distance) (© Robert Goodison).

Wood Crane's-bill in Askrigg Bottoms (Wensleydale) hay meadow, cut annually but not grazed (© Robert Goodison).

1.5   Specialised semi-natural dry grassland types

1.5.1   Calaminarian grasslands

Calaminarian grasslands are restricted to outcrops of veins of heavy metals, such as lead, zinc, chromium and copper, in calcareous rocks, mine workings and spoil heaps, river gravels rich in heavy metals; and unusual minerals, such as serpentine exposures. The NVC recognises one vegetation type, Sheep's-fescue–Spring Sandwort grassland (OV37), which is known from the Mendip Hills, Derbyshire, the Yorkshire Dales, North Pennines and Wales (Rodwell 1992; Stevens *et al.* 2010). Other related assemblages are not adequately covered by the NVC, for example, the calaminarian grasslands on the serpentine of sites such as the Keen of Hamar in Shetland. Calaminarian grassland has an open structure, in which tussocks of Sheep's-fescue, patches of Common Bent and scattered Spring Sandwort are prominent. Although herb diversity is low, metal-tolerant species include Alpine Penny-cress, Sea Campion, Bladder Campion and Thrift. Scarce or rare metal-tolerant species may also be present, such as Young's Helleborine (*Epipactis helleborine* var. *youngiana*) and Forked Spleenwort. The endemic Shetland Mouse-ear-hawkweed occurs only in serpentine grassland on Unst, in the company of boreal species, such as Arctic Sandwort. Some calaminarian grassland sites also support rare lichens and bryophytes.

Calaminarian grassland is also recognised in Annex I habitat 6130 (Calaminarian grasslands of the *Violetalia calaminariae*) of the EU Habitats and Species Directive. Although widely distributed in Britain, its extent is very restricted, and estimates for the 2013 UK Habitats Directive Report indicate 200 ha in England, 80 ha in Scotland and 49 ha in Wales.

A JNCC report found that only 46% of calaminarian grassland sites were in favourable condition (Williams 2006). The major reasons cited for the deterioration of habitat was under-management and successional change, largely resulting from undergrazing and a gradual reduction in soil toxicity, allowing succession to proceed.

Near-natural calaminarian grassland forms part of serpentine debris habitat at the Keen of Hamar, Unst, Shetland. Closed serpentine heath is also present where soil has formed on the slopes.

Several nationally scarce species, such as Shetland Mouse-ear-hawkweed (left) and Northern Rock-cress (right) are found in the sparse calaminarian grassland habitat of the Keen of Hamar.

1.5.2 Machair

Machair is a distinctive grassland community of the coastal plains of north and west Scotland, and western Ireland. It is found only on calcareous sands blown inland by prevailing westerly winds, from beaches and sand dunes. The sand has a high shell content, and supports both dry and seasonally wet grasslands typical of calcareous to neutral substrates. The short turf grasslands that develop can be complex, and most are believed to have a long history of management by local communities, involving seasonal grazing and low-input rotational cropping. No NVC communities are restricted to machair systems, but two communities which are most characteristic of the vegetation are the Daisy–Meadow Buttercup subcommunity (SD8d) and Selfheal subcommunity (SD8e) of Red Fescue–Lady's Bedstraw fixed dune grassland. These communities are dominated by Yorkshire-fog, Creeping Bent and Cock's-foot, and by herbs such as Daisy, Meadow Buttercup, Selfheal, Creeping Buttercup, eyebright species, Harebell and Red Clover, which give the communities their distinctive

Left: Machair grassland with Kidney Vetch and spikes of Common Spotted-orchid (ssp. *hebridensis*), Cnip, Isle of Lewis. Right: Wet machair with Northern Marsh-orchid, Northton Bay, Isle of Harris (Peter Buckley).

character (Rodwell 1992). Few rare plants are confined to machair communities, but they support a rich invertebrate fauna, and important populations of Corn Crake and breeding waders such as Lapwing.

Machair was designated as a priority habitat in the UK BAP. Its conservation status is also recognised in Annex I habitat 21A0 (Machairs) of the EU Habitats and Species Directive. Estimates for the 2013 UK Habitats Directive Report indicate 14,500 ha in Scotland. Williams (2006) did not distinguish machair grassland in his report on the condition of dunes, shingle and machair SSSIs. Within this group, approximately 51% of SSSIs were in favourable condition. The major reason cited for sites in unfavourable condition was under-management, largely resulting from undergrazing.

1.5.3   Limestone pavements

Limestone pavements are a rock exposure with a complex pattern of crevices or 'grikes' separating large blocks of limestone or 'clints', formed by the scouring action of ice and percolating water during the last ice age. Pavements are typically horizontal or gently inclined. They are widely scattered in Britain, found on Carboniferous limestone in Wales and northern England, and on the Dalradian and Durness limestones of Scotland. Limestone pavements support a range of vegetation types, including grassland, heath, scrub, woodland and rock ledge communities. Grazing pressure plays a key role in determining the vegetation type, which may be essentially open, shrubby or wooded.

Open pavement is characterised by a rich diversity of plant species, with herbs and ferns growing in the grikes, scattered trees and shrubs and abundant insect populations. Many rare species are found in these open pavement communities, such as Rigid Buckler-fern, Downy Currant, Dark-red Helleborine and Solomon's-seal. Rare species normally found in montane environments include Green Spleenwort and Baneberry. Open pavement communities may also be rich in lichens and bryophytes.

Limestone pavement includes five NVC communities; CG9, OV38–40 and MG5 (Table 1.6), although peat-filled grikes and hollows support mire communities, and grike flora is more reminiscent of a W9 Ash–Rowan–Dog's Mercury field layer (Rodwell *et al.* 2000). Limestone pavements are also included in Annex I habitat 8240 (Rocky habitats and caves) of the EU Habitats and Species Directive. Estimates for the 2013 UK Habitats Directive Report indicate 1,978 ha in England, 300 ha in Scotland and 753 ha in Wales, with the largest areas in North Yorkshire and Cumbria (Maddock 2008). England holds 77% of the British resource, and the UK supports a significant proportion of European limestone pavement (Backshall *et al.* 2001).

A JNCC report found that only 27% of limestone pavement habitats were in favourable condition (Williams 2006). The unfavourable condition of rocky habitats, such as the higher altitude limestone pavement that supports open vegetation communities, is often due to overgrazing, which leads to a loss of vegetation structure, poor reproduction of more vulnerable and palatable species and the spread of unpalatable, rank species. Rarer plants may become restricted to the deeper, narrow grikes out of reach from grazing sheep. Those that favour the top of grikes, such as Bloody Crane's-bill and Rigid Buckler-fern are particularly vulnerable.

Table 1.6 Summary of main limestone pavement communities (Rodwell 1992)

NVC type	Description
CG9	Blue Moor-grass–Limestone Bedstraw grassland (see Table 1.4).
OV38	Limestone Fern–False Oat-grass: confined to sunny exposures of calcareous bedrock. Characterised by open stands of vegetation dominated by Limestone Fern and False Oat-grass, and occasionally Red Fescue or Sheep's-fescue, and False Brome. Supports a range of other vascular plants and bryophytes.
OV39	Maidenhair Spleenwort–Wall Rue: an open crevice community characterised by diminutive ferns such as Maidenhair Spleenwort and Wall Rue, and bryophytes. With the exception of Sheep's-fescue, Crested Hair-grass, Biting Stonecrop, Wild Thyme and Common Rock-rose; other vascular plants are infrequent.
OV40	Green Spleenwort–Brittle Bladder-fern: a community of shaded crevices and ledges, particularly at high altitudes, often characterised by a variety of ferns. Few other vascular plants are common, but in less-shaded crevices Sheep's-fescue, Herb-Robert and various hawkweeds occur.
MG5	Crested Dog's-tail–Common Knapweed: see Table 1.3.

Limestone pavement on Great Ormes Head (left), North Wales and Bheinn Shuardail, Isle of Skye (right).

Field Gentian growing on limestone pavement at Bheinn Shuardail, Isle of Skye.

1.6   Dry grassland scrub communities

Scrub is a complex habitat: some types are temporary, seral stages in a successional series from open ground to mature woodland, while others are more stable and permanent. The conservation value of scrub can be remarkably high, but is frequently underrated. Dry scrub communities may be found on acid, neutral or calcareous soils. Five scrub types and two underscrub communities are recognised in the NVC. In addition, its conservation status is also recognised in Annex 1 habitat H5130 of the EU Habitats and Species Directive (*Juniperus communis* formations on heaths or calcareous grasslands). Scrub is also a component of Annex 1 habitat H6210 (Semi-natural dry grasslands and scrubland facies) on calcareous substrates.

Scrub is often regarded as a problem in grassland management, and there are circumstances where it can become invasive and ultimately damage a site if not properly managed. However, in many cases scrub represents an important element of native grassland ecosystems, often contributing significantly to species richness, particularly on lowland calcareous soils. The conservation value of scrub communities has been described in detail elsewhere (Mortimer *et al.* 2000; Day *et al.* 2003). Management issues which might arise include interrupting succession and invasion to prevent the loss of open grassland to scrub encroachment, maintaining a balance between scrub and grassland, and conserving wildlife interest associated with scrub (Section 3.4).

1.6.1   Lowland communities

The most common scrub types in lowland Britain are Hawthorn–Ivy (W21), Bramble–Yorkshire-fog (W24) and Blackthorn–Bramble (W22). Bramble–Yorkshire-fog scrub, which frequently occurs on abandoned farmland, often develops into Hawthorn–Ivy scrub on neutral or base-rich soils, and to Gorse–Bramble scrub (W23) on acidic soils.

Scrub communities on deeper, more fertile calcareous soils are mainly composed of Hawthorn, Blackthorn and Bramble. On shallow, less fertile calcareous soils, species-rich scrub communities can develop, including shrubs such as Common Whitebeam, Box, Yew, Dogwood, Juniper and Wayfaring-tree (W21c and W21d). Box scrub (subcommunities of W12 and W13) occurs very locally in southern England. Scrub is widely recognised to support many rare or threatened plants and invertebrates, in addition to a number of birds and mammals. Several communities are important for the shrub species themselves, namely Box, Juniper and, on some more inaccessible limestone sites in the north and west, rare endemic whitebeams. Where grazing is reduced or abandoned, scrub will expand, and gradually develop into woodland without intervention.

Hawthorn scrub is the dominant community on neutral soils, often including Blackthorn, Elder and elms (W21). On deeper, moist and richer neutral soils, Blackthorn–Bramble (W22) scrub may dominate, accompanied by Privet and Hazel on soils with a higher base status, and Gorse on base-poor soils. On abandoned semi-natural grasslands such as Crested Dog's-tail–Common Knapweed grassland (MG5), a mosaic may develop with patches of Hawthorn and Blackthorn scrub, interspersed with Bramble–Yorkshire-fog (W24). Although common, scrub on neutral grassland can provide an important habitat for declining species such as farmland birds in otherwise intensively farmed landscapes (Mortimer *et al.* 2000). Elder occurs particularly in areas of disturbance, caused, for example, by Rabbit burrows or Badger setts. As with calcareous grassland, scrub will expand and gradually develop into woodland without intervention, except on very exposed sites.

Scrub on the calcareous grassland slopes of Box Hill, Surrey.

Gorse and Bramble scrub (W23), sometimes dominated by Broom, is found on free-draining, acidic brown earths where it occurs in mosaics with acid grassland, particularly Sheep's-fescue–Common Bent–Heath Bedstraw grassland (U4), or more acidic forms of MG5. It has a low botanical diversity, but may be important for populations of other species such as the Dartford Warbler, particularly in a mosaic of acid grassland and lowland heath. Heavy grazing can also encourage the development of Bracken–Bramble underscrub. These communities may also be found in association with calcareous grassland, where drift deposits of acidic soils occur. Gorse scrub will also develop into woodland without grazing or burning, but too much soil disturbance can result in the invasion of trees such as birch and pine.

1.6.2 Upland communities

The dry scrub forest zone is dominated by Hawthorn, with species such as Blackthorn, Hazel, Rowan, Crab Apple and Gorse, which colonise Common Bent–Sheep's-fescue upland acid grassland communities. This community is not described in the NVC. It has a low botanical diversity, but is important for some upland birds such as the Stonechat and Tree Pipit. Communities of Ash, Rowan and birches and Atlantic Hazel woods are poorly understood.

Two other types of scrub are largely confined to mountainous parts of northern Britain. Juniper heath (W19) occurs in association with a range of grassland heath and mire communities, including Blue Moor-grass–Limestone Bedstraw grassland (CG9) and Sheep's-fescue–Common Bent–Wild Thyme grassland (CG10). It is found in the eastern and central Scottish Highlands and in more isolated stands on hills south to the Lake District. High-altitude stands of dwarf willows, such as Downy Willow (W20) represent one of the rarest habitats in the UK, and hence are of considerable conservation value (Mortimer *et al.* 2000). They occur as isolated stands, in a mosaic with CG10, and montane heath, dwarf-herb and ledge communities, located mainly in the southern and central Highlands.

Upland scrub is generally controlled with browsing and grazing, but it is more likely to require measures to expand the area, rather than to reduce it.

1.7 Semi-improved dry pastures and meadows

Semi-improved dry pastures and meadows encompass a wide range of communities, from the richer swards of False Oat-grass grassland derived from pasture where grazing has been neglected (MG1c, d and e) to the ubiquitous Perennial Rye-grass swards (MG7) improved by seeding with cultivated varieties of grasses and White Clover (Table 1.7). However, within this category there are communities with some botanical value. MG1 grassland can be quite diverse, and not simply a rank, eutrophicated community of roadside verges and neglected pasture. The composition of the sward depends to some extent on the condition of the soil, including its moisture status, trophic state or base-richness. Of particular interest are the more species-rich MG1 subcommunities: Common Knapweed (MG1e), Meadowsweet (MG1c) and Wild Parsnip (MG1d), the latter more calcareous in character. Richer swards are also found within Perennial Rye-grass–Crested Dog's-tail grassland (MG6). Upland hay meadows of this community have already been mentioned (Table 1.5), but there are also botanically interesting examples of the MG6b Yellow Oat-grass subcommunity and meadows rich in Yellow Iris (a variant of the typical MG6a subcommunity) (Rodwell *et al.* 2007).

Ubiquitous False Oat-grass grassland (Tone Blakesley).

Table 1.7 Summary of the characteristics of improved pasture and meadows (Rodwell 1992)

NVC type	Description
MG1	False Oat-grass grassland: dominated by False Oat-grass and other tussocky coarse grasses such as Cock's-foot and Yorkshire-fog. An ungrazed lowland grassland of circumneutral soils, found on neglected meadows and pastures, roadside verges, railway embankments and industrial sites. Without mowing, scrub encroachment will ensue. Ubiquitous across Britain's lowlands, with communities generally reflecting differences in management or neglect. The Red Fescue (MG1a) and Common Nettle (MG1b) subcommunities represent ranker grasslands, dominated by grasses, supporting few herbs, with Common Nettle a constant in the latter. The Meadowsweet subcommunity (MG1c) is slightly richer, and characteristic of moister soils. Soil conditions restrict the distribution of the richer and varied Wild Parsnip subcommunity (MG1d) to more calcareous soils of the south and east, and the Common Knapweed subcommunity (MG1e) to mesotrophic soils across Britain. The Burnet-saxifrage variant of MG1e is particularly rich, although only found on moist calcareous soils of rocky limestone slopes in the Mendip Hills, Derbyshire, Yorkshire and Durham.
MG6	Perennial Rye-grass–Crested Dog's-tail grassland: improved permanent dry grassland with a short sward, usually species-poor and dominated by grasses, although richer swards can occur within the MG6b Yellow Oat-grass subcommunity (see also Table 1.5). Typical on moist, but free-draining permanent lowland pasture on mesotrophic brown soils. Perennial Rye-grass is often abundant, with frequent Crested Dog's-tail, Red Fescue and Common Bent; Yorkshire-fog and Cock's-foot may also be frequent. Derived from a wide range of precursors, such as MG5, CG2 and U4 communities, through continued agricultural improvement.
MG7	Perennial Rye-grass leys: specialised, species-poor swards, ubiquitous throughout lowland Britain. Many have been sown with agricultural grass varieties, together with improved strains of White Clover. Derived from a wide range of precursors, sometimes through seeding into meadows and pastures. Some examples of the Perennial Rye-grass–Meadow Foxtail–Meadow Fescue (MG7c) flood meadow communities are now known to be richer than first thought (Rodwell et al. 2007).

2. Grassland wildlife

Following the discussion of plant communities in Chapter 1, here we introduce some of the other wildlife inhabiting or using dry grassland habitats in Britain, focusing primarily on species of conservation concern. Some consideration of key management issues for invertebrate and bird communities is also included.

2.1 Invertebrates

Semi-natural dry grassland at various successional stages provides habitat for a wide variety of invertebrates, including some of the rarest to be found in Britain. All grassland types are valuable for invertebrates, with some supporting species found only in that grassland type. Invertebrate diversity is influenced by a wide range of factors, such as topography, soil, vegetation structural diversity and the presence of decaying organic matter (Table 2.1). Notably, sward structure can be as important as a diverse plant flora for maintaining the richest invertebrate assemblages. Essentially, invertebrates use semi-natural grassland to breed, feed and overwinter or hibernate.

Topography is important for invertebrates, and a semi-natural grassland site with varied topography is likely to support a more diverse assemblage than a level site. Aspect and associated variations in soil depth influences vegetation structure, and the degree of

The Roman Snail found in tussocky grassland including calcareous grassland, was added to Schedule 5 of the Wildlife and Countryside Act, 2008; in addition to pressures on habitat, these molluscs are threatened by collectors.

Table 2.1 Summary of key dry grassland features important for invertebrates (Kirby 2001; © RSPB)

Features	Characteristics
Topography	
South-facing slope	Warm and sunny; variation in soil depth and vegetation structure, including short turf and bare ground.
Slopes of other aspects	Including banks, ditches, anthills, and so on, which may be sheltered and cooler; slopes offer variation in soil depth and vegetation structure.
Soil structure	
Free-draining soils	Tend to drain and warm up more quickly, usually loose and crumbly; support many burrowing invertebrates at the northern edge of their range.
Sandy base-rich	Friable, loose, easily warmed soils, particularly near the coast, e.g. Breckland; good for burrowing invertebrates.
Vegetation structure	
Patches of bare ground	Hunting, basking and nesting sites (including paths and tracks).
Short turf	Open structure.
Grass tussocks	Microclimate providing shelter and protection, feeding and hibernation.
Taller swards	Diverse species, structure and stages of plant development important, with flowers from March to October providing pollen and nectar, and seed heads; retention of dead seed heads and hollow stems important for provision of overwintering and hibernation sites.
Thatch layer	Nesting and hibernation sites, and sites for detritus feeders.
Patches of scrub	Shelter; nesting; basking; feeding on foliage, pollen, nectar and honeydew; hunting.
Decaying matter	
Litter	Hibernation and feeding for detritus feeders.
Dung	Rich and unique habitat for species contributing to the nutrient cycle of grasslands.
Carrion	Attract carrion specialists and their predators.

shelter it provides. South-facing slopes are particularly important for invertebrates that require sunshine and warmth, depending on the degree of slope, the height of the sward and the presence of bare ground. On thinner soils, such as chalk downland, for example, shorter turf on these slopes may be maintained by a combination of grazing and natural desiccation in the summer months. In contrast, north-facing slopes contribute cooler and damper conditions to the structural diversity of a grassland site.

Grassland types differ considerably in the diversity of invertebrate species, size of populations and the number of species of conservation concern they support (Table 2.2). Soil structure, soil composition and vegetation and management all influence invertebrate

Species-rich calcareous grassland can support a rich assemblage of insects.

communities. Calcareous grassland supports rich assemblages of invertebrates (Alexander 2003), more so than other semi-natural dry grassland types. It is characterised by often thin, free-draining soils which tend to warm up quickly, and grassland with a diverse and species-rich vegetation structure. Lowland dry acid grassland is also considered to be moderately species-rich, particularly on sites with a range of vegetation types. Lowland meadows cut in late June to late July support fewer invertebrates, particularly those of conservation concern. Meadows cut annually also support fewer invertebrate species, although less is known about the rarer species using this habitat.

In any semi-natural grassland site, structural diversity is one of the key factors influencing invertebrate diversity. The most diverse assemblages of invertebrates are likely to be found on sites exhibiting a variety of successional stages, from bare ground through to patches of scrub. Diverse swards might include short turf, tussocks and areas of taller grass, which are likely to support plants at different growth stages, particularly flowers, fresh seed heads and the dead seed heads and stems which persist through the winter. The merits of managing a site to accommodate such structural diversity will

The Black-veined Moth is now restricted to a few sites in Kent, where it requires large areas of chalk downland with variable structure, with Tor-grass, Marjoram and Common Knapweed. Overgrazing is an important threat.

Table 2.2 Suitability of semi-natural dry grassland types for invertebrates

Grassland type	Invertebrate fauna	Examples of species of conservation concern[a]
Lowland calcareous grassland	Very good for invertebrates; scrub also richer than other grassland types.	Wart-biter Bush Cricket Hazel Pot Beetle Adonis Blue Black-veined Moth Shrill Carder Bee Phantom Hoverfly
Lowland dry acid grassland	Generally poor for invertebrates, but supports many specialists, particularly ground-dwelling and burrowing invertebrates in loose, sandy soils.	Four-banded Weevil-wasp Banded Mining Bee Brush-thighed Seed-eater Field Cricket
Lowland meadows	Some specialist invertebrates, particularly in grazed pasture, but generally less species-rich than calcareous grassland.	Large Garden Bumblebee Hornet Robberfly
Upland calcareous grassland	Very good for invertebrates, including a number of species of conservation concern.	Wall Mason Bee Round-mouthed Whorl Snail Geyer's Whorl Snail
Upland hay meadows	Invertebrates can be numerous, but relatively few, often common species.	Weevils, e.g. *Omiamima mollina* Click beetles, e.g. *Ctenicera pectinicornis* Flies, e.g. *Nanna brevifrons*

[a]More comprehensive lists can be found on the Buglife website (https://www.buglife.org.uk/).

depend to some extent on the grassland type and the topography. Structural diversity might be beneficial for some lowland pastures, for example, but not on south-facing calcareous grassland slopes. Relatively short downland turf with an open structure is now rare. Therefore, where this occurs, the development of extensive areas of tall grass or even scrub should be avoided.

For any dry grassland site, its primary conservation value must always be considered, and the site managed accordingly.

2.1.1 Butterflies

Butterflies are selected for special attention here because they are considered to be representative of invertebrate populations more generally and they are good indicators of long-term changes in biodiversity, responding to changes in management and the environment (including climate change). Long-term data sets are available that continue to be closely monitored by a large number of amateur recorders. Butterflies are also relatively easy to identify and monitor. Hence they can be used as one measure of the success of habitat restoration programmes.

Semi-natural grassland supports more species of butterfly than any other single habitat in Britain. Eleven of the 13 wider countryside species which might be considered characteristic of semi-natural grassland form colonies (Table 2.3), while the Brimstone and Painted Lady tend to range widely across the countryside, breeding in suitable habitat when they come across it. The majority of wider countryside species are widespread, with either stable or expanding populations. However, the Wall and Small Heath butterflies

Table 2.3 Wider countryside butterflies characteristic of semi-natural dry grassland, including light scrub (Asher *et al.* 2001; Fox *et al.* 2006; 2011; 2015; © Butterfly Conservation)

Species	Notes on range, population and habitat
Small Skipper	Widespread and range expanding but significant decline in abundance.
Essex Skipper	Southern and eastern England. Range expanding but significant decline in abundance. Favours acid grassland.
Large Skipper	Widespread in England and Wales. Range expanding but significant decline in abundance. Widespread and common on range of soil types, usually with shrubs, tall herbs and grasses.[a]
Brimstone	Widespread in England and parts of Wales. Range expanding. Favours scrubby grassland, particularly those with Buckthorn or Alder Buckthorn.
Small Copper	Widespread. Range stable, but has suffered severe decline. Wide range of habitats including calcareous grassland.[a]
Brown Argus	Widespread in southern and eastern England. Expanding range may be due to climate change and wider range of food plants. Favours calcareous grassland, but also found in range of other habitats.
Common Blue	Widespread and relatively stable. Range of soil types, especially if Common Bird's-foot-trefoil is present.
Painted Lady	Regular migrant. Breeds in any open habitat with thistles.
Wall[b,c]	Once widely distributed in England, but declined severely in distribution and abundance since 1970s, particularly in inland areas. Found on a range of soil types.
Marbled White	Range expanding, though many colonies lost throughout twentieth century. Favours calcareous grassland.
Gatekeeper	Widespread in southern, central England and Wales. Range expanded in 1980s and 1990s, though reported to be in severe decline[a] colonies lost through agricultural intensification, and loss of hedgerows and unimproved grassland. Favours grassland with tall grasses close to scrub and hedgerows.
Meadow Brown	Widespread. Range stable. Common and widespread, but declining through agricultural intensification. Found on a range of soil types.
Small Heath[b,c]	Widespread but declined severely in distribution and abundance since 1970s Many colonies lost through loss of native grasses to arable or improved grassland; now much less abundant locally, and causing considerable concern. Found on grassland with fine grasses, particularly downland.

[a]See www.gov.uk/government/uploads/system/uploads/attachment_data/file/389401/agindicator-de6-18dec14.pdf.
[b]Country-level priority species.
[c]Red List species (Fox *et al.* 2010).

Small Heath is a country-level priority species that has suffered long-term declines.

are classified as species of conservation concern, having suffered long-term declines; both are country-level priority species, i.e. species of principal importance (formerly UK BAP priority species) and included on the UK red list (Fox *et al.* 2010).

Thirteen habitat specialist butterflies are also characteristic of semi-natural grassland, such as the Duke of Burgundy and Chalkhill Blue (Table 2.4). A further seven habitat specialist butterflies, including Heath Fritillary and Small Pearl-bordered Fritillary also use dry grassland, although this is not their main habitat type (Table 2.5). With the exception of the Green Hairstreak and Dark Green Fritillary, all habitat specialists using grassland are included on the butterfly red list for Great Britain, representing more than 50% of the species on this list (Fox *et al.* 2010). Most are also classified as country-level priority species, highlighting the serious extinction risk facing many grassland species in Britain. The decline in habitat specialists, such as the Duke of Burgundy, Adonis Blue and Marsh Fritillary, has been linked to changes in semi-natural grassland management, such as reduced grazing leading to taller swards and the invasion of coarse grasses (Bourn

Duke of Burgundy, a country-level priority species in England has been the focus of concerted efforts in recent years to stabilise populations.

Table 2.4 Habitat specialist butterflies characteristic of semi-natural dry grassland, including light scrub (Asher *et al.* 2001; Fox *et al.* 2006; 2011; 2015; © Butterfly Conservation)

Species	Notes on range, population and habitat
Lulworth Skipper[a,b]	Range restricted to south of Dorset, where population is fairly stable. Calcareous grassland (including chalk downland) and coastal undercliff grassland maintained by erosion and soil slippage.
Silver-spotted Skipper[b]	Range declined dramatically during the twentieth century, but some recovery of range and large increase in abundance since 1980. Short, sparse turf of thin-soiled chalk grassland in southern England. Has responded to site conservation management, but needs extensive networks of habitat patches. Adults have reasonable powers of colonisation.
Dingy Skipper[a,b]	Long-term decline due to habitat loss and changes in land management such as scrub encroachment on chalk downland. Increase in abundance and occurrence over past decade. Wide range of open, sunny habitats including chalk downland.
Grizzled Skipper[a,b]	Range and population declined severely due to loss of semi-natural grassland, and management changes in remaining fragments. Stronghold remains in southern England. Range of habitats including chalk downland and other calcareous soils including clays.
Green Hairstreak	Widespread but declining, possibly due to shading. On chalk grassland, short turf (< 10 cm) important, with light scrub beneficial. In other habitats, strongly associated with scrub. Suitable ant colonies required to bury pupae.
Small Blue[a,b]	Substantial decline in the twentieth century throughout range, extinct across much of northern England, due to habitat loss and lack of management. Local throughout Britain, with stronghold in the south. Wide range of sheltered grasslands with abundant Kidney Vetch, including calcareous grassland and coastal grassland.
Chalkhill Blue[b]	Widespread decline in colonies over last 100 years, particularly in the 1950s with the reduction in Rabbit numbers; positive signs in the past decade. Found in southern England on dry calcareous grassland, tolerating sward up to 10 cm. Some individuals recolonised sites up to 20 km.
Adonis Blue[b]	Major decline in range over last 200 years, particularly in 1950s and 1960s, but recent substantial increase in its stronghold areas in southern England. Dry calcareous grassland with abundant food plants growing in short turf. Some individuals recolonised sites over 10–15 km.
Large Blue[a,b]	Formerly scattered colonies across southern England, extinct in 1979. Reintroduction programme started in 1983. Prefers warm, dry, predominantly acidic coastal or limestone grassland with abundant Wild Thyme and colonies of host red ant, *Myrmica sabuleti*.
Duke of Burgundy[a,b]	Steady decline throughout the twentieth century, becoming substantial since 1950. Population levels increased over the past decade and range decline halted. Found in scattered colonies across southern England with isolated colonies in northern England. Occurs in open areas in ancient woodland (where most substantial losses recorded) and scrubby calcareous grassland.

Table 2.4 continued

Species	Notes on range, population and habitat
Dark Green Fritillary	Widespread, but substantial decline over past 100 years in north and east Britain, which accelerated since 1950s, but may now be stabilising. Main habitats are: calcareous grassland; damp grassland, flushes and moorland; grassland with Bracken; and coastal grassland, dunes and scrub. Occasionally found in woodland open space. Relatively mobile. Thrives in taller, transitional habitat until driven out by scrub. With loss of herb-rich grassland, may be more reliant on woodland open space.
Marsh Fritillary[a,b]	Once widespread in Britain, but has declined severely across whole range over past 100 years, becoming extinct in eastern Britain; decline continued into the twenty-first century. Occurs on dry calcareous grassland (usually west- or south-facing slopes) in England, and also on damp or heathy tussocky grassland, usually with Devil's-bit Scabious. Has responded locally to conservation measures, but threatened by both under- and overgrazing.
Glanville Fritillary[a,b]	Virtually restricted to coastal landslips and chalk downland on the Isle of Wight, where population stable in recent years, but still vulnerable. Requires early successional habitats with abundant Ribwort Plantain, where susceptible to increase in Rabbit population and drought. Habitat restoration and creation may be required to support its natural pattern of site extinction and colonisation.

[a]Country-level priority species.
[b]Red List species (Fox *et al.* 2010).

and Thomas 2002). This trend has been recognised across Europe, where the European Grassland Butterfly Indicator showed a strong negative trend between 1990 and 2011, declining by almost 50%, suggesting a considerable loss of grassland biodiversity (European Environment Agency 2013).

There is clearly an urgent need to increase the populations of butterflies of conservation concern on grassland sites through more appropriate conservation management practices.

Silver-studded Blue (Tone Blakesley).

Table 2.5 Habitat specialist butterflies for which semi-natural dry grassland is not the main habitat (Asher *et al.* 2001; Fox *et al.* 2006; 2011; 2015; © Butterfly Conservation)

Species	Notes on range, population and habitat
Silver-studded Blue[a,b]	Severe decline in range and abundance; extinct across much of central and northern England, parts of Wales and the North Downs, due to heathland destruction and loss of early successional habitat. Increase in occurrence over the past decade. Found on lowland heathland, sand dunes and calcareous grassland in North Wales and Dorset. Adults extremely sedentary.
Northern Brown Argus[a,b]	Small, scattered colonies in northern England and Scotland, declining due to habitat loss and lack of management. Also threatened by climate change. Sheltered, semi-natural grassland on thin, base-rich soils not heavily grazed. Also found on neutral and occasionally acid soils in Scotland where Common Rock-rose present. Adults sedentary.
Small Pearl-bordered Fritillary[a,b]	Widespread in western and northern Britain but severe decline in the south. Main habitats include damp grassland, woodland open space and grassland with Bracken; also dry calcareous grassland, especially Carboniferous limestone/scrub mosaics (Mendip Hills and Morecambe Bay).
Pearl-bordered Fritillary[a,b]	Widespread in Scotland but long-term decline in England and Wales ; increase in abundance over past decade. Three main habitats include woodland open space, open woodland in Scotland and sparse dry grassland with scattered scrub (such as the Morecambe Bay limestone grasslands) or dense patches of Bracken.
Heath Fritillary[a,b]	Once widespread in southern England, but already in decline at the start of the twentieth century, which continued until early in the twenty-first century; concerted conservation action has been required to ensure its continued survival in Britain. Primarily found in heathland combes (South-West) and coppiced woodland (South-East), but breeds on a few semi-natural grassland sites in the South-West.
Mountain Ringlet[a,b]	Found in the mountains of the Scottish Highlands and Lake District, long-term decline in occurrence, and may be threatened by climate change. Predominantly a species of damp grassland, but also uses drier 'grass/heath' vegetation (CG11) on rocky outcrops.
Grayling[a,b]	Widespread coastal species, particularly in the south and west of Britain, but a decline in both abundance and distribution in southern England recorded in recent decades. Occurs on calcareous grassland and heathland inland, but mainly on coastal dunes, saltmarshes and clifftops. Few calcareous grassland colonies survived the outbreak of myxomatosis in the 1950s.

[a]Country-level priority species.
[b]Red List species (Fox *et al.* 2010).

Dry grassland management and restoration should also aim to improve connectivity between existing colonies to facilitate dispersal and maintain metapopulations. This may allow butterflies to recolonise former sites, or colonise new sites where grassland habitat

Small Pearl-bordered Fritillary.

has been created. This will require implementation of management practices such as scrub control, cutting and grazing regimes designed specifically to meet the requirements of those species most likely to benefit on a particular site, while not compromising the wider conservation objectives of that site. This is important because management practices implemented for one species, such as the maintenance of short turf on downland for the Silver-spotted Skipper might compromise other threatened species, such as the Small Blue, which has different habitat requirements.

2.1.2 Implications for management

The key management issues affecting grassland invertebrate populations are common to many semi-natural grassland communities, although any site should be assessed and management plans developed accordingly. Sites should also be carefully monitored to ensure that management is providing high-quality habitat throughout the year, and that this is maintained in the long term. Changes such as increases or decreases in grazing, for example, can have significant impacts on invertebrate communities in just a few years, and if species are lost, recolonisation may not happen without human intervention, particularly where semi-natural grassland is highly fragmented. Consistency in management is the key message for invertebrate conservation, and changes in management regimes should generally be avoided on sites in good condition, which have historically been managed in particular ways.

2.1.2.1 Lowland calcareous grassland

This represents one of the richest grassland habitats for invertebrates, supporting a wide range of species of conservation concern, such as the Wart-biter Bush Cricket, Phantom Hoverfly and various leaf beetles, butterflies and moths. The characteristics supporting diverse assemblages of insects include: a rich flora; diverse vegetation structure with areas of bare ground; thin, well-drained soils which are friable and warm quickly; a range of aspects; optimum grazing regimes; and species-rich scrub, which is also richer for invertebrates than scrub developing on other grassland types. While maintaining historical areas of scrub is valuable for invertebrates, allowing this to expand is likely to be highly detrimental to the calcareous grassland sward; also allowing scrub to develop into woodland will greatly reduce its value for invertebrates. Invertebrates in lowland calcareous grassland are also threatened by: fertiliser application or reseeding; changes in management practices, such as undergrazing leading to damage to the sward or dominance by rank grasses, such as Tor-grass; higher stocking densities resulting in trampling and the loss of some plant species; and in some cases, abandonment of grazing leading to rank grassland and scrub. In any management plan, care must be taken to ensure that grazing (or mowing) results in structural diversity in the sward, and does not lead to a uniform, closely grazed summer sward, with an absence of flowers.

2.1.2.2   Lowland dry acid grassland

These communities usually have fewer plant species than calcareous or neutral grassland, and this is reflected in poorer invertebrate communities. However, many of the invertebrates characteristic of acid grassland are specialists, not found in neutral or calcareous grassland, such as Brush-thighed Seed-eater and species which prefer loose, sandy soils, such as ground-dwelling and burrowing invertebrates, e.g. weevil wasps and mining bees. Consequently, areas of dry, sandy grasslands with south-facing slopes are important habitats. Some species, such as solitary bees, also require patches of taller vegetation for sources of pollen and nectar. Other specialists require tussocky grasses to complete their life cycles. Acid grassland is usually managed as pasture, benefitting most from light or rotational grazing regimes that promote structural diversity, including areas where plants can flower and set seed. Overgrazing is one of the most serious threats to semi-natural lowland acid grassland, leading to rank grassland, Bracken and possible scrub encroachment. Conifer plantations are a threat in some areas, while agricultural improvement involving the use of chemicals or reseeding is also a significant threat. Some dung-feeding invertebrates might be threatened if grazing animals have been recently treated with broad-spectrum antiparasitic avermectin medicine. If grazing is no longer possible, cutting may be introduced, but care is needed to avoid damaging delicate sandy soil habitats by heavy machinery; it is also important to cut on rotation to preserve the structural diversity of vegetation across a site. Artificial disturbance might also be necessary to maintain dry, open areas on sites that have historically been managed in this way.

2.1.2.3   Lowland meadows

These include enclosed neutral hay meadows and pasture across the lowlands of Britain. Due primarily to the timing of the summer hay cut, hay meadows support relatively few Nationally Scarce species, and have restricted invertebrate assemblages. Traditional management practices can help to maintain and expand invertebrate populations in these meadows if refuges are created in the immediate vicinity. Semi-natural dry pasture supports a wider diversity of invertebrates, although still not as rich as calcareous grassland. This is due to a range of factors, including lower plant species richness, higher soil fertility in some cases, heavier soils, less short turf and less bare ground. Neutral pastures are often on level sites, but varied structural diversity is maintained by grazing. Invertebrates in lowland meadows and pasture are threatened by improvement through inorganic fertiliser application or reseeding, changes in management practices, such as a shift from hay to silage, earlier and more synchronised annual hay cuts, higher stocking densities on pasture, and in some cases, abandonment leading to rank grassland and scrub. For hay meadows, maintenance of traditional management is important, preferably with some uncut margins where plants can flower and seed. If cutting is introduced to replace grazing, invertebrate populations are likely to suffer. Maintaining structural diversity in the sward is key. Mowing different parts of a site on rotation and varying cutting height or leaving uncut margins will maintain structural diversity in the sward.

2.1.2.4   Upland calcareous grassland

Upland calcareous grassland represents a rich habitat for invertebrates. Although the diversity of species of conservation concern is generally lower than lowland calcareous grassland, diverse assemblages may be found in association with seepages and streams. Invertebrates benefit from habitat mosaics similar to lowland calcareous grassland and also face a similar range of threats and management issues (see earlier). In addition, quarrying for limestone and other rocks is a continuing threat to upland calcareous grassland communities. Many of the rare and scarce invertebrates relying on seepages and flushes are vulnerable to changes in the water table, reductions in water flow or damage to associated vegetation. Overgrazing can have a particularly detrimental impact on sensitive wetter areas.

2.1.2.5   Upland hay meadows

Upland hay meadows support relatively few Nationally Scarce species, but there has been little research in this area. Like their lowland counterparts, the vegetation in upland hay meadows is allowed to grow tall, but is cut in midsummer, which restricts invertebrate assemblages to the more common species, whose life cycles fit the management regime. However, with the loss of so much semi-natural grassland, even these communities are important for invertebrate conservation. Cutting is unavoidable, but a wider variety of invertebrates might be more sustainable in upland hay meadows if there were refuges in the immediate vicinity, such as uncut margins or fields designated for a late cut each year. Invertebrates in upland hay meadows are also threatened by more efficient management practices, such as increased grazing pressure in the spring, earlier and more synchronised annual cutting and increases in the application of farmyard manure. Grazing benefits dung-feeding invertebrates in particular, providing farm animals have not been recently treated with avermectin. Maintenance of traditional management is important, because in some cases abandonment of hay cutting is a threat.

2.1.2.6   Scrub

Scrub is clearly an important habitat for invertebrates in semi-natural grassland communities in some circumstances. It can provide sources of nectar, food plants, a nesting habitat, shelter and warmth, important for different species at various stages of their life cycle. Structural diversity, particularly on the interface with grassland can attract a rich diversity of species. Shaded, damper areas are also important for some species. For more information on scrub management, see Section 3.4 and *The Scrub Management Handbook: Guidance on the Management of Scrub on Nature Conservation Sites* (Day *et al.* 2003).

2.2 Birds

Semi-natural dry grassland types vary considerably in their associated bird communities, but all provide important habitats for both common birds and some of the most threatened species in Britain. It is perhaps the Corn Crake, more than any other bird, which symbolises the decline in semi-natural dry grassland birds over the past 100 years or so. Historically, Corn Crakes bred regularly in almost every English county, with their rasping call commonly heard in traditionally managed hay meadows. With the introduction of mechanised and earlier cutting and habitat loss, Corn Crakes had virtually disappeared as a breeding species in England by the 1980s.

2.2.1 Lowland birds

Most species which nest in lowland dry grassland, either on the ground or in grassland scrub are of conservation concern, with the Skylark being numerically the dominant species (Table 2.6). Relatively few species make extensive use of the interior of lowland dry grassland fields for nesting (Brown and Grice 2005). Rarer species that formerly nested locally in dry grassland, but now do so only rarely include the Nightjar, Wheatear, Woodlark and Whinchat. Quail and Montagu's Harrier now make more use of arable fields for nest sites (Brown and Grice 2005). Chalk downland, where the sward is close-cropped and there are bare soil patches, is locally important for the Stone-curlew in Britain, and populations are increasing, for example, on Salisbury Plain (see Box 4.1). The other core breeding area for the Stone-curlew is the Breckland. Although birds nest on arable fields, the grass/heaths, where heathland grades into semi-natural lowland dry acid grassland, are especially important.

The presence of scrub in lowland grassland increases bird species diversity and supports the nesting of species of conservation concern such as the Dunnock, Common Nightingale, Grasshopper Warbler, Willow Warbler, Linnet, Yellowhammer and Reed Bunting. The Common Nightingale has become strongly dependent on scrub in England

Common Nightingales have become strongly dependent on scrub in England (Tone Blakesley).

Kestrels hunt in grassland, but nest in other habitats (Tone Blakesley).

(Fuller 2012). The Turtle Dove, Tree Pipit and Whinchat have become very rare, more typically absent in these habitats in recent years. Scrub also supports two warblers not classed as birds of conservation concern, the Garden Warbler and Lesser Whitethroat. The number of 'grassland' breeding species is approximately one third that of the number of 'scrub' species and the overall diversity of breeding species in grass/scrub mosaics increases with scrub cover and thickness (Fuller 1982).

Species of conservation concern which nest in adjacent woodland or other habitat and feed on grassland, include the Kestrel, Stock Dove, Mistle Thrush and Yellow Wagtail. The Chough is a rare species mainly found in the coastal counties of Wales and western Scotland; it nests on cliffs and inland rock faces, but feeds on short dry grassland turf. Other species that feed a lot on grassland, all year round, include the Common Buzzard, Common Starling, Rook, Jackdaw and Magpie. Species visiting grassland to feed tend to be insectivorous, favouring grazed fields, sometimes attracted by insects associated with dung. Many birds feed around the margins of meadows and pasture, where safety can be sought in nearby hedgerows and trees. A wide range of factors determine the extent to which a field might be used by feeding birds, including soil type, size, the timing of mowing and grazing regimes (type of animal, stock density) and so on (Vickery *et al.* 2001). Aerial feeders such as Barn Swallows and Swifts also feed over grassland; uncut grassland can attract large numbers of Swifts and Barn Swallows, which take emerging insects. Mown grassland can also create a flush of accessible invertebrates that attract Barn Swallows, wagtails, pipits, thrushes, and sometimes waders such as the Eurasian Curlew and Northern Lapwing. Post-breeding flocks of Golden Plover may also be attracted under such conditions.

Barn Swallow (left) and other insectivorous birds such as Yellow Wagtail (right) may be attracted to uncut grassland to feed on emerging insects (Tone Blakesley).

The Skylark is numerically the dominant species of lowland dry grassland (Andy Hay, http://www.rspb-images.com/).

In winter, dry grassland attracts relatively few species, although gulls, Redwing, Fieldfare, Rook and Common Starling may often be found. In some parts of lowland Britain, dry grassland provides major feeding opportunities for the Northern Lapwing and Golden Plover, particularly semi-natural grassland rich in earthworms (Fuller 1982). Other winter visitors which feed on grassland, although not exclusively, include Bewick's Swan, Whooper Swan, Wigeon and several goose species: Bean, Pink-footed, Greylag, Barnacle and Brent Geese. Wildfowl tend to favour improved grassland, due to the higher nutritional value of the forage. Grassland scrub supports feeding thrushes and roosting passerines in the winter. Larger concentrations of feeding and roosting birds also attract raptors, such as the Hen Harrier, Sparrowhawk, Merlin and Peregrine.

Table 2.6 Birds of conservation concern (BoCC) (Eaton *et al.* 2015) that make use of lowland semi-natural dry grassland (pink areas signify use of the resource)

	Nesting		Feeding		BoCC[a]
	Ground	Grassland scrub	Summer	Winter	
Kestrel			■	■	Amber
Northern Lapwing[b]	■		■	■	Red
Stone-curlew	■		■		Amber
Grey Partridge[b]	■		■	■	Red
Corn Crake	■		■		Red
Stock Dove			■	■	Amber
Turtle Dove		■	■		Red
Skylark[b]	■		■	■	Red
Meadow Pipit	■		■	■	Amber
Yellow Wagtail[b]	■		■		Red
Dunnock		■	■	■	Amber
Common Nightingale		■	■		Red
Song Thrush			■	■	Red
Mistle Thrush			■	■	Red
Fieldfare				■	Red
Redwing				■	Red
Grasshopper Warbler		■	■		Red
Willow Warbler		■	■		Amber
Common Starling			■	■	Red
Tree Sparrow			■	■	Red
Linnet[b]		■	■	■	Red
Cirl Bunting[c]		■	■	■	Red
Reed Bunting		■	■	■	Amber
Yellowhammer		■	■	■	Red

[a]In the BoCC, red-listed birds are of the highest concern, amber moderate and green of low conservation concern.
[b]Also associated with upland grasslands.
[c]Formerly more widespread, now confined to Devon.
Birds that have become very rare or are usually absent from this habitat include the Nightjar, Woodlark, Whinchat and Tree Pipit.

2.2.2 Upland birds

The uplands are dominated by unenclosed heather moorland, acid grassland and blanket bog, with some enclosed semi-natural grassland. Their breeding bird communities are quite distinct from those of the lowlands (Table 2.7). Although semi-natural upland hay meadows represent only a small proportion by area, they are notable for their bird populations, and support a high density of breeding waders, mainly Northern Lapwing, Eurasian Curlew and Oystercatcher. Grey Partridge and passerines such as the Skylark, Wheatear, Yellow Wagtail and Meadow Pipit also breed. Upland hay meadows are also a very important source of seed for foraging House Sparrow, Linnet and Twite, while Golden Plovers feed in meadows following the hay cut. In the summer months, plentiful small mammals and birds can attract predators of conservation concern such as the Hen Harrier, Merlin and Short-eared Owl. In contrast, far fewer species breed in upland calcareous grassland, although Pennine limestone grasslands support relatively high densities of Golden Plover.

Upland hay meadows support important populations of breeding waders, such as the Northern Lapwing (left) and Eurasian Curlew (right) (Andy Hay, http://www.rspb-images.com/).

Uplands are far less important for birds during the winter months due to the harsh climate, with many birds migrating, or moving to coastal or lowland locations. One exception is the Black Grouse, which use hay meadows for feeding during the winter months, and occasionally to rear their chicks in the summer (Brown and Grice 2005).

2.2.3 Conservation

Many declining grassland birds may be found on the list of Birds of Conservation Concern (BoCC) (Eaton *et al.* 2015), which is a non-statutory approach to classifying breeding and wintering British birds depending on their conservation status. Red-listed birds are of the highest concern, amber moderate and green of low conservation concern. These lists are periodically reviewed to take account of recent trends and issues. The country-level priority species lists include birds for which the UK has international importance, including those that are endemic species or races, and which have undergone large declines; most red-listed grassland birds and some amber-listed species are included.

The Wild Bird Indicator is part of the UK Government's Quality of Life approach, first introduced in 1999. It is compiled annually in conjunction with the British Trust for Ornithology (BTO), Royal Society for the Protection of Birds (RSPB) and the JNCC, from

Table 2.7 Birds of conservation concern (BoCC) (Eaton *et al.* 2015) that make use of upland semi-natural dry grassland (pink areas signify use of the resource)

	Ground-nesting	Feeding		BoCC[a]
		Summer	Winter	
Hen Harrier[b]	(pink)	(pink)		Red
Merlin[b]	(pink)	(pink)		Red
Whimbrel[c]	(pink)	(pink)		Red
Eurasian Curlew	(pink)	(pink)		Red
Black Grouse			(pink)	Red
Short-eared Owl[b]	(pink)	(pink)	(pink)	Amber
Whinchat[d]	(pink)	(pink)		Red
Twite		(pink)		Red

[a]In the BoCC, red-listed birds are of the highest concern, amber moderate and green of low conservation concern.
[b]Nest only in unenclosed uplands areas, more commonly in heather.
[c]Nest in Shetland.
[d]Nest in hay meadows and rank vegetation, occasionally in lowlands.

bird counts made using standard methods to assess change over time. Many grassland species are included in the farmland bird category of the Wild Bird Indicator, which was 55% lower in 2013 than its 1970 level. Although there has been an overall decline in this assemblage, the indicator masks a more complex picture of increases in some more common species and more dramatic declines in others. For example, Wood Pigeon and Jackdaw populations have more than doubled since 1970, whereas Tree Sparrow and Turtle Dove numbers have fallen by 90 and 96% respectively (UK Biodiversity Indicators 2014). Other species, including the Northern Lapwing, Skylark and Yellowhammer also continue to decline. The reasons for such dramatic declines are likely to be complex, and have been linked to factors such as the loss or improvement of semi-natural dry grassland (Section 4.1), increased pesticide use, field boundary removal, increased nest predation due to a reduction in gamekeeping, more mechanised and earlier cutting, and the intensification of grassland management. In addition, some of the declining species, including the Turtle Dove and Yellow Wagtail, are sub-Saharan migrants, and their declines may be related to conditions on migration or in their African wintering grounds.

There is clearly a need to increase BoCC populations on dry grassland sites through more appropriate management practices and to extend their ranges where possible. This will require the implementation of a suite of management practices aimed at the birds most likely to benefit on a particular site, while not compromising wider conservation objectives.

2.2.4 Implications for management

Many of the dramatic declines in farmland bird populations in the lowlands have resulted from the loss of semi-natural grassland and the intensive management of improved grassland, particularly for silage. For example, early cutting often destroys Skylark nests, and birds that have a second clutch may fail to raise their young before the next cut (see Wilson *et al.* 2009).

Lowland grassland provides nesting and feeding for a range of birds including residents such as the Northern Lapwing and Song Thrush, and winter visitors such as the Golden Plover, Redwing and Fieldfare. Waders prefer larger areas of grassland where they may be safer from predators, whereas thrushes tend to feed in smaller fields where they can find cover in hedgerows and trees. In mixed farming environments, grassland provides invertebrates for foraging birds, with different species favouring different sward structures (Table 2.8). Management practices that increase invertebrate populations generally fall into two categories:

- *Grazed pasture*: Birds requiring shorter (often complex) swards for nesting and feeding; also a number of birds feed on invertebrates associated with dung, including waders, Swifts, Barn Swallows, wagtails, Common Starlings and corvids (and also many bat species); and
- *Meadows*: Preferred by birds requiring taller grassland cover for nesting; traditional hay meadows can provide an abundance of seed for finches such as Linnets, which feed on sorrels and dandelions in particular. Areas of shorter turf are often beneficial for birds on these sites. Uniquely, recently cut swards can also create a flush of invertebrates; this attracts a range of species, as described earlier.

Resources that have become scarce in the lowland countryside are seeds and large, long-lived invertebrates. For feeding, pastures with structurally diverse swards are particularly valuable for birds (Table 2.8): short swards provide access to soil invertebrates; open areas allow birds to see approaching predators; while tussocks provide cover, offer a richer supply of invertebrates and may provide seed. Birds may also feed in areas where animals congregate, such as drinking troughs and supplementary feeding areas, attracted by bare ground where they can forage for seeds and invertebrates.

Use of farmyard manure on semi-natural hay meadows and semi-improved grassland boosts soil invertebrate numbers, such as earthworms, which are important food for birds (see Section 3.2). Solid manure also brings invertebrates closer to the surface, making them more accessible for birds. The occasional application of manure on hay meadows is also associated with more diverse plant communities, and hence richer invertebrate populations, in contrast to grassland receiving more regular applications of inorganic fertiliser.

Scrub is also important for several bird species, particularly scrub with good diversity of structure and age that is managed on rotation. This provides denser areas for nesting and more open areas for foraging in the breeding season, and may provide a winter supply of berries (Section 3.4). For more information on bird habitats, see Fuller (1982); for scrub/grassland mosaics, see Fuller (2012); and for birds of lowland grassland and farmland, consult the RSPB's *A Management Guide to Birds of Lowland Farmland* (Winspear and Davies 2005).

Table 2.8 Sward structure and use of dry grassland by some lowland birds of conservation concern (grey areas indicate where the resource is not used) (modified from Winspear and Davies 2005; © RSPB)

Species	Nesting	Summer foraging	Winter foraging
Lowland			
Stone-curlew	Tightly grazed lowland swards (calcareous grassland in Wessex (south-western England); acid grassland in the Brecks) with sparse vegetation and bare areas		
Northern Lapwing	Short sward (5 cm) with scattered taller tussocks (< 10% of area) in larger fields (5 ha), often wet grassland		Short, open vegetation such as pasture (< 5 cm) and bare earth for soil invertebrates
Meadow Pipit	Swards of 10–50 cm, possibly including sparse, open scrub		Moderately tall, gappy, grazed swards
Skylark	Large fields, sparse vegetation (20–50 cm), e.g. lightly grazed pastures or hay meadows		Large, open fields with seed heads
Song Thrush		Short, open vegetation (< 5 cm) and bare ground for soil invertebrates, including cattle-grazed pasture	
Common Starling		Short, open vegetation (< 5 cm) and bare ground for soil invertebrates	
Linnet		Swards with seed heads, e.g. dandelions and sorrels	Fields with seeding herbs
Yellow-hammer		Structurally diverse swards with tussocks, short grass and bare ground	Fields with seed heads
Upland			
Eurasian Curlew	Plenty of cover, e.g. upland rough pasture and semi-natural hay meadows; now rarely breeds in lowlands		
Golden Plover		Hay meadows on higher ground, early in the season and after the hay cut	
Black Grouse			Local distribution; forages for herbs and grasses in upland hay meadows (notably in the North Pennines)
Whinchat	Formerly bred on lowland downland and in hay meadows; nests in open scrub and Bracken in upland grassland habitat mosaic		

2.3 Reptiles and amphibians

Semi-natural dry grassland supports the majority of Britain's indigenous reptiles and amphibians, thus representing an important habitat for this group. Reptiles use grassland for all aspects of their life history, while amphibians reproduce in aquatic environments, using grassland for hunting, shelter and hibernation.

2.3.1 Reptiles

Semi-natural dry grassland is an important habitat for the Adder, Grass Snake, Viviparous or Common Lizard and Slow Worm, each being widely distributed in Britain (Table 2.9). All have been designated as priority species in recognition of declines resulting from habitat loss and degradation. Reptiles are ectothermic, which means that they require the warmth of the sun to raise their body temperature. This has a strong influence on their behaviour and hence their habitat requirements. Reptiles prefer sites on well-drained soils, with south-facing slopes or varied topography that provides good opportunities for basking. Conversely, they must also avoid overheating, so a good reptile habitat must have good structural diversity. Taller vegetation provides humid areas for the animals to cool down on hotter days, and also to seek cover from predators. Consequently, over-management of grassland areas, for example, by excessive grazing, can be very damaging to reptile populations. When management work is undertaken, the structural complexity of the sward should not be damaged. Hibernation usually takes place underground, for example, in Rabbit burrows, but also in above-ground structures, such as rotting logs or large grass tussocks. Maintaining a structurally diverse habitat suitable for reptiles usually requires some form of regular management, such as grazing or scrub control.

Two other reptiles are found in Britain, both of which are priority species and also European Protected Species (EPS). The Smooth Snake is exclusively a species of lowland heathland in southern England, but it is included in this account because it will hunt in dry acid grassland that is part of a lowland heathland habitat mosaic (Table 2.9). The Sand Lizard is the final member of the British reptile fauna. It is exclusively a lowland heathland inhabitant, with a very restricted distribution, and unlikely to be found in dry grassland habitats.

Viviparous Lizard (Tone Blakesley).

2.3.2 Amphibians

Semi-natural dry grassland is an important habitat for the Common Frog, Common Toad and Britain's three species of newts (Table 2.9). The Natterjack Toad is very rare, but it is also included because it can be found in acid grassland that forms part of lowland

Table 2.9 Terrestrial reptiles and amphibians found in semi-natural dry grassland habitat

	Notes on range, population and habitat
Reptiles	
Adder[a]	Widespread but patchy distribution across Britain, with extensive declines in recent decades (Edgar *et al.* 2010) due to habitat loss and land use changes. Prefers undisturbed, sunny open glades, or slopes, usually close to thicker cover. Favours chalk downland, heathland, moorland and woodland rides, on light chalk or sandy soils. Primary food is small mammals, but also takes invertebrates, reptiles and small birds.
Grass Snake[a]	Widely distributed in the lowlands of England and Wales, though patchy in northern England, with severe declines in some areas (Edgar *et al.* 2010). Highly aquatic, but uses a range of wet and dry grassland, including chalk grassland, which offers cover and opportunities for basking. Primary food is amphibians, but also takes fish and occasionally small mammals and nestling birds.
Slow Worm[a]	Widespread in Britain, particularly in southern and eastern England, though probably declining nationally through habitat loss (Baker *et al.* 2004). Found in a variety of habitats including most types of grassland. Diet includes soft-bodied invertebrates.
Viviparous or Common Lizard[a]	Widespread but patchy distribution in the British uplands and lowlands, with large declines in recent decades due to habitat loss (Edgar *et al.* 2010). Found in a variety of habitats, including most dry grasslands, with diverse topography and opportunities for basking. Feeds on invertebrates.
Smooth Snake[a,b]	Serious range decline since the mid-twentieth century through habitat loss and degradation; now restricted in southern England. Found almost exclusively on lowland heath, although hunts in acid grassland if immediately adjacent.
Amphibians	
Common Frog	Widespread in the British uplands and lowlands, but becoming much less common due to the loss of terrestrial habitat and breeding ponds. Forages exclusively on land for invertebrates in rough grassland and pasture, and a wide variety of other habitats in the vicinity of breeding ponds. Chalk grassland generally too dry to support large populations.
Common Toad[a]	Widespread in Britain, but significant declines in recent years due to the loss of terrestrial habitat and breeding ponds. Terrestrial catchments can be large; avoids grazed pasture and improved grassland, preferring to forage for invertebrates in rough grassland, scrub and woodland.
Natterjack Toad[a,b]	Serious range decline during the twentieth century through habitat loss and degradation, distribution now highly restricted in England. Predominately a species of sand dunes, saltmarsh and lowland heath (including acid grassland in the habitat mosaic).

Table 2.9 continued

	Notes on range, population and habitat
Great Crested Newt[a,b]	Lowland species with wide distribution in England, and parts of Wales and Scotland. Major decline in Britain over the last century due to loss of pond foraging habitat. Uses taller grassland and scrub in vicinity of breeding pond for foraging and shelter.
Palmate Newt	Patchy distribution in the English uplands and lowlands, more common in Wales and Scotland. Forages for invertebrates in grassland and other terrestrial habitats in the vicinity of breeding ponds.
Smooth Newt	Wide distribution in England, and parts of Wales and Scotland. Forages for invertebrates in grassland and other terrestrial habitats in the vicinity of breeding ponds.

[a]Country-level priority species.
[b]European Protected species.

heathland habitat mosaics and sand dunes. The Natterjack Toad and Great Crested Newt have EPS status, and together with Common Toad, are also priority species, in recognition of major declines in their populations resulting from habitat loss and degradation. All amphibians require water bodies for breeding, although the amount of time they spend in water and their specific requirements differ between the species. Great Crested Newts, for example, use ponds that are generally devoid of fish, which would predate their larvae. Common Toads breed in larger ponds, including slow moving rivers, usually with good vegetation cover. They will breed where fish are present, because toad tadpoles are thought to be unpalatable to fish. In contrast, Smooth and Palmate Newts breed in a wide range of ponds. All amphibians spend a considerable part of their lives on land, so the terrestrial habitat surrounding water bodies is of considerable importance. In general, the landscape within a kilometre or so of breeding ponds should comprise a mix of semi-natural pasture, with scrub, hedgerows and woodland. Structural diversity in semi-natural grassland is important for shelter, and for invertebrates for foraging amphibians.

Common Toads avoid grazed pasture and improved grassland, preferring to forage for invertebrates in rough grassland, scrub and woodland.

Management operations should avoid damaging the structural diversity; adult Common Frogs, for example, feed entirely on land.

There are many opportunities in grassland management or restoration schemes to create excellent habitat features for both reptiles and amphibians, such as the creation of cover or the provision of additional structure through grassland management. If the presence of reptiles or amphibians is suspected on a site where management or restoration operations are being planned, surveys based on current best practice should be undertaken, with a view to implementing a mitigation programme if necessary. For further information, the *Herpetofauna Workers Manual* (Gent and Gibson 2012) provides comprehensive guidance to all aspects of reptile and amphibian conservation and management, including site assessment, species translocation and the law. Natural England's *Reptiles: Guidelines for Developers* (English Nature 2004) and Amphibian and Reptile Conservation's *Reptile Habitat Management Handbook* (Edgar *et al.* 2010) should also be consulted, together with the latest environmental management guidance on the GOV.UK website (https://www.gov.uk/).

2.4 Mammals

Mammals returned to Britain after the last ice age, but flooding of the land bridge with Europe brought the migration to an end (Yalden 1999). In the post-glacial wildwood, carnivores such as Wolves and Lynx would have hunted herbivores such as Aurochs. Smaller carnivores included the Wildcat, Red Fox, Badger, Polecat and Pine Marten. The Lynx and Brown Bear survived in Britain at least until Roman times, while the Wolf finally became extinct in Britain in the seventeenth century. Of their prey, Red Deer and Roe Deer survive to the present day, despite intense hunting pressure in the Middle Ages.

Today some mammals are associated with semi-natural dry grassland (Table 2.10 and Table 2.11), but none are confined to this habitat. Very few British mammals have specialised requirements; most could be considered as generalists, although many are adapted to woodland. Two exceptions might be the Mole and Field Vole. Moles are confined to the topsoil, while field voles occupy the surface soil of the ground zone,

Hedgehogs have suffered a long-term decline in Britain.

Pine Martens forage in rough grassland close to woodland, and may be reintroduced to parts of Britain in the future.

although both species occur in a wide variety of habitats. Semi-natural grasslands are particularly important for shrews, voles and other small rodents. Most of these are not designated as priority species (with the exception of the Harvest Mouse), but they are an important source of prey for raptors, owls and carnivorous mammals, such as the Polecat, other small mustelids and the Red Fox. Mammalian carnivores that hunt in grassland tend to seek shelter and breed in woodland, scrub or hedgerows.

Semi-natural grasslands are often managed for plants and invertebrates, with less attention given to mammals. While the most serious mammalian declines in Britain are in species that use other habitats, such as the Red Squirrel and Dormouse, the Polecat is still rare in parts of Britain, the Pine Marten and Wildcat are rare, and in the case of the Wildcat, critically endangered. The Polecat and Wildcat are also threatened by hybridisation with feral animals (with ferrets and polecat-ferrets in the case of the Polecat and domestic feral cats in the case of the Wildcat), and Red Deer by hybridisation with the non-native Sika Deer.

2.4.1 Bats

Bats have a close association with woodland, which they use for both foraging and roosting. However, many species regularly use semi-natural grassland for feeding in the summer months (Table 2.12). Grazed pasture in particular may support flies and beetles associated with dung degradation, which provide an important food source for bats. A landscape that includes mature woodland, together with a mosaic of hedgerows, riparian woodland, scrub, semi-natural grassland, ponds and rivers (both slow- and fast-flowing) would suit the feeding and commuting activities of all British bats. All of these habitats can provide insect food in abundance, which is one of the major factors governing the suitability of a habitat for foraging bats. Such a landscape may once have

Table 2.10 Native mammals found in semi-natural dry grassland habitats in Britain (Harris *et al.* 1995 and cited reports (© Joint Nature Conservation Committee))

Species	Notes on range, population and habitat
Hedgehog[a]	Widespread in Britain, locally common, but long-term decline (Tracking Mammals Partnership (TMP) 2009), though more stable in recent years. Prefers woodland, hedgerows and gardens, in close proximity to grassland. Diet of invertebrates, but also eggs and chicks of ground-nesting birds. Protected under the Wildlife and Countryside Act 1981, but is a problem on offshore islands.
Common Shrew	Widespread but patchy distribution in Britain; common, but past declines due to loss of semi-natural grassland (Battersby and TMP, 2005). Found in dry grassland and a variety of other habitats. Forages for invertebrates on surface but lives in burrows. Protected under the Wildlife and Countryside Act 1981.
Pygmy Shrew	Widespread in Britain, common, but past declines due to loss of semi-natural grassland (Battersby and TMP, 2005). Forages for invertebrates on surface, making use of surface runways, but lives in burrows. Protected under the Wildlife and Countryside Act 1981.
Mole	Widespread in Britain, common. Present in most habitats where soil deep enough to burrow, thriving in pasture. Diet of earthworms predominantly, but also insect larvae in the summer, so its presence is a good indicator of soil invertebrate abundance.
Field Vole[b]	Widespread but patchy distribution, locally common. Threatened by increased grazing pressure and loss of rough grassland (Battersby and TMP, 2005). Found in undisturbed semi-natural grassland with tussocks, but also other habitats, such as scrub and woodland; uses underground network of burrows and surface runways to evade predation; nests in base of tussocks or underground. Feeds primarily on fine grasses. Conservation important as it is an important prey animal.
Wood Mouse	Widespread, very common and populations stable. Found primarily in grassland and woodland, where it lives in burrows. Less susceptible to grazing than the Field Vole. Diet includes seeds, fruits, leaves and invertebrates.
Harvest Mouse[a]	Widespread in eastern and southern England, with scattered colonies further north, locally common and possibly introduced. Declining due to changes in habitat management and agricultural methods (Battersby and TMP, 2005). Nests of woven grass above ground in dense grassland or cereals, Bramble or grassy hedgerows. Diet a mix of seed, berries and insects. IUCN Red List (Least Concern category as of September 2015).
Red Fox	Widespread in Britain, common. Found in most habitat types in Britain, often hunting in grassland. Little legal protection.
Weasel	Common and increasing in Britain (TMP 2009). Found in most habitat types in Britain where there is sufficient cover and small rodents, which is its main food. Threatened by loss of linear features, poisoning and competition with foxes.

Table 2.10 continued

Species	Notes on range, population and habitat
Stoat	Common and increasing in Britain (TMP 2009). Found in most habitat types in Britain where there is sufficient cover, hunting mainly for small mammals; animals do not like to be out in the open. Threatened by loss of linear features, poisoning and competition with foxes.
Polecat[a]	Locally common, increasing and spreading across England in recent years. Hunts in farmland, particularly in the lowlands, with abundant Rabbits. Uses Rabbit burrows as dens.
Pine Marten[a]	Locally common in parts of Scotland, very rare England and Wales. Heavily persecuted in the nineteenth century. Population trends unclear. Avoids grazed pasture but forages in rough grassland close to forests, where voles occur in high densities. Protected under the Wildlife and Countryside Act 1981.
Badger	Widespread in England and Wales but thinly distributed in Scotland, common but no clear trend in populations. Nocturnal, diet of mainly earthworms, but a variety of other foods. Protected by a number of laws.
Wildcat[a]	Confined to the north of Scotland, critically endangered. Hunts Rabbits, Hares and small mammals on the edge of moorland, in pasture, scrub and forests. IUCN Red List.
Red Deer	Scotland and parts of England, particularly the Lake District, East Anglia and South West. Common, long-term increase in numbers (TMP 2009). Grazes in woodland and moorland, but also semi-natural dry grassland. Dense populations can have severe environmental impacts. Protected under the Wildlife and Countryside Act 1981.
Roe Deer	Widespread in Scotland, northern and southern England. Common, long-term increase in numbers (TMP 2009), and spreading. Typically found in woodland, particularly at the edges, but moves out into grassland to graze. Dense populations can have severe environmental impacts. Protected under the Wildlife and Countryside Act 1981.

[a]Country-level priority species.
[b]Bank Voles are more dependent on cover, but may also frequent grassland.

Table 2.11 Introduced mammals found in semi-natural dry grassland habitats in Britain

Species	Notes on range, population and habitat
Brown Hare[a]	Widespread with the exception of north-western Scotland, common; no clear trend in numbers over recent years (TMP 2009). Diet of young grass shoots, including cereals. Little legal protection.
European Rabbit	Widespread over much of Britain, common. Significant declines in recent years (TMP 2009). Pest in agriculture and forestry, conservation threat to machair.
Fallow Deer	Locally common in parts of Britain. Popular in parks. Dense populations can have severe environmental impacts. Protected under the Wildlife and Countryside Act 1981.

[a]Country-level priority species.

Table 2.12 Bats that forage over semi-natural dry grasslands; their status, foraging habitat and food (Harris *et al.* 1995; Bat Conservation Trust; Vincent Wildlife Trust; and cited reports)

Species	Notes on range, status, habitat and diet
Greater Horseshoe Bat[a]	South Wales and the South-West, very rare and endangered (Battersby and TMP 2005). Forages over pasture and woodland. Diet predominantly beetles and moths.
Lesser Horseshoe Bat[a]	South Wales, Welsh borders (as far west as Oxfordshire) and the South-West, rare and endangered (Battersby and TMP 2005). Primarily a woodland and scrub foraging species. Diet of flies and small moths.
Natterer's Bat	Widespread, relatively scarce. Forages in woodland, wet woodland, linear woods, woodland edge, open grassland, over water; commutes/forages along hedgerows. Diet includes wide range of insects and spiders.
Daubenton's Bat	Widespread, population estimates poor. Forages along riparian woodland, over water, other woodland at certain times of year, and may forage over the periphery of dry grassland. Diet of insects with aquatic larvae.
Whiskered Bat	Widespread but local; population estimates poor. Predominantly aerial hawkers in riparian habitats, woodland, woodland edge, open country including grassland. Diet of moths, spiders and wide range of other insects.
Brandt's Bat	Common in northern and western England; population estimates poor. Believed to have similar habits to Whiskered Bat. Forages in damp upland areas but is more closely associated with woodland. Diet of moths, spiders and wide range of other insects.
Alcathoe Bat	Recently discovered in Yorkshire, Sussex and the West Midlands. Foraging habitat and diet unknown in Britain, but probably similar to Brandt's Bat and Whiskered Bat; considered to be a woodland specialist, particularly in areas with watercourses.
Serotine Bat	Appears to be declining in south-east England, with a population shift west and north, but population estimates poor. Forages in pasture, lowland parkland, woodland edge, hedgerows. Diet of beetles, moths and other insects.
Noctule Bat[a]	England and Wales, uncommon. Forages in wide range of habitats, including open grassland. Diet of beetles, flies, moths.
Leisler's Bat	England, scarce (Battersby and TMP 2005). Forages in wide range of habitats, including open grassland. Diet of flies and other small insects.
Common Pipistrelle	Widespread; population estimates poor. Forages in wide range of habitats, including grassland where linear features close by. Diet of flies and other insects.
Soprano Pipistrelle[a]	Widespread; population estimates poor. Forages along woodland edge and rides, more associated with wet areas than Common Pipistrelle and may forage over the periphery of dry grassland. Diet of flies and other insects.
Nathusius' Pipistrelle	Rare, mostly forages in richly structured forest habitat, riparian woodland, wet woodland, park landscapes (grassland). Diet includes flies.

Table 2.12 continued

Species	Notes on range, status, habitat and diet
Grey Long-eared Bat	Southern England, very rare. Gleans from foliage/other surfaces over semi-natural grassland. Diet includes moths.
Barbastelle[a]	Widespread, very rare. Forages where its prey is common; in woodland canopy, woodland edge, orchard, downland and pasture. Diet includes moths.

[a]Country-level priority species.

Pipistrelle species forage over grassland, especially when linear features such as hedgerows are close by.

been commonplace, but is less so today. Consequently, the management and restoration of semi-natural dry grassland could provide an important foraging habitat for bats and make an important contribution to bat conservation in the wider landscape.

Most bats have suffered significant declines in the last 100 years or so (Stebbings 1988; Hutson 1993; Harris *et al.* 1995; Bat Conservation Trust 2014), due in part to agricultural intensification resulting in habitat loss, including roost sites (Hutson 1993). All species and their roosts are protected by legislation and are listed in Annex IV of the EU Habitats Directive (species in need of particularly strict protection); of these, four species (including the Barbastelle) are given additional protection under Annex II. Seven British bats are country-level priority species, including several species that forage over semi-natural grassland; the Barbastelle is also on the IUCN Red List of Near Threatened Species.

2.5 Fungi

Semi-natural dry grasslands on nutrient-poor soils represent a very valuable habitat for fungi, but this is rarely considered in grassland management plans, as far less is known about the British mycota than flowering plants. Pastures are particularly important for macrofungi, as grazing (or sometimes mowing) creates a short sward, which many macrofungi appear to be associated with. Hay cutting followed by aftermath grazing – which maintains a relatively short sward – is conducive to abundant fruiting of grassland fungi (Griffths *et al.* 2012). The maintenance of a short sward after hay cutting is also compatible with maintaining the diversity of grassland plants and invertebrates. In contrast, longer swards inhibit the fruiting of macrofungi (Spooner and Roberts 2005), but the mycelium (which is very long-lived) may persist in the soil for long periods.

Large numbers of microfungi are also found in dry grassland, particularly in ungrazed meadows. Many of these species are saprobes, responsible for recycling nutrients from decaying vegetation. Others include parasites of vascular plants (which are usually minor in their effects on the host, and some are of conservation concern (Woods *et al.* 2015)) and bryophytes, and arbuscular mycorrhizae such as species of the genus *Cortinarius*.

Lichens are very colourful, and familiar to most people. They are fungi which host photosynthetic algae or cyanobacteria (sometimes both together); this is a stable, self-supporting or 'symbiotic' relationship. One often reads the phrase 'lichens and fungi' whereas the truth is that lichens are fungi, though lichenised fungi occur in several groups across the fungal kingdom (most are cup fungi (ascomycetes) but some mushroom-forming basidiomycetes are also lichenised). Lichens are not usually associated with grassland, though in low-nutrient dune grasslands they often represent a significant proportion of the ground cover. The richest communities of terricolous lichens are found on calcareous grasslands and the acidic Breckland inland dunes (SD11b). Even on semi-natural grassland, lichen-rich swards may be very localised, in areas where short turf has been disturbed to some extent, often by humans, allowing lichens to enter (Gilbert 2000).

2.5.1   Macrofungi

Most of the fungi found in dry grasslands could be considered 'grassland specialists', as relatively few of these species occur in other habitats. The macrofungi most characteristic of dry grasslands include the mushrooms and toadstools (agarics) such as the Field Mushroom, waxcaps (family Hygrophoraceae), and species of the genera *Dermoloma* and *Entoloma*. Club and coral fungi (family Clavariaceae) represent another major macrofungal group of dry grasslands; despite their unusual appearance these too are closely related to the mushroom-forming fungi. Other frequent grassland agarics include members of the genera *Conocybe*, *Mycena* and *Stropharia* genera, although they are not restricted to grassland. Some puffballs, such as the Meadow Puffball, are also found in semi-natural dry grassland. Many of these are saprotrophs, which may be found in the litter layer, soil/humus, and dung.

Fungal communities are believed to be dependent to some extent on soil characteristics, such as type, pH and mineral content, water relations, climate and environment, in a similar way to plant communities. Low fertility is particularly important. Some waxcaps such as *Hygrocybe calciphilia* and *H. spadicea*, and some *Entoloma* species are calcareous grassland specialists, while others such as *H. laeta*, *H. turunda* and *H. miniata* prefer acid grassland (Evans 2003).

Waxcaps are particularly charismatic, and are some of the best-known macrofungi of semi-natural dry grasslands. They are characterised by their thick waxy gills, moist flesh and usually brightly coloured fruiting bodies. Some, such as the Pink Waxcap have pointed 'pixie caps'. Waxcaps are easy to observe, but most are difficult to identify, and require expert knowledge. They are usually found on unimproved semi-natural grasslands that have come to be known as 'waxcap grasslands'. Up to 20 species or more may be found on the better sites, which include lowland pasture (MG5), lowland

Vermillion Waxcap growing on semi-natural limestone grassland.

calcareous grassland (CG1 and CG2) and upland acid grassland (U4), although most semi-natural dry grassland is likely to contain at least one or two species. However, even a single dressing of fertiliser may be sufficient to destroy the waxcap mycota of a site. Consequently, these fungi are considered to be good indicators of 'unimproved' semi-natural grassland in Britain, together with several other agaric genera, such as the pink-spored species of *Entoloma* (pink gills) and *Dermoloma*, fairy clubs (clavarioid fungi) and earth tongues (ascomycetes of the family Geoglossaceae). Once lost, it may take many decades before waxcaps begin to recolonise, but this can happen, as demonstrated by the diverse and sometimes rare fungal communities found on older lawns, and in churchyards and cemeteries, which have some of the best assemblages of waxcaps in Britain. A high diversity of macrofungi can also be found on some semi-improved grassland, such as Perennial Rye-grass–Crested Dog's-tail communities (MG6) (Rotheroe 2001). However, on most improved grassland and parks, the fungi present tend to be common grassland species able to tolerate higher levels of N, such as the Fairy-ring Toadstool, Parasol Ink-cap and *Panaeolina foenisecii*.

A provisional Red List for Threatened British Fungi including nearly 400 species (Evans *et al.* 2006) was followed by the publication of the Red List for *Boletaceae* (Ainsworth *et al.* 2013). Many of the species listed in the provisional Red List are found in semi-natural grassland. Some of these, including many waxcaps and other grassland species are designated as priority species. Despite the sensitivity of these species to changes in management, only a handful of sites have been designated as SSSIs for their mycological importance (Plantlife 2008). Threats to these fungi include ploughing, grassland 'improvement' through the application of fertiliser (particularly N) and changes in grazing regimes (usually a reduction leading to taller swards) (Griffith *et al.* 2004). It is important that areas of fungal diversity are recognised, and that there is improved awareness among grassland managers; this will allow these sites to be managed sympathetically.

2.6   Assessing the conservation value of a site

Biodiversity should be evaluated before any plans to change the management of a site are implemented. If species of conservation concern are using the habitat, the effects of management changes on the species present must be considered. Simple baseline surveys of biodiversity may be sufficient to establish the wildlife value of a site and its immediate surroundings and inform a future restoration strategy. For larger sites, or where there is interest in monitoring the changing status of the flora and fauna over a number of years, more sophisticated methods may be needed, and advice should be sought from professional ecologists or conservation organisations. Other sources are local naturalists, Biological Records Centres and conservation organisations for records of species of conservation concern, such as priority species, Red Data species, Nationally Scarce species or any species protected under Schedule 5 of the Wildlife and Countryside Act 1981. The National Biodiversity Network (NBN) Gateway (https://data.nbn.org.uk/) contains biodiversity data from a wide range of sources, provides national and regional species distribution maps, and allows searches for all species recorded in SSSIs and 10-km squares; a useful starting point when considering the conservation value of an area.

2.6.1 Phase 1 habitat survey

The Phase 1 habitat survey is the most widely used general survey technique. It focuses on vegetation structure, and topographic and substrate features, to identify areas of potential conservation value. Practitioners should consult the comprehensive field manual *Handbook for Phase 1 habitat survey* (JNCC 2010). Surveyors may also consult the Higher Level Stewardship (HLS) *Farm Environment Plan (FEP) Manual* (Natural England 2010a), free to download (with official updates) from the Natural England website (https://www.gov.uk/government/organisations/natural-england), which is designed to provide the guidance needed to complete a FEP.

Phase 1 habitat classification and field surveys are relatively simple; they are standardised techniques which map semi-natural communities and other wildlife habitats. Surveyors record site topography, elevation and the structural habitat category, such as grassland and heathland. Qualitative descriptions may relate, for example, to substrate type (acid, neutral or basic) or wetness (wet, dry) and to management status (e.g. improved and unimproved). Broad habitat types, such as unimproved neutral grassland and semi-improved neutral grassland, do not distinguish vegetation at the community level. Distinguishing between semi-improved and improved grassland and amenity grassland at Phase 1 level can be challenging. Phase 2 surveys focus specifically on detailed vegetation surveys of areas found to support semi-natural habitats; they usually follow the NVC, which is widely accepted as a robust classification of semi-natural vegetation in the British Isles.

2.6.2 Wildlife surveys

2.6.2.1 Vegetation

For arable land, sophisticated survey techniques may not always be necessary, unless rare or endangered plants, such as cornfield annuals, are present, or remnants of unimproved grassland remain along field margins. In other areas, such as existing grassland, an NVC survey might be appropriate. Consult the *National Vegetation Classification: Users' handbook* (Rodwell 2006), which provides details of the methodology for sampling and describing vegetation communities. Refer to the keys given in Rodwell (1991 *et seq.*) to determine the most appropriate NVC community classification. Computer programs are also available to calculate the 'goodness of fit' of data collected from quadrats to the expected species composition of semi-natural woodland communities and subcommunities recognised by the NVC. One such program, TABLEFIT (Hill 1996), is available freely from the Centre for Ecology & Hydrology, as is another in the Modular Analysis of Vegetation Information System (MAVIS) software (Centre for Ecology & Hydrology 2015).

2.6.2.2 Birds

Bird surveys may be necessary on known sites, or those suspected to support scarce breeding birds, such as Skylark, Corn Bunting, Tree Sparrow and Grey Partridge. Grassland that might support ground-nesting birds, including breeding waders such as the Northern Lapwing, should also be surveyed. In exceptional cases, species on Schedule 1 of the Wildlife and Countryside Act 1981 or Annex 1 of the European Commission (EC) Birds Directive may be found. A winter bird survey should also be considered if the area might include important feeding or roosting habitat for waterfowl or raptors.

Bird surveyors should be familiar with UK legislation, which protects all wild birds, their nests and eggs, with limited exceptions. Nesting is considered to have started as soon as nest building starts. Some rare species are given special protection; in Britain it is a criminal offence to disturb, at or near the nest, a species on Schedule 1 of the Act. In Scotland it is also an offence to disturb Capercaillie and Ruff at their leks. For more detailed information, it is advisable to consult the Act itself.

Various survey methods might be considered (Hill *et al.* 2005), although in many cases, surveys may be based on the BTO's Breeding Bird Survey. This is a national project aimed at monitoring changes in breeding populations of widespread bird species across the UK. Typically, two visits are made to a site: the first between early April and mid-May, to coincide with the main period of breeding activity of resident birds, although this may be later in the north; the second between mid-May and late June, during the main breeding period for migrant birds, which again will vary depending on the location. Further details on this and other survey methods can be obtained from the BTO website or from *Common Standards Monitoring Guidance for Birds* (JNCC 2004) which can be downloaded from the JNCC website (http://jncc.defra.gov.uk/pdf/CSM_birds_incadditionalinfo.pdf). Additional information on rare and declining farmland birds can be found on the RSPB *Advice for Farmers* web pages.

2.6.2.3 Invertebrates

Invertebrate surveys can be both expensive and time-consuming; an alternative is the identification of habitats of high invertebrate biodiversity value. Natural England's HLS *Farm Environment Plan (FEP) Manual* (Natural England 2010a) provides useful guidance. If four or more microhabitat features (summarised below) are present, the site is likely to have a high potential for invertebrates; seven or more indicate that it is a high biodiversity grassland 'habitat for invertebrates':

- variable topography or near vertical areas of exposed soil;
- free-draining light soils;
- some species-rich semi-natural vegetation;
- frequent patches of bare ground (0.01–0.1 m^2);
- anthills;

- patches of scrub, ancient trees, natural springs and flushes, or other water bodies;
- variable vegetation structure;
- abundant seed or flower production throughout the year;
- dry stone walls;
- fibrous dung attractive to beetles.

In addition, specialists could be engaged to survey selected groups, such as butterflies or hoverflies. These can act as 'indicators' of invertebrate diversity and habitat health. Butterflies are highly visible and most are relatively easy to identify. The requirement of adults for abundant flowers for nectaring can be indicative of potential feeding sources for a range of other invertebrates. The diversity of species and the presence of colonies will also give some indication about the quality of existing vegetation in surrounding open areas. Details of the methodology for transect recording can be found on the UK Butterfly Monitoring Scheme website (http://www.ukbms.org/).

2.6.2.4   Mammals

If species of conservation concern are known to be present on or adjacent to the site, then surveys should be undertaken. These include priority species and species protected under Schedule 5 or 6 of the Wildlife and Countryside Act 1981. If protected species are suspected, then advice from the statutory conservation agency must be sought prior to undertaking surveys appropriate to the species concerned (for a summary, see Institute of Environmental Assessment, 1995). The services of a professional ecologist will almost certainly be required if surveys are necessary.

2.6.2.5   Amphibians and reptiles

All native reptiles and amphibians are protected to varying degrees, and the majority use semi-natural dry grassland. Four species have European Protected Species status (Section 2.3), of which the Great Crested Newt is the species most likely to be encountered. For this widely distributed species, the NBN Gateway is a useful starting point to gather information on local populations, although absence of records does not mean that the species is not present in the area. Local Biological Records Centres and local amphibian and reptile groups are also useful sources of information. Newts use the area surrounding breeding ponds for foraging, dispersal, resting and hibernation, so a survey may need to be commissioned if a pond used by Great Crested Newts is within 500 m of a site.

The Adder, Grass Snake, Slow Worm and Viviparous Lizard are found in a range of habitats, including grassland and brownfield sites, although they are less abundant in intensively farmed landscapes and upland areas. Activities that could harm reptiles include land clearance, cutting vegetation to a low height, removal of woodpiles or other debris and unsympathetic grazing. Where their presence is suspected, and they are likely to suffer from operations involved with site preparation for grassland creation or restoration, surveys based on current best practice should be undertaken by professional ecologists, with a view to implementing a reptile mitigation programme if necessary. Reptile activity is highly seasonal; animals hibernate between October and March, and their activity during the summer months is dependent on the weather. Reptiles also become less active (aestivate) in prolonged periods of hot weather. The *Herpetofauna Workers Manual* (Gent and Gibson 2012) provides comprehensive guidance to all aspects of reptile and amphibian conservation and management, including site assessment, species translocation and the law. Natural England's *Reptiles: Guidelines for Developers* (English Nature 2004) should also be consulted.

2.6.2.6 *European Protected species*

European Protected Species are protected under Annex IV of the European Council Directive on the Conservation of Natural Habitats and Wild Fauna and Flora (92/43/EEC), implemented in Britain through the Habitat Regulations 1994 (amended many times since). European Protected Species have full protection under the Wildlife and Countryside Act 1981 (as amended); EPS which use dry grassland include: all bats; Wildcat; Great Crested Newt; Natterjack Toad; Large Blue butterfly; and several rare plants including Early Gentian and Lady's-slipper. A project which 'carries out operations on, over or under land', or involves a 'material change in use of land' that might adversely affect an EPS requires a special Development Licence, issued by the Department for Environment, Food & Rural Affairs (Defra).

3. Semi-natural dry grassland management

This chapter provides site managers with a range of information on the management of semi-natural dry grassland, including grazing, cutting, weed control, scrub management and fertiliser application. Although the discussion relates specifically to semi-natural grassland, the information provided should also be useful for semi-improved grassland management.

Management plans for semi-natural grassland should focus on the basic techniques that have given rise to the plant and animal communities characteristic of a particular site. Traditionally, grasslands were maintained either as pastures that used grazing animals exclusively, or meadows (removing a summer hay crop to produce winter feed for livestock, usually accompanied by grazing periods in the spring, autumn or winter). Plant communities that evolved under continuous grazing differed substantially from those associated with cutting and aftermath grazing. If management has changed in the recent past, for example, from hay meadow to pasture, or grazing has been abandoned, the conservation interest is likely to have been affected. In such cases, careful evaluation of the grassland community is important, to see whether a return to 'historical' management practices would benefit wildlife. Farm records, such as grazing regimes, farmyard manure applications, plant species diversity in the sward, and so on might be useful in planning management strategies for the future. The restoration of neglected semi-natural grassland, semi-improved grassland, and the creation of new grassland habitats are considered in later chapters.

3.1 Grazing

Grazing results in the gradual removal of plant material, allowing invertebrates to move and persist in the sward, in contrast to the immediate effect of mowing. It also provides habitat for dung communities. Grazing is often chosen as the best management option for semi-natural dry grassland, as it can maintain diversity in the sward and provide bare ground for regeneration. It is also possible to carefully manage grazing intensity to ensure a site remains in pristine condition, for example, by rotational grazing, or the use of temporary fencing to restrict grazing to certain parts of a site.

Livestock selectively graze, generally preferring younger foliage and species they find more palatable. The ability of highly competitive species to dominate is usually limited. Cattle, for example, may graze preferentially on finer grasses, but will turn to coarse and tussocky grasses when stocking densities are high, or more palatable material becomes scarce. Some less palatable species may not be grazed at all. Grazing also helps to prevent scrub encroachment. Trampling has beneficial effects on the composition and structure of grassland communities; for example, bare patches allow seeds to germinate and provide habitat for some invertebrates; layers of thatch and leaf litter are broken up. However, bare patches can also provide regeneration niches for thistles and other noxious weeds. In wetter areas, or along regular pathways, poaching can occasionally lead to soil compaction and impede drainage. Grazing generally results in the development of an uneven sward, with the community composition and species abundance reflecting the livestock species, breed and stocking density, as well as soil moisture, season and grazing intensity. Cattle, sheep, horses and ponies are most commonly used to manage semi-natural grassland.

In the past, the dispersal and exchange of propagules between areas of semi-natural grassland was effectively achieved by wandering herds of grazing animals. The historic grazing practice of moving animals from hill pastures to valley paddocks overnight may also have once been effective by impoverishing hill land while manuring the valley fields. Nowadays, animals tend to be left on semi-natural grassland, with the possible build up of nutrients in areas where animals prefer to defecate, such as paths and laying up areas. However, on some modern farms, grazing can actually cause nutrient enrichment, if stock are alternated between semi-natural grasslands and nutrient-rich improved pastures, or if supplementary feeding is given in winter.

Undergrazing is now a key concern for semi-natural dry grassland in the lowlands. In recent years, decreasing grazing pressure, along with other factors such as increasing fertility and habitat fragmentation have contributed to decreasing species richness. Undergrazing is also a concern in the uplands, where land abandonment is becoming an issue. Overgrazing due to increased numbers of sheep is also a major concern for a large number of SSSIs in the uplands (English Nature 2005). Overstocking has led to the replacement of some upland plant communities with close-cropped uniform grassland; upland calcareous grassland, for example, may be more palatable than adjacent heaths or more acidic grassland. Upland hay meadows are also vulnerable to overgrazing, causing changes in community composition and structure.

3.1.1 Livestock species and breeds

Choice of animals depends to some extent on the quality of grazing; nutritionally poor grassland, for example, requires animal breeds able to thrive in such conditions. Many modern breeds of commercial livestock have been selected for grazing high-quality improved pasture in Britain. These animals may not do well on poor-quality grassland. Conservation graziers often use more traditional, hardy breeds, which thrive on semi-natural grassland. A grazier's local knowledge and experience can be invaluable in selecting appropriate breeds. Alternatively, consider Stock Keep, a database matching livestock to grazing across Britain (http://www.stockkeep.co.uk/); or contact the Grazing Animals Project (GAP), part of the Rare Breeds Survival Trust (RBST), who maintain a register of stock-keepers and their animals (http://www.grazinganimalsproject.org.uk/). The GAP website also includes useful leaflets on topics such as cattle handling, animal welfare and equine handling.

The GAP's free online services for registered users include:

- Stock Keep: it links stockholders seeking grazing to those who require livestock for grazing;
- Ready Reckoner: it allows graziers to calculate their grazing system budgets easily.

All species graze selectively, and their effects on the sward can be difficult to predict (Table 3.1). Species preference may vary from one year to the next, even with animals of the same breed. Using a variety of species and breeds might result in the best grazing outcome, resulting in a range of sward types and habitat diversity, but in many cases this will be impractical. And as sward height is reduced, and preferred species are heavily grazed, graziers must monitor the welfare of their animals, irrespective of breed.

Table 3.1 Summary of grazing characteristics of cattle, sheep, horses and ponies on semi-natural dry grassland

Characteristics	Cattle	Horses and ponies	Sheep
Feeding method	Bulk grazer, tears off vegetation, maintains longer swards (typically 5 cm) Ruminant	Bulk grazer, also selective, nips vegetation very close to the ground (swards typically 2 cm) Non-ruminant	Selectively graze vegetation close to the ground (minimum sward height 3 cm) Non-ruminant
Diet	Less variable	Variable	Variable
Grazing patterns	Use tongues to bite and pull vegetation into their mouths; unable to manipulate vegetation like sheep	Forward pointing teeth result in grazing like Rabbits; very selective, but do not manipulate vegetation or uproot plants	Thin, mobile lips, able to manipulate vegetation and push plants aside, and also uproot plants
Species selectivity	Wide mouths results in low selectivity; prefer grasses and forbs	High, prefer grasses, favouring the development of herb-rich swards; some breeds eat Bracken in late summer	High, often target flowering plants
Selectivity of vegetative structures	Avoid flowers, grass stems eaten	Some flowers and grass stems are eaten	Flowers and grass stems are eaten
Dead plant material and litter	Consume	Consume	Avoid
Coarse, tall and tussocky grasses	Graze	Graze, but some areas become rank grassland 'latrines'	Avoid, often trampled
Browsing trees and shrubs[a]	Sometimes browse trees and shrubs	Browse a range of trees and shrubs	Frequently browse trees and shrubs
Browsing brambles	May browse in winter	Native ponies will browse	Browse in winter

[a]See also Table 3.7.

3.1.2 Cattle

Cattle use their tongues to pull tufts of vegetation into their mouth, which they then tear off. This results in longer swards than grassland grazed by sheep and horses, with tussocks that are good for invertebrates and small mammals. Cattle are sometimes preferred for semi-natural dry grassland because their wide mouths mean that they are less selective than other animals. They are particularly suited to summer grazing because they do not target flower heads; this supports species diversity in the sward. Cattle avoid grazing within 20 cm or so of the edge of dung, which also supports structural variation in the sward. They can effectively control tall, coarse grasses and areas of rank grassland through grazing and trampling and they can also control invading scrub, particularly pioneer willows. The animals tend to tear leaves and shoots, which can have a significant impact on young saplings. They may be less effective in controlling thorns such as Hawthorn. Some species of rushes are grazed, and even Wood Small-reed may be grazed, particularly in the winter months. Their heavier weight causes more substantial poaching than sheep, offering more opportunities for regeneration, although this can cause problems in extensive periods of wet weather.

The behaviour of cattle varies between age groups, as well as breeds. Younger stock, suckler cows or store stock are most commonly used to graze semi-natural dry grasslands. Dairying is usually impractical, except for the smaller breeds such as the Jersey, due primarily to the higher nutritional requirements of the animals. Most conservation grazing therefore uses beef or dual-purpose cattle, often those breeds that have not been improved for the market (Table 3.2). These include: upland beef cattle such as Highland, Welsh Black, Galloway and Beef Shorthorn; lowland breeds such as Traditional Hereford, Sussex and White Park; and dual-purpose cattle such as British White, Red Poll and Shetland. More details can be found in the *Breed Profiles Handbook:*

Table 3.2 Summary of the characteristics of cattle for conservation grazing (after GAP 2001a; © Rare Breeds Survival Trust/Grazing Animals Partnership)

Breed of cattle	Suitability for conservation grazing
Upland beef	Hardy and thrifty breeds More limited damage to sensitive swards due to small to medium size Can be difficult to handle Slow-growing breed Carcass quality moderate to good
Lowland beef	Moderately hardy and thrifty More limited damage to sensitive swards due to medium size (UK breeds) Placid, easier to handle than upland beef (UK breeds) Moderately fast-growing and early-maturing Carcass quality good
Lowland dual purpose	Moderately thrifty and hardy Tend to have more milk for a growing calf Suitable for conservation grazing Adapted to handling, placid Moderate growth rate, with little supplementary feeding Carcass quality moderate to good

British Whites (rare breed) – a lowland dual purpose breed, winter grazing on calcareous grassland on the North Downs.

Beef Shorthorn, a traditional native breed, grazing year-round in neutral meadows in Hainault Forest, Essex.

A Guide to the Selection of Livestock Breeds for Grazing Wildlife Sites (GAP 2001a), which can be downloaded from the GAP website.

3.1.3 Sheep

Sheep have thin, mobile lips that enable them to selectively graze much closer to the ground than cattle, potentially creating a more uniform, shorter sward. They are highly selective, and will target flowers and buds of a range of herbs, which can affect species diversity if they are present in the summer months, or for long periods. Uniform short swards, resulting from high stocking densities, are poor for most invertebrates and can cause difficulties for ground-nesting birds. However, lower stocking densities are likely to result in much more structural variation in the sward. Sheep are generally less able to cope with taller, ranker grassland than cattle, so they will be less effective on sites that have been neglected. Poaching is less of a problem than with cattle because of their relatively small feet and light weight, although they will create some bare ground and cause compaction. Sheep browse on trees and shrubs, although this is dependent to some extent on breed (Section 3.4). They can be used in summer scrub management programmes, but there is a risk of overgrazing vulnerable herb communities.

Browsing and grazing behaviour varies with breed and age. Three categories of sheep are generally considered most suitable for conservation grazing: upland breeds, such as the Beulah Speckled Face; primitive breeds, such as the Shetland; and hill breeds, such as the Herdwick (Table 3.3). Detailed profiles of these, and many more breeds can be found in *The Breed Profiles Handbook* (GAP 2001a). Southdowns may also be used, although other lowland breeds are more suited to improved or more fertile sites, and are less likely to browse.

Sheep are especially vulnerable to dogs, and may be unsuitable for sites that are heavily used by dog walkers.

Table 3.3 Summary of the characteristics of sheep breeds for conservation grazing (after GAP 2001a; © Rare Breeds Survival Trust/Grazing Animals Partnership)

Breed of sheep	Suitability for conservation grazing
Hill (mountain)	Hardy to extreme weather Thrive on poor quality vegetation Browsing habits help to control scrub Require less assistance at lambing time Slow-maturing Moderate carcass size
Upland	Most are hardy Graze on semi-natural grassland and coarser vegetation More limited browsing than hill breeds Require less assistance at lambing time Moderate- to slow-maturing Moderate to good carcass size
Primitive	Hardy to extreme weather Thrive on poor quality vegetation Requirement for browsing, so good for invasive scrub control Unlikely to have problems at lambing Slow-maturing Small carcass size
Lowland	Less hardy, particularly heavy breeds Require more fertile semi-natural or improved grassland, and winter food supplements, but some breeds do well on rough downland (e.g. Southdown) No significant browsing Ewes often require assistance during lambing Moderate- to fast-maturing Good carcass size

Southdown sheep (rare breed) are used for conservation grazing on lowland semi-natural dry grassland and rough downland.

Badger Face Welsh Mountain sheep (rare breed) are a hardy upland sheep of the Welsh mountains (Tone Blakesley).

3.1.4 Horses and ponies

Horses and ponies are bulk grazers like cattle, but their teeth enable them to be selective, and to graze very close to the ground. They can deal effectively with rank vegetation, particularly if used in combination with cattle. They also tend to be used in areas where: horse grazing is traditional; grazing is too poor to support commercial cattle or sheep grazing; a mosaic of habitat is required to support a wide range of important invertebrates; or short vegetation is required. However, if stocking density is not carefully controlled, horses can cause serious damage to semi-natural grassland. Overgrazing results in: patches of grassland with a very tight sward or 'lawns'; large patches of bare ground which are ideal for Common Ragwort establishment (Section 3.3.3); and 'latrine' areas of tall, rank grasses. Grazing with other species can be considered in this case. Poaching can also be a significant problem in grassland overgrazed by horses, and supplementary winter-feed may cause eutrophication. Grazing with horses is less common in upland areas.

Table 3.4 Summary of the characteristics of equines for conservation grazing (after GAP 2001a; © Rare Breeds Survival Trust/Grazing Animals Partnership)

Equine type	Suitability for conservation grazing
Native ponies	Hardy, adapted to difficult environmental conditions May require supplements to avoid mineral deficiency when grazed year-round Adapted to range of food types
Non-native primitive breeds	Hardy, adapted to difficult environmental conditions Show no signs of requiring supplements for mineral deficiency Adapted to range of food types
Domesticated horses	Best suited to meadows and some calcareous grassland Less hardy than native and primitive breeds Thin-skinned horses susceptible to insects
Donkeys	Not usually hardy May require supplements to avoid mineral deficiency when grazed year-round Adaptable to range of food

Four categories of equines may be considered for conservation grazing: native ponies, such as Shetland and New Forest; non-native primitive breeds, such as Konik; domesticated horses; and donkeys (Table 3.4). Native ponies and primitive horses are particularly hardy, and can be used in both the management of semi-natural grassland and the restoration of neglected areas, and will trample and graze on coarse grasses. Some breeds also graze on Bracken later in the summer and early autumn, when toxicity levels are lower, but they are reported to require grass to avoid being poisoned (Oates and Bullock 1997). Some will also push through areas of scrub, and browse on the shoots of trees and shrubs, helping to control invasive scrub. Detailed profiles of these, and many more breeds can be found in *The Breed Profiles Handbook* (GAP 2001a). Surrey County Council's *Horse Pasture Management Project* (http://www.surreycc.gov.uk/environment-housing-and-planning/countryside/looking-after-the-countryside/countryside-advice/horse-care-and-pasture-management) has information that is more generally applicable across Britain.

Exmoor ponies (left) and Fell ponies (right) are highly effective for conservation grazing; the Exmoor ponies are owned by Moorland Mousie Trust (© Juliet Rogers and Nicola Evans).

3.1.5 Stocking density

A grazing management plan should prescribe the species, breed, age and the number of animals required to achieve the desired effect on vegetation structure and species diversity. The effects of trampling, poaching and dunging must also be considered. Discussions of conservation grazing sometimes refer to livestock units (LSUs), which use one dairy cow as the base unit. Graziers use LSU coefficients to calculate the capacity of grassland in terms of the number of animals it will support, and when supplementary feeding might be necessary.

Unfortunately there is no easy formula for determining stocking densities, and land managers may need to rely on the experience of a grazier and careful monitoring of the effects of grazing on the structure and composition of the sward, adjusting numbers of animals accordingly. In contrast to commercial agricultural systems, stocking densities will generally be lower than the carrying capacity of the site, to preserve the characteristics of the grassland community. Some of the more important factors to consider are listed in Table 3.5, although these are by no means

Table 3.5 Factors to consider when determining stocking density

Factor	Considerations
Site fertility	Less fertile sites will support fewer animals; on more fertile sites, summer grazing may be essential to control growth and maintain structural variability.
Soils	Semi-natural grassland with light, well-drained soils, such as calcareous soils, may support fewer animals, particularly in the summer months; faster growth rates on neutral grassland might require higher stocking densities to maintain structural variation in the summer.
Vegetation	Species composition and community type are an important consideration in determining numbers, whether other species/breeds might be more appropriate, and what seasons to graze.
Productivity	Light grazing may not keep up with productivity in the summer months, particularly on more fertile soils, but animals may 'catch up' later in the year.
Species	LSUs differ between species, e.g. a 1 yearling beef cow may be equivalent to four or five adult ewes.
Duration	Grazing all year or seasonally; grazing over shorter periods requires higher stocking densities (e.g. twice the number of LSUs can be grazed for half as long on a given area) but this can be highly detrimental, particularly for invertebrate communities; consider also that plant growth in summer is much higher than in winter.
Accessibility	If the site has variable vegetation communities, consider how different areas support animals, and whether all are equally accessible; also note that the placement of water points will affect grazing and poaching density, as animals congregate at watering points, especially in the summer.
Other grazers	Consider grazing by wild herbivores, such as deer and Rabbits.
Flexibility	Is it possible to increase grazing intensity at certain times or move animals to lay back areas at others?

LSU, livestock unit.

prescriptive. If the stocking density is inappropriate for high-quality semi-natural grassland, then the habitat may be adversely affected. High stocking densities may quickly reduce tall, rank vegetation, but this can be very damaging for invertebrates and ground-nesting birds. If supplementary feeding is required, this is best provided away from the core conservation area if possible, due to the likely effects of poaching and eutrophication. Remember also that inappropriately sourced hay might inadvertently introduce unwanted species to a site. Further information on sustainable management schemes can be obtained from Defra (https://www.gov.uk/guidance/grazing-and-pasture-sustainable-management-schemes).

Grazing pressure is also determined by the duration of grazing, in addition to the number of animals. Increasing the stocking density by a factor of two, for example, might halve the duration of grazing (Figure 3.1). However, duration will also be influenced by the season and plant growth rates, the desired amount of defoliation and the sensitivity of plant and animal communities. It is therefore important to consider any deleterious effects of rapid defoliation on grassland communities (Table 3.5). In many instances, low stocking densities that gradually change the vegetation structure are more beneficial for vulnerable invertebrate communities. In the summer months, there may be a temptation to increase stocking rates during periods of more rapid plant growth, but animals are likely to compensate for this as growth rates decline in the autumn.

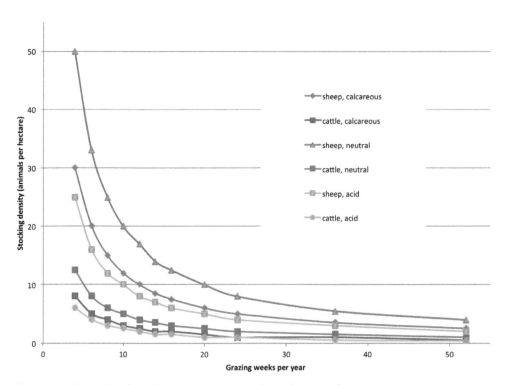

Figure 3.1 Example of stocking densities (animals per hectare) for conservation management of major lowland grassland types, based on medium-sized animals (after Crofts and Jefferson 1999; © Natural England).

3.1.6 Timing and duration of grazing

Graziers may be keen to turn animals out as early as possible in the spring, but this may not always suit conservation interests on semi-natural dry grassland. Ground-nesting birds can delay conservation grazing, for example, breeding Stone-curlew on the chalk downland of Salisbury Plain. In some parts of the uplands, grazing regimes may also need to be modified to accommodate breeding waders and species, such as the Skylark, Wheatear, Yellow Wagtail and Meadow Pipit. In general, conservation managers will need to consider how much grazing is required, how long a site is likely to sustain a herd and the sensitivity of different communities and individual species (plant and animal). Generic grazing advice may need to be modified to suit the requirements of endangered species most likely to benefit on a particular site, without compromising other threatened species. This can be problematic, as some species that require grazing, such as Marsh Fritillary butterflies, can be threatened by both over- and undergrazing.

Grazing regimes are generally seasonal or rotational, but in some instances can be continuous. Seasonal grazing is the main form of grazing on lowland semi-natural dry grassland. Each season carries its own advantages and disadvantages, and implications for wildlife (Table 3.6). Aesthetically, many people like to see ungrazed summer grassland, with orchids and tall flowering herbs, and abundant insects such as butterflies, bees and hoverflies. But this may also allow scrub and coarser grasses to gradually dominate a

Table 3.6 Effects of grazing semi-natural dry grassland in particular seasons

Season	Advantages	Disadvantages
Spring (April–May)	Useful for controlling unpalatable dominant species such as Tor-grass Control growth of scrub seedlings where scrub invasion is problematic	Avoid grazing grassland with early flowering rarities, such as Pasque-flower and Green-winged Orchid Yellow-rattle susceptible to heavy spring grazing Heavy grazing can damage invertebrate populations and cause local extinctions. Risk of trampling ground-nesting birds Can inhibit flowering of plants whose seeds are eaten by birds
Summer (May–September)	Sward productivity at its highest Traditionally practised on lowland neutral pasture Young shoots of woody species, such as willows, Sycamore and Ash palatable Control growth of competitive species Drier sites suffer less poaching Faster microbial activity reduces nutrient build-up from dung in soil In some areas, produces shorter swards favoured by wintering wildfowl	If stocking density is too high, adversely affects species requiring flowers, fruits and seeds Some invertebrates may be lost from sites with heavily grazed short summer swards Annuals may be lost from a site with repeated heavy summer grazing having a high impact on seed set Weather conditions need to be carefully monitored – in drought summers, animals may need to be moved

Table 3.6 continued

Season	Advantages	Disadvantages
Autumn (September–October)	Most plants have flowered and set seed Invertebrates least prone to disturbance Animals may assist seed dispersal and establishment Traditional, hardy breeds may do well on vegetation of lower nutritional value	Palatability is lower, animals may be forced to graze less palatable species, but these may be rejected, lowering grassland quality Woody species are less palatable Late flowering species such as Devil's-bit Scabious are more vulnerable.
Winter (October – April)	Trampling can disturb leaf litter and create regeneration niches for spring-germinating annuals Sheep may cause less damage than heavier animals, though cattle and horses more useful for breaking up scrubby sites Less damage to most invertebrates and dormant herbs Particularly suited to sites with light, well-drained soils, such as calcareous grassland	On wetter ground, reduce stocking density to avoid damage by poaching with the risk of invasive weed regeneration Consider condition of animals (traditional, hardy breeds may fare better) Supplementary feeding may be necessary Unlikely to be sufficient to control scrub growth and the development of rank grassland Animals may need to be removed when excess growth has been grazed
Year-round	Simpler to manage Practised over large areas of upland grassland Lowland parks and small horse paddocks Benefits dung fauna	Moderate to high stocking density can be very damaging, particularly on small sites Seasonal disadvantages can be problematic in year-round grazing if stocking densities are inappropriate

site. Grazing through the spring and summer months may require careful adjustment of stocking levels to maintain the desired outcome, which might be structural diversity and healthy populations of flowering plants. Autumn grazing has the particular advantage of following flowering and seed set, but animals will have less palatable and less nutritious herbage to graze. Consequently, some plant material will be left uneaten, and trampled. On larger sites stock could be moved around on rotation, using temporary fencing. For smaller sites, animals could be moved from one site to another. However, care should be taken not to introduce too many animals to a relatively small area too quickly, which might cause serious damage to invertebrate populations through sudden changes in the vegetation structure.

In winter, low stocking rates will ensure structural diversity is maintained, and some tall seed heads survive. In some cases, grazing may be undertaken in different seasons on the same site. For example, spring may be combined with autumn grazing, or late summer grazing with winter grazing. Continuous low-intensity grazing can also maintain diverse swards and good structural diversity, but great care has to be taken to avoid overgrazing during the flowering period, thus preventing seed set. Continuous grazing is usually practised in upland areas or larger sites in the lowlands, using stocking levels

Summer grazing with cattle on species-rich calcareous grassland at Wye National Nature Reserve (left), Southwick Hill (middle) and Park Gate (right) has resulted in a very short sward, with flower stems restricted to occasional inaccessible hollows.

just sufficient to maintain a varied structure. Without careful monitoring, continuous grazing might represent too great a risk for semi-natural dry grassland in many lowland situations.

3.1.7 Animal management

Cost is as important in conservation management as in commercial farming. However, cost implications can be more difficult to manage for conservation grazing, because the operation cannot be judged purely as a profit or loss exercise. The success of conservation grazing is dependent on the outcome for the habitat, in addition to the financial aspects of stock production. Pursuit of profits can compromise habitat management in the short term, while poor financial management might result in grazing becoming non-viable, compromising habitat management in the longer term.

For individuals and conservation organisations contemplating the purchase of animals, there are many factors to be carefully considered. There are considerable advantages to owning your own stock: conservation can be the main priority for grazing; animals can graze less productive semi-natural grassland for as long as required, with less pressure to move them to more nutritious swards or provide supplementary food; profits can be made in certain circumstances; and the purchase of traditional, rare breeds can enhance the work of the organisation. On the other hand, considerations conservation managers might find challenging, include:

- Purchasing animals and the legal considerations for moving stock need to be considered.
- Managing animals requires considerable time and expertise.
- Labour requirements can be high, for example, shepherding.
- Grazing management can be quite complicated if animals are moved from site to site.
- Transportation is required.
- Handling areas and buildings will be required.
- Health problems, such as fly strike of sheep or parasites of sheep and cattle, can be costly to treat in terms of veterinary bills and time.
- Some health problems are preventable; most sheep keepers now use pour-on products rather than dips.
- Stock ownership can tie up large amounts of capital.
- Returns on investment may not achieve the desired levels.
- Public liability for damage or injury caused by animals needs to considered.
- Purchase or rental of lay-back land, when animals are not required on the conservation grassland, may be required.

Garden staff at Wakehurst Place had to learn about animal husbandry to manage a flock of Southdown sheep brought in to graze restored grassland around the Millennium Seed Bank.

An alternative approach is to find a good local grazier who will rent conservation grassland for grazing, particularly if traditional or rare breeds are available. Conservation sites might be eligible for agri-environment schemes, which might offset any costs to the conservation site manager's budget. This should minimise the input of resources and expertise required, while maximising the management objectives for semi-natural grassland. The GAP register of stock-keepers and their animals, managed by the RBST may be helpful. Also contact other conservation organisations who lease land to farmers, or who might have purchased their own animals; some of these organisations might be interested in forming a grazing partnership. Details of existing partnerships can be found on the GAP website. Farmers who have not leased conservation grassland before may be unfamiliar with the constraints of conservation grazing, and their priorities for maximising profits from stock management may be quite different to those of the conservation site manager. Consequently it is important to maintain close contact with the grazier to ensure that conservation objectives are properly met. A written licence/ agreement is important, particularly with respect to animal welfare. This might include:

- Clauses confirming responsibility for the health and welfare of the animals to the grazier. (Consider an Animal Health Plan, as outlined in the GAP information leaflet *Animal Health Plans* (Gap 2007a).);
- A named individual responsible for the animals (including daily welfare checks) who is familiar with Defra's *Code of Recommendations for the Welfare of Livestock*;
- Clauses relating to fees and payments;
- Details of public access to sites being grazed. There is a responsibility to people using a site with public access under the Countryside and Rights of Way Act 2000, and to sites crossed by public rights of way (PROW);
- Confirmation that the grazier follows the Welfare of Farmed Animals (England) Regulations 2000, or the equivalents in Wales/Scotland.

Owners and graziers must also satisfy themselves that their stock is not exposed to the risk of poisonous plants such as Common Ragwort (Section 3.3.3). For additional daily stock checks, some organisations use members of the local community or 'lookers'. GAP's leaflet *Grazing Stock on Sites with Public Access* (GAP 2007b; http://www.grazinganimalsproject. org.uk/stock_management.html) includes an example of a Stock Checkers Report Form and a more detailed Stock Checking Procedure, and the RBST runs courses to train potential stock checkers. Owners or registered keepers of animals should also be responsible for livestock movement, disease and biosecurity. More information on animal welfare and the completion of a livestock risk assessment can be found in *A Guide to Animal Welfare in Nature Conservation Grazing* (GAP 2001b). If you are considering reintroducing grazing to common land, consult *Grazing Our Commons* (Natural England 2010b).

Irrespective of ownership, there are a number of infrastructure considerations common to most sites, which should be discussed with the grazier before any animals are introduced. These include:

- Fencing – permanent fencing will usually be required to keep animals in the designated areas, avoiding sensitive areas if possible, as poaching can be particularly heavy along fence lines.
- Temporary or permanent handling facilities – important when animals need to be confined in a safe, secure environment for work, such as drenching sheep for worms or trimming their feet; temporary hurdles can suffice for sheep, but more robust structures will be required for larger animals. (See GAP leaflets *Cattle Handling Facilities* (GAP 2007c) and *Equine Handling Facilities* (GAP 2007d).)
- Shelter – some animals require shelter, although in some circumstances trees can naturally provide this.
- Water – all animals need access to water, so a water trough is essential if there is no natural source.
- Designated area for supplementary feeding may be needed, away from more sensitive grassland communities.

In addition, it is important to consider whether a site is heavily used by dogs. If so, there is a risk to sheep from dogs off-lead, and dog walkers can be worried by cattle and horses. Signs and advice for dog walkers should seriously be considered. (See GAP leaflet *Dogs and Grazing* (GAP 2007e), which raises awareness of the dangers to livestock.)

3.1.8   Grazing by Rabbits

Where grazing has been withdrawn from lowland semi-natural grassland, usually for economic reasons, Rabbits may be the only way in which the sward is controlled. In some sites, Rabbits may have been responsible for preventing serious declines in habitat quality, particularly on some semi-natural acid and calcareous grassland communities. Where numbers have fallen, particularly since the outbreak of myxomatosis in the 1950s, sites have suffered from the development of rank grassland and scrub encroachment. In contrast, some sites, particularly lowland calcareous grassland have been damaged as Rabbit numbers recovered, with serious damage done to plant and animal communities. A good example is the semi-natural chalk grassland at Seven Barrows described in Box 5.1. In any stock grazing exercise, the impact of Rabbit grazing should always be a consideration, and may require modifications to grazing regimes if Rabbit numbers cannot be controlled, which is often the case. Coupled with an annual cut, Rabbits may be able to maintain some interest in sward diversity, although controlled grazing by livestock is much more desirable.

BOX 3.1 Case study: Upland hay meadow restoration through the Yorkshire Dales Millennium Trust's Hay Time Project

Topic: Restoring upland hay meadows
Location: Yorkshire Dales National Park (Yorkshire Dales Millennium Trust)
Area: 370 ha
Designation of sites: Semi-improved meadows being managed within agri-environment schemes
Management and community use: Privately owned and managed within working farms; public access is limited to PROW or, for some sites, by arrangement
Condition of sites: Species-poor meadows with potential for restoration
Website: www.ydmt.org/programme-details-hay-time-14609

Muker Meadows SSSI, Swaledale (© Don Gamble; YDMT). SSSI, Site of Special Scientific Interest.

Background

Upland hay meadows are a priority habitat and also an Annex 1 habitat of the EU Habitats Directive (see Section 1.4.5). The very richest upland hay meadows contain more than 30 plant species/m^2 and up to 120 species per field. They are of high habitat value for a range of fauna, many of which are also priority species: they provide feeding areas for invertebrates, bats and other mammals, and feeding and nesting sites for breeding waders and passerines.

However, there has been a dramatic decline in the extent and quality of upland hay meadows over the last 70 years, primarily due to changes in agricultural practice. The area of the habitat is fragmented, with sites isolated within a more intensively managed landscape. Less than 1,000 ha of upland hay meadows now survive in the UK, of which about half is found in the upland dales of North Yorkshire, Cumbria and County Durham. A small but significant proportion is confined to roadside verges.

The Hay Time Project was set up in 2006 by the Yorkshire Dales Millennium Trust (YDMT) and the Yorkshire Dales National Park Authority, with the aim of making a significant contribution towards UK BAP targets for the expansion/ restoration of upland hay meadows at that time. The project uses sustainably harvested seed from species-rich donor meadows to increase the species diversity of suitable receptor meadows. The project also runs events and activities to increase public awareness, enjoyment and understanding of hay meadows. A sister project was set up at the same time in the North Pennines Area of Outstanding Natural Beauty (AONB), and YDMT has subsequently helped establish similar projects in the Forest of Bowland AONB and Nidderdale AONB.

Seed donor and receptor meadows must be identified, surveyed and matched, including the analysis of soil samples from receptor meadows. Implementation plans are prepared by YDMT, liaising with farmers, contractors and Natural England. Receptor meadow preparation and seed harvesting has to be co-ordinated with spreading, donor and receptor meadows require monitoring, and the YDMT provides meadow management advice and training, and implements community and education initiatives. The project has a range of specialist seed harvesting and spreading machinery available. To date, approximately 370 ha at 67 farms have had seed added to them.

Upland hay meadow restoration methods

Where possible, seed is harvested from nearby donor meadows (either SSSIs or undesignated species-rich sites). Preferably the donor and receptor meadows are in the same dale, but due to the scarcity of good quality donor meadows seed is sometimes transported between dales. Where a local seed source is not available, British-provenance processed seed (as mixes or individual species) is sourced from reputable seed companies.

Four seed harvesting methods are used: green hay; hay concentrate; brush harvesting; and vacuum harvesting and hand-collecting. All methods have their pros and cons and no single method is suitable for all schemes.

Green hay

Soon after the receptor meadow has been cut, cleared and harrowed, a tractor-pulled Amazone flail mower is used to cut and collect green hay from an agreed area of the donor meadow. The green hay is transported to the receptor meadow and spread by a tractor-pulled Millcreek muck spreader. Green hay is spread at a ratio of between 1:3 and 1:5 (that is, 1 ha of green hay is spread on 5 ha of receptor meadow).

Green hay has also been collected using a conventional forage harvester. This method is the quickest of the field-scale seed introduction methods but it can only be used where the slope of the donor meadow is shallow enough for a muck spreader or trailer to be towed alongside for the green hay to be sprayed into. Most of YDMT's donor meadows are too steep to use a forage harvester.

Green hay is bought from the donor farmer at a rate of £505/ha. This rate was agreed with Natural England to reflect the value of the seed and to enable the farmer to buy in replacement hay. As well as the seed cost there is the cost of the contractor to harvest, transport and spread the seed. (There is no charge for use of

Tractor-pulled flail mower cutting and collecting green hay (© Don Gamble; YDMT).

YDMT specialist machinery.) Green hay schemes cost around £550/ha of receptor meadow.

Green hay is the preferred method, as a large quantity of local provenance seed from the widest range of plants is collected. It is the most flexible of the large-scale seed introduction methods, as the donor farmer can cut the rest of the meadow before or after harvest of the green hay from the agreed area. Crucially, green hay is the least affected by wet weather; if the ground is not too wet for tractors, green hay can be cut and spread during light rain. The disadvantage is that a large volume of material has to be transported and spread within an hour or so of being collected, which means that the donor and receptor sites have to be within about 40 minutes of each other.

Hay concentrate

This method is similar to green hay except that only the top third to a half of the standing herbage from the agreed area is cut and collected, leaving the rest to be made into hay as usual. Obviously seed has to be collected before the donor farmer cuts the meadow. The seed also has to be collected during dry weather. The seed is collected by a quadbike-pulled hay concentrate harvester and spread by either a quadbike-pulled spreader or the tractor-pulled Millcreek muck spreader. Hay concentrate is used when the donor farmer wants to reduce hay loss, or the donor site is too far away for green hay to be used, or where tractor access is difficult.

Brush harvesting

The Logic quad-towed brush harvester collects ripe seed but does not cut the grass, so hay can be made afterwards (although the sward is slightly flattened). Brush harvesting must be carried out in dry weather and once the morning dew has evaporated. Harvested seed is emptied on to a tarpaulin and spread out to

Quadbike-pulled hay concentrate harvester (© Pippa Rayner; YDMT).

dry. (This also allows invertebrates to escape.) Once dry, the seed can be cleaned using a sieve to remove most of the stalky material and then stored. Seed is spread by hand on prepared plots but it can also be broadcast using a push spinner. A key advantage of this method is that the seed spreading 'window' can be extended into September.

Vacuum harvesting and hand-collecting
Seed harvested from one or two plots marked out in the donor site is spread on a tarpaulin in an airy barn, usually at the receptor farm, to dry. When the receptor meadow has been cut and cleared the seed can be spread by hand on harrowed plots or on patches of bare ground created during haymaking. To enable future

Quad-towed brush harvester collecting ripe seed without cutting the grass (© Don Gamble; YDMT).

monitoring the location of the plots is recorded, either by measuring the distance and bearing of the centre of the plots from a fixed feature, such as a gatepost, or by recording the national grid reference of the plot centre using a hand-held GPS device. This method can be used when the donor farmer does not want to lose any hay; the donor site is too far away for green hay to be used; the receptor site is already fairly species-rich and only needs seed of targeted species introduced; or where the receptor site is too small to justify the expense of field-scale seed addition. Although cost-effective, this method collects the least amount of seed and seed can only be collected in dry weather.

Seed of appropriate species in roadside verges can be collected by hand. Ripe seed heads are removed by hand or using scissors and dried before being stored in marked envelopes. Hand-collected seed is used to supplement seed collected by leaf vacuum and hay concentrate.

Vacuum harvesting by hand (© Don Gamble; YDMT).

Monitoring

All receptor meadows are surveyed before seed addition and again in the following summer. The survey method involves a W-shaped transect across the field, stopping ten times to record the frequency and abundance of all vascular plant species within 1 m^2.

In 2011, all of the receptor meadows worked on in the previous 4 years were resurveyed and the results were statistically analysed (Perry and Gamble 2012). The key findings were as follows:

- All seed addition methods have led to statistically significant increases in species richness, diversity and composition.
- Green hay addition is associated with increased abundance or the introduction of a large number of species.
- The vegetation at most sites is, with time, moving towards the target NVC (MG3) and away from that associated with semi-improved grassland (MG6).

For more resources produced by the project, visit http://www.ydmt.org/resources. Information provided by Don Gamble (YDMT).

3.2 Cutting

Hay meadows were traditionally managed by a combination of spring and autumn grazing with hay cuts in late summer for winter forage, and occasional top dressing with farmyard manure and applications of lime. Cutting reduces the abundance of coarse grasses (Parr and Way 1988), reducing shading and altering the competitive balance to allow more stress-tolerant species, especially low-growing forbs, to survive. Unlike grazing, cutting is unselective and produces a relatively uniform sward height and structure, which is poor for invertebrates. It removes biomass while returning ripe seeds to the soil, but only if the vegetation is left long enough to flower: early silage cuts mean reduced numbers of species regenerating from seed, and smaller quantities of seed shed. Furthermore, there will usually be much less bare and disturbed ground caused by trampling by animals, little redistribution of nutrients in dung patches, and a greater build-up of litter and thatch at the soil surface, particularly where the cuttings are not collected. Cutting at the same point in the growing season favours a suite of species producing ripe seed at that time, while increasing the frequency of cutting may prevent flowering of some species altogether.

Hay meadows provide an important source of forage for livestock feed in winter, when the growth rate of grass is low. Where agricultural production is the main objective, farmers manage meadows to maximise yields, which has led to the 'improvement' of large areas of semi-natural grassland, through reseeding with high-yielding species, such as varieties of Perennial Rye-grass, and regular dressing with fertilisers. Silage is the highest quality of forage, and represents the vast majority of forage harvested in lowland Britain. It is cut in the spring, before most species flower, with repeat cuts every six weeks or so. Hay is still widely cut from improved grassland, usually in June or early July.

Semi-natural meadows that have not been improved are now rare, and most of these are managed for conservation, although some are still managed by farmers for a 'lower' yield hay crop. Conservation management objectives usually aim to maintain or restore the conservation value of a site and to prevent the deterioration of a site through the development of areas of rank grassland or invasive scrub.

Neutral meadows are usually harvested for hay, coupled with grazing at different times of the year, including: lowland Crested Dog's-tail–Common Knapweed grassland (MG5) and upland hay meadows such as Sweet Vernal-grass–Wood Crane's-bill (MG3). Herb-rich flood meadows of the Meadow Foxtail–Great Burnet community (MG4) and Crested Dog's-tail–Marsh Marigold flood pasture (MG8) are also managed for hay production. Following traditional management systems, hay meadows are cut in late summer, followed by aftermath grazing. They may also be grazed in the spring, and receive occasional top dressing with farmyard manure and applications of lime (Section 3.5). These meadows often have species-rich swards; diversity can also result from varying the time of cutting, or missing some years altogether.

3.2.1   Cutting methods

Drum or disc mowers towed by tractors are most commonly used for cutting semi-natural grassland, although forage harvesters and toppers may also be used. Pedestrian-operated cutters and balers can be used on smaller sites. Heavy machinery can damage the sward and create ruts after periods of very wet weather, so mowing under these conditions should be avoided. Care must also be taken when setting cutting heights, to avoid excessive soil disturbance and creating significant areas of bare ground, which can be invaded by coarse grasses and noxious weeds. Bales are generally removed within three weeks or so to avoid damage to the sward. If the hay is not being used for forage, for example, because of contamination, it should still be removed from the field, to avoid nutrient enrichment of the soil, and to prevent the development of a mat of leaf litter and dead vegetation. Litter layers prevent seed from reaching the soil surface, reduce light levels and promote the growth of mosses. Repeated removal of cuttings maintains the low nutrient status of semi-natural grassland, thus inhibiting the growth of more aggressive grasses.

3.2.2   Timing

Mowing is dependent on the weather and geographical location. Mowing of upland hay meadows will be typically later than lowland meadows in southern England, for example. Farmers will be concerned about yield and quality; conservation managers will be concerned about the impact of mowing on biodiversity, especially plants, and in some cases, invertebrate and bird communities.

Later-flowering Betony and Dyer's Greenweed thrive along the margins of a semi-natural lowland hay meadow; the margins of this meadow in East Sussex are cut later in the summer than the main meadow to ensure that later-flowering species have an opportunity to flower and set seed, thus allowing for some regeneration.

Mowing should be timed to allow key species in the plant community to flower and set seed, mimicking the effects of traditional management that allowed diverse grassland communities to develop and flourish. It is important that annuals, biennials and short-lived perennials, in particular, are allowed to set seed before cutting. However, within the window of cutting, which is usually between late June and late July (depending on weather and location), some species will not have flowered or set seed. Many of these species are perennials, but the long-term effects on their populations may vary in different situations.

For sites supporting ground-nesting birds such as waders, mowing should not commence until eggs have hatched, and adults and their broods are able to escape the mowing machinery. Mowing should commence in the centre of the field and work outwards, to allow any broods to escape. For these species and in the case of species such as Skylarks, delaying mowing until the young have fledged is important (see Section 6.4.3).

Due to the long history of summer mowing, hay meadows tend to have evolved invertebrate communities of more common species, whose life cycles tolerate cutting at this time. Management plans could be modified to develop the invertebrate interest in hay meadows, but this may well have adverse impacts on plant communities, so is not recommended as a general practice.

Fertility also influences cutting time: on more fertile sites, where the growth of competitive grasses is more vigorous, an earlier cut might be necessary; on less fertile sites, later cutting may enhance species diversity.

With mechanisation and the imposition of more detailed management plans, there is a tendency for managers and contractors to start mowing close to the start of the 'official' cutting period, and to complete the task as rapidly as possible. This leads to uniformity, and risks a reduction in species richness. The introduction of some variety in cutting date between patches of grassland or between farms would reduce the risk of losing species from meadows. Late cutting every few years would also support later-flowering species such as Betony and Meadowsweet.

3.2.3   Cutting and aftermath grazing

Traditional hay meadow management which combines spring (usually in upland meadows) and autumn grazing with hay cuts in late summer promotes maximum flowering, seed set and opportunities for germination in gaps (Figure 3.2). For this reason, this is also popular in grassland restoration (Section 5.3). In practice the hay cutting date varies considerably from July to September, depending on weather conditions in that particular year and the accessibility of fields from the farmhouse, the nearest tending to be cut first. This strongly influences species composition and only in late-cut meadows will later-flowering species, such as Common Knapweed and Wood Crane's-bill, thrive. The most common seasonal alternatives are *spring* and *summer meadows*. In spring meadows, the sward is briefly grazed or mown in early spring (March and April), or grown unchecked until late June, then cut at a height of approximately 7.5 cm (occasionally 10 cm, to allow the hay to dry more quickly on tall stubble). The regrowth is mown or grazed to 7.5 cm every 3–4 weeks until growth ceases in November, or a single cut is taken in September. In summer meadows, spring grazing or mowing is optional and slightly later, from March to May, with the hay cut in late summer (July and August). The aftermath is grazed or cut from September to October, and occasionally grazed over winter. Lammas meadows were traditionally grazed between Old Lammas day (12

August) and Lady Day (25 March), then 'shut up' at the beginning of the new growing season to allow growth for a hay cut in July/early August.

In their surveys of farm practices in the North Pennines and Yorkshire Dales, Smith and Jones (1991) noted that while average starting dates for the hay harvest were clustered around 1 July, the finishing dates varied widely and depended on the harvesting sequence of different meadows. Finishing dates also tended to occur much earlier in the season since the 1950s, due to advances in agricultural efficiency, especially with the advent of large baling machinery. This altered timing of cutting, together with the application of mineral fertilisers on some fields, profoundly affected the botanical composition of the meadows, the earlier finishing dates compromising many late-flowering species which were harvested before they could set ripe seed. Cutting species-rich hay meadows on the Somerset Levels, either in May or September, rather than at the optimum time in August has also been shown to significantly reduce botanical diversity (Kirkham and Tallowin 1995). Possibly the best way to optimise diversity is to have a rotation of mowing dates among hay meadow tracts, creating a mosaic of mown, unmown and regrowing patches, following the traditional agricultural practice still employed in parts of Eastern Europe (Babai and Molnár 2014).

Early hay harvests, applications of farmyard manure/fertiliser and trampling by cattle and sheep all have the potential to damage the nests and fledglings of ground-nesting birds, especially when operations are carried out during the breeding season early in spring. For example, farm stock may cause considerable disturbance and mortality when meadows are grazed well into May. Similarly, added nutrients in the form of fertilisers or farmyard manure will gradually reduce plant diversity and the foraging ability of

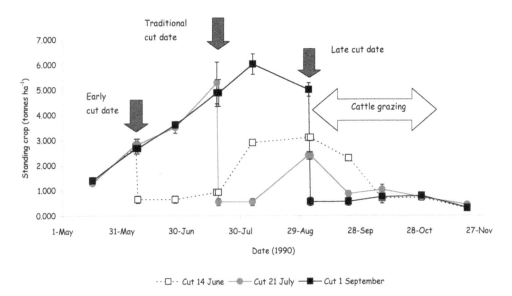

Figure 3.2 Above-ground standing crop of a species-rich MG3b meadow under three different cutting regimes in an otherwise traditionally managed meadow, Upper Teesdale (County Durham) 1990. The June hay cut gives a low yield with the aftermath grazing producing an equivalent yield; a traditional July cut date increases yield, with a slight reduction in aftermath grazing; an early September cut has a standing crop equivalent to that at the July cut date, but results in little aftermath regrowth for cattle (© Roger Smith, using data collected by Helen Buckingham).

young birds. With moderate applications of farmyard manure, however, the abundance and accessibility of prey items, such as soil-dwelling invertebrates, may be increased by bringing them closer to the surface. Late spring grazing intensity also affects the way in which the sward is used by different bird species. Fuller (1996) reported that Lapwing, for example, preferred moderate-to-high levels of spring grazing which produces shorter vegetation; other species such as Snipe, Redshank, Eurasian Curlew, Whinchat and Skylark preferred lightly grazed, tussocky vegetation.

Changes in the management regime of grazed hay meadows can cause significant changes to rare plant or invertebrate communities, which may have developed as a result of management. For example, replacement of cutting by grazing might threaten populations of early flowering species such as Yellow-rattle. If aftermath grazing following a hay cut is stopped, this may affect species diversity, for example, through loss of regeneration niches caused by animal hooves, or by allowing competitive grasses to become better established. Traditionally, supplementary feed comprising hay from semi-natural grassland may have helped to maintain species diversity in the past; the opposite may be true if hay from improved grassland is spread on to semi-natural hay meadows. Supplementary feeding should be confined to areas away from the most sensitive parts of the sites, due to increased levels of poaching and dunging.

3.2.4   Cutting to replace grazing

Most calcareous and acid semi-natural dry grassland would normally be grazed. However, there may be circumstances where grazing is no longer practical. In other situations, pastures may be too small to establish a grazing regime, or it may not be possible to undertake aftermath grazing on hay meadows. In these cases, regular mowing can be used as a substitute, although the outcome for species diversity may change. For sites that have previously been managed by grazing, the introduction of mowing could have a significant impact on both plant and invertebrate communities, and should be carefully evaluated.

Like grazing, mowing will remove plant bulk and biomass; it will help maintain low nutrient status (assuming cuttings are removed); and it can also lead to a dense sward if done regularly. However, there are a number of outcomes from grazing which do not result from mowing alone:

- Creation of bare patches – mowing does not create the essential patchwork of regeneration niches resulting from poaching.
- Creation of structural diversity in the sward – mowing produces a uniform sward, unlike the structural mosaic resulting from grazing at low stocking densities, which supports more diverse invertebrate communities.
- Selective control – mowing is non-selective, whereas some animals are highly selective, favouring unpalatable species and low growing herbs.
- Seed production – the successful production of seed is dependent on the timing of mowing, whereas animals may be more selective.
- Gradual removal of plant material – mowing is immediate, and does not allow invertebrate populations to adjust as they would on grassland with a low stocking density of animals.

Any conservation site that is to be managed primarily by cutting will therefore need some imagination to maintain structural diversity and hence healthy invertebrate populations.

This cannot be prescribed, but should be implemented as appropriate on a site-by-site basis. Options to consider include:

- For a single cut, make this as late in the season as possible, to minimise adverse effects on invertebrate communities.
- Vary the height of the cut across the site.
- Cut parts of the site on a rotational basis.
- Cut at different times.
- Alter the frequency of cutting.
- Maintain small areas of scrub or rank grassland scattered across a site.
- Disturb the soil surface to replicate the effects of poaching. (Note that some heavy mowing machinery may cause damage through soil compaction.)

Provided that cuttings are removed, steps like these should ensure that some flowers and seed heads survive each year, together with the invertebrates they support.

3.2.5   Combined cutting and grazing regimes

Traditional meadow management combining spring and autumn grazing with hay cuts in late summer are popular in restoration, to promote maximum flowering, seed set and opportunities for germination in gaps. In practice, hay cutting depends on weather conditions and the accessibility of fields. Dry matter production of hay from unfertilised, species-rich meadows typically ranges between 20 and 80% of that produced from a fertilised Perennial Rye-grass ley, cut for silage. With increasing application levels of N fertiliser, up to 400 kg/ha, there is often a linear increase in dry matter productivity, assuming no other major limiting nutrients, but at the cost of species diversity as the less responsive species are competed out (see Section 3.5.1).

During conventional haymaking, losses in overall production of 10–25% may occur due to respiration and leaching of the harvested material, areas left uncut and material not gathered by the harvester. The nutritional quality of hay is highest when cut early in the season, especially when the fields have been shut up for short periods of growth after a late spring grazing. Hay quality deteriorates if cut later, after the peak of flowering, when the proportion of lignin-carbohydrate complexes increases in the herbage. This reduces its digestibility in ruminant animals to about 20% below that of silage from agriculturally improved grassland. Similarly, when haymaking is delayed by wet weather, forage quality deteriorates further, exacerbated by soluble nutrient losses through leaching.

3.3   Weeds and herbicides

The principal weed species recognised as being injurious in semi-natural dry grassland (and other land) are Creeping Thistle, Spear Thistle, Common Ragwort, Broad-leaved Dock and Curled Dock; invasive weeds include Japanese Knotweed, Giant Hogweed and Himalayan Balsam, all of which are specified under the Weeds Act 1959. Other species that can be problematic in some circumstances include Bracken and rushes. While all reasonable steps may be taken to prevent the spread of injurious and invasive weeds, some are also valuable to wildlife as: food plants for invertebrates; sources of seed for farmland birds; natural constituents of the wider plant community; and because they contribute to the structural diversity of the sward. Creeping Thistle, Spear Thistle and Common Ragwort, for example, support several rare invertebrates, especially on sites where they have been present for long periods of time. On well-managed semi-natural grassland, these species are not usually a problem. Here, if necessary, site managers might consider control rather than eradication if numbers increase. On other sites, the presence of these species may not meet with agricultural objectives, for example, grazing with horses on pasture with Common Ragwort, or they may become a threat to agricultural land. On any semi-natural grassland site, improved habitat management would be the ideal method of choice for controlling weeds, such as modified grazing regimes. This may be supplemented in the first instance by various non-chemical control treatments, to comply with Defra's code of good agricultural practice (Defra 2009). Site managers should also try to identify potential weed problems as early as possible, to reduce the time and cost of subsequent control measures.

3.3.1 Non-chemical control

Prevention is preferable to cure where pernicious weeds are concerned. If some form of control is required, management of the site needs to be reappraised to avoid further infestations in the future. Some of the reasons why weeds can become a problem on semi-natural dry grassland are:

- excessive trampling and poaching;
- bare ground in areas where supplementary feed is supplied regularly;
- use of forage contaminated with weed seeds, particularly if there is excessive bare ground;
- overgrazing in general;
- burning;
- excessive fertiliser application.

Non-chemical control methods (by hand or mechanical) should be used wherever possible. Recent research has contrasted environmentally sustainable weed control with herbicide application, including combinations of different grazing regimes, cutting and herbicide wiping (Pywell *et al.* 2010). The results of a six-year trial showed that thistles can be controlled without herbicides, if lenient grazing is practised in spring and autumn. Severe infestations may need herbicide treatment, but the authors point out that these are best prevented through appropriate management.

Non-chemical control may be by hand, or mechanical:

- Weeding by hand should be undertaken before plants flower. Use a spade or hoe just below ground level (not for Common Ragwort). Alternatively use a thistle hoe (not for Common Ragwort) or hand weed by pulling (useful for Common Ragwort if numbers are manageable). Note that even dead Common Ragwort may be palatable to livestock, so all plant material should be removed.
- Mechanical control may employ repeated cutting on a tractor-pulled topper for thistles and docks. Cut just before flowering for thistles and repeat a month later. This should prevent seeding and reduce plant vigour, but it is unlikely to eliminate the plant. It is important to consider ground-nesting birds when planning Common Ragwort control, so where birds are nesting, it may not be possible to start control before midsummer.

If these methods are not successful, or impractical, then chemical control might be considered.

3.3.2 Chemical control

Although chemicals are potentially dangerous, and they can harm the environment, specified species in particular may require chemical control in certain circumstances. Under the Food and Environment Protection Act 1985 and the Health and Safety at Work etc. Act 1974, a new code of practice for using plant protection products came into force in 2006 (Defra 2006a). More recently, the Plant Protection Products (Sustainable Use) Regulations came into force in 2012. However, anyone using pesticides professionally should refer to the 2006 Code of Practice and check the Health and Safety Executive website for updates (http://www.pesticides.gov.uk/guidance/industries/pesticides/topics/using-pesticides/codes-of-practice). The Code of Practice provides practical guidance to professionals on the safe use of plant protection products that meet the legal conditions covering their use. It also provides information on training and certification requirements in the use of pesticides and the Control of Substances Hazardous to Health (COSHH) assessment and assessment of risks to the environment, which must always be completed. Following the advice in the Code will keep you within the law, although practitioners are permitted to work in a different way, providing that way is just as safe. Practical guidance is also available for those who sell, supply and store pesticides for sale by way of the *Code of Practice for suppliers of pesticides to agriculture, horticulture and forestry*, available on the Pesticide Safety Directorate (PSD) website (http://www.pesticides.gov.uk/guidance/industries/pesticides).

Consent may be required from the Environment Agency if herbicides are to be used near watercourses. In addition, it is important to follow the guidance on the product label. Practitioners should also consult Defra's *Code of good agricultural practice* to help protect the environment (Defra 2009), available on the GOV.UK website. We have not attempted here to include the large volume of information required to meet legal and operational guidance, but refer our readers to *The Herbicide Handbook* (English Nature 2003) when making choices about techniques and appropriate site conditions and requirements. Also, we encourage readers to avoid disturbing breeding birds in spring and early summer.

3.3.2.1 Herbicide selection and application

There is a wide range of chemicals available, so it is important to choose the chemical which best fits the species requiring control and to ensure that the right growth stage is targeted. Narrow-spectrum, selective herbicides minimise the effects on other plants; broad-spectrum, non-selective herbicides, such as Roundup (glyphosate), can also be used, but great care is required to avoid spraying other species. Herbicides and uses must be approved under the Control of Pesticides Regulations 1986 (as amended) or the Plant Protection Products Regulation. All herbicides have standard or 'off-label' approval for use in specific situations. Approved products are listed on the PSD website and details of all herbicides can be found in the *UK Pesticide Guide*, jointly published annually by CABI and the British Crop Protection Council. In addition to the guide, and of course the label itself, useful information sheets for selected herbicides, including environmental risks, can be found in *The Herbicide Handbook* (English Nature 2003). For off-label approval, read the relevant Notice of Approval on the PSD website. Refer also to the herbicide Environmental Information Sheets produced under the Voluntary Initiative (http://www.voluntaryinitiative.org.uk/en/home) and *Grassland Weed Control* published by the Voluntary Initiative and the Agricultural Industries Confederation (http://www.voluntaryinitiative.org.uk/importedmedia/library/1510_s4.pdf).

The method of application is important, as are the weather conditions, to ensure that target species are controlled, while avoiding damage to the environment. Knapsack sprayers or weed wipers are most frequently used to spot-treat pernicious weeds. Knapsack spraying allows precise application of herbicides to the target species (providing narrow nozzle tips are used), and is particularly suited to relatively small areas of semi-natural grassland, sites with steeper slopes (if it is safe to do so) and sites with anthills. Even with knapsack sprayers, buffer zones of several metres should be allowed to avoid damage to sensitive plants. For small-scale, localised applications, other hand-held sprayers may be used. COSHH assessments will be required for any chemicals that do not carry knapsack recommendations, provided that they are not prohibited from use with knapsack sprayers.

Weed wipers also allow targeted herbicide application, and may be hand-held or tractor-mounted. They have several distinct advantages, including delivery of a controlled dose of herbicide to the underside of plants where it is less likely to be washed off by rain, while avoiding ground contamination and spray drift onto the sward. However, there are also disadvantages:

- Weeds need to be at least 10 cm higher than the surrounding vegetation, possibly requiring temporary changes in a grazing regime to achieve this.
- It may not be possible to use weed wipers at the ideal growth stage of the weeds because of height issues.
- Fewer herbicides have approval for use with weed wipers.
- Herbicides are used at higher concentrations.
- There is a risk of some 'volatile' herbicides reaching non-target vegetation in warm temperatures (> 20 °C).

Herbicides without approval would require comprehensive COSSH assessments and would need to meet Defra's Code of Practice (Defra 2006a).

Following treatment, grazing restrictions will apply, depending on the specified grazing interval of the particular herbicide. This specifies the time for the chemical to act, not necessarily when it is safe for animals to return in the case of Common Ragwort control.

Finally it is important to re-emphasise that herbicides should only be used if other measures fail to tackle the cause of the problem. Furthermore, the text here is a general introduction to the use of herbicides for weed control on semi-natural dry grassland, and therefore no liability is accepted.

3.3.3   Common Ragwort control

Common Ragwort (*Senecio jacobaea*) is the only member of the huge genus *Senecio* that is a specified weed under the Weeds Act 1959. It is a natural component of some semi-natural dry grassland communities, where it can support rare invertebrates. Continued good management is the best way to prevent serious Common Ragwort problems. However, populations can increase dramatically on sites where management has changed, for example, overgrazing can create regeneration niches that Common Ragwort can readily exploit. This is of concern to graziers because the plant contains toxins that can be fatal to horses and other animals. For semi-natural dry grassland sites in general, populations of Common Ragwort and other specified weeds should be kept to a minimum if the land is being grazed.

High population densities on nearby sites can also result in colonisation by windborne seed of Common Ragwort, so risk should be assessed (after Defra 2004):

- High risk: Present, flowering/seeding within 50 m of grazing land or land used for feed/forage;
- Medium risk: Present within 50–100 m of grazing land or land used for feed/forage;
- Low risk: Present > 100 m from grazing land or land used for feed/forage.

The distances cited should be used as guidelines only, with other environmental factors taken into consideration, such as topography, natural barriers, and so on. Ragwort contamination of hay and silage also presents a hazard to animals, as they are less likely to reject dried ragwort, which remains highly toxic. Contaminated forage must be declared unfit as animal feed and safely disposed of.

Defra's Code of Practice states that owners of livestock and forage producers should practise good management, undertake regular inspections, and safely control and dispose of Common Ragwort. In addition, livestock owners must 'satisfy themselves that their stock is not exposed to the risk of ragwort poisoning' (Defra 2004), and may need to move animals if necessary. Other landowners should also follow these guidelines, and in addition assess the risk:

- High risk: immediate action required using appropriate control, taking account of land status.
- Medium risk: establish control policy to identify and deal with likely changes from medium to high risk, using appropriate control, taking account of land status.
- Low risk: no immediate action required, but continue monitoring and prepare contingency plans.

3.3.3.1 *Control*

Many semi-natural dry grassland sites will have a conservation designation, and on such sites control methods might damage plant or invertebrate communities. On these sites, preventative management is important, and any action should be cultural rather than chemical. For SSSIs in England, managers must consult their local Natural England Area Team to agree appropriate control methods; in Wales, consult Natural Resources Wales (NRW); and in Scotland, consult Scottish Natural Heritage. Risk assessments may need to be carried out on grazing land, but outcomes will vary, depending on local factors, such as the susceptibility of the particular grazing animals, and the abundance of alternative palatable herbage. For sites that are not grazed, Common Ragwort may be tolerated unless it is a threat to grazing animals on neighbouring land, or meadows used for forage, in which case the risk would need to be assessed. Similar advice is recommended for sites with non-statutory designations. In addition, control of weeds on land covered by agri-environment schemes, such as Environmental Stewardship in England, Glastir in Wales and the Rural Stewardship Scheme in Scotland should be included in individual agreements.

There are a number of control methods that may be advised, all of which are only likely to be effective if accompanied by improved management. The main control methods on sites with conservation designations are cutting, levering out and pulling (Figure 3.3).

Pulling by hand, or levering out before flower heads mature is effective in relatively small areas, but it is labour intensive, and larger areas might require mechanical pulling, unless large numbers of volunteers are available. It may be necessary to repeat later in

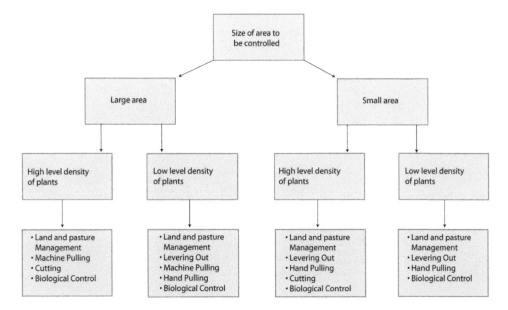

Figure 3.3 Decision tree to select the most appropriate non-chemical control method (after Defra 2004, *Code of Practice on How to Prevent the Spread of Ragwort*. Crown copyright).

the season and over successive years; this may avoid the need to use herbicides. Plants should be disposed of safely, to avoid any risk of animal consumption, or of seeds being dispersed. Gloves should always be worn, as poisoning is a risk to humans through direct contact.

Cutting is less labour intensive, and can be effective in preventing seed dispersal, particularly over large areas. However, it is not a substitute for pulling (or herbicide treatment) because the plants are not killed, and are likely to produce vigorous new shoots capable of flowering later in the season. Cut stems must again be disposed of safely.

In the past, *grazing* has been considered as a possible means of controlling Common Ragwort (English Nature 2003), but the risk to animals is too great, and grazing should not be undertaken (Defra 2004).

On some sites, herbicide treatment (Figure 3.4) might be appropriate, particularly sites of lower conservation value.

Annual treatment with *herbicides* can effectively control the spread of Common Ragwort. As with any herbicide treatment, COSHH risk assessment and risks to the environment should be completed; only use approved chemicals (Section 3.3.2.). Herbicides may be applied by:

- *Spot treatment* at the early rosette stage, or just before flowering, is the most labour intensive and relies heavily on finding all the plants, particularly at the rosette stage; products containing citronella oil may also be used for spot treatment.
- *Weed wiping* just before flowering is less labour intensive than spot treatment and better suited to heavier infestations of ragwort, but it will only be effective against taller plants.
- *Selective spraying* at the rosette stage is also less labour intensive, and suitable for large areas, and dense colonies of ragwort; however, most herbicides are broad-spectrum, and will kill other herbs.

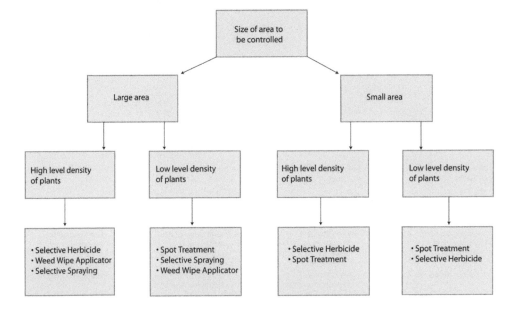

Figure 3.4 Decision tree to select the most appropriate herbicide treatment (after Defra 2004, *Code of Practice on How to Prevent the Spread of Ragwort*. Crown copyright).

Anyone concerned about the risk posed by Common Ragwort should consult the appropriate national code of practice, published in England (Defra 2004), Wales (Welsh Government 2011) and Scotland (Scottish Government 2008); and the *Guidance on the Disposal Options for Common Ragwort* (Defra 2005).

3.3.4 Bracken control

Bracken is a natural component of woodland flora, but is now characteristic of some open habitats in Britain, where it may be a valuable component of a habitat, supporting a wide range of birds, mammals and invertebrates. However, it is a vigorous fern, with extensive underground rhizomes that help the plant to spread very quickly across a site if not controlled. Resulting single species stands may threaten other, more valuable habitats, as well reducing grazing areas. For example, Bracken can be a major problem on heathland and moorland. Encroachment can also threaten lowland acid grassland (including enclosed acid grassland in the uplands), lowland meadows and in some cases, upland calcareous grassland. For this reason it may be necessary to control Bracken, and much has been written about this (e.g. Backshall *et al.* 2001; SEARS 2008; Natural England 2008a; 2008b).

The primary objective when assessing the need to control Bracken is the protection and continuing management of habitats of the highest conservation value. Where Bracken encroachment is a problem on semi-natural dry grassland, Bracken is unlikely to be the feature of highest conservation value on that site; it is more likely that grassland plant and animal communities would be threatened. If the ecological interests of Bracken and semi-natural grassland are similar, then a management plan must aim to benefit both. Consent may be required on SSSIs. Management options are essentially as follows (after Brown and Robinson 1997):

- No control – steep slopes; heavily grazed areas; sites where wildlife considerations are important, or eradication is unlikely to lead to the restoration of dry grassland of conservation value; control may damage archaeological features;
- Conservation management – limited control to maintain habitat mosaics including patches of Bracken for conservation value;
- Limited control – to reduce area or severely limit spread without eradication;
- Eradication – often impractical on dry grassland sites.

Prevention is better than cure; for example, Bracken invasion might result directly from changes in grazing regimes. Overgrazing can lead to excessive areas of bare ground allowing Bracken to invade. The control of invasive scrub might also provide regeneration opportunities for Bracken.

The options for herbicide control have been severely limited by the ban in 2011 on the use of Asulam. This chemical was preferred because of its relatively high specificity. Glyphosate herbicides, such as Roundup Pro Biactive can also be used to control Bracken, but extreme care has to be taken to avoid damage to other plants if this broad-spectrum herbicide is used. Application rates of 5 L/ha have been recommended (Natural England 2008b). They should be applied when fronds have expanded, but before tips die back, to maximise translocation to the rhizome. Glyphosate is most suitable for patches of Bracken with deep litter, with other plants absent. Elsewhere, it may only be appropriate to use glyphosate as a spot treatment, with a knapsack sprayer or tractor-mounted weed wiper to create selectivity. Treated Bracken can poison livestock, so animals should be excluded from a treated area until Bracken has completely senesced or been removed, to reduce the risk of poisoning. In any case, some form of follow-up treatment will be necessary, involving either further herbicide spot treatments or mechanical treatments.

Bracken was traditionally cut or rolled by hand, as a means of controlling, but not eradicating the plant. Mechanical methods now used for Bracken control include cutting and crushing, and for larger sites, ploughing. Ploughing, which damages rhizomes by

Bracken on Butterfly Conservation's Bentley Station Meadow in Hampshire is cut in late June with an Allen Scythe – a finger-bar mower – to manage Bracken that outcompetes lower-growing herbs; Bracken clearance also creates areas of short turf on which butterflies can bask.

exposing them to drought and frost can be effective, but it is unsuitable for semi-natural grassland sites. It is most appropriate for monocultures of Bracken, where no damage to an underlying sward will result from the treatment.

Bracken should be *cut* twice each year for 2–3 years, followed by annual cuts when Bracken is under control. This has the overall effect of weakening the rhizome and inhibiting frond production. The first cut should be in mid-June/July as the plants reach maturity, or later if nesting birds might be present. This should be followed by a second cut in August (about six weeks after the first cut) to remove new fronds stimulated by the first cut, thus further weakening the rhizome. Control should be achieved, but eradication is unlikely.

Crushing is suitable for smaller sites; it involves flattening and bruising Bracken stems in midsummer, which weakens rhizomes in a similar manner to cutting. The treatment may need to be repeated later in the first year, and in subsequent years, to keep on top of the Bracken. Lightweight rollers or specialised 'Bracken bruisers' can be towed behind tractors.

Grazing can also be an effective preventative measure, keeping Bracken under control following successful mechanical treatment. Heavy livestock in winter will also control Bracken to some extent through trampling and bruising stems, and breaking up litter layers. Animals should be removed in spring to avoid poisoning.

For more information on Bracken control, see *Bracken Control: A guide to best practice* (SEARS, 2008), *Bracken Management and Control* (Natural England, 2008b) and the RSPB's Bracken management advisory sheets (http://www.rspb.org.uk/forprofessionals/).

Highland cattle have been used to control scrub, Brambles and trample Bracken on an orchid-rich meadow in East Kent.

3.3.5   Control of other species

Other specified weeds that may require control include Creeping Thistle, Spear Thistle, Broad-leaved Dock and Curled Dock.

3.3.5.1 Thistles

Creeping Thistle can be abundant and withstands grazing, so it can become severe in pastures. It produces relatively few viable seeds, spreading primarily by rhizomes to effectively colonise a site. In contrast, Spear Thistle is a biennial species with a deep taproot and establishes well on overgrazed swards. As with other specified weeds, prevention is the primary means of control, avoiding excessive areas of bare ground that might facilitate colonisation, and maintaining balanced grazing. Cultural and chemical control methods can be used.

Hand control can be effective for both species, including pulling and hoeing, although if pieces of Creeping Thistle rhizome are left in the ground, they will produce new shoots. Repeated cutting at ground level will also weaken plants, but not eradicate them. On the other hand, Spear Thistle is a biennial, so will not withstand repeated cutting to prevent flowering and seed dispersal; it should be cut prior to flowering in summer, with a second cut to prevent flowering later in the season.

Chemical control may be considered for more severe infestations, where rapid control is required. Thistles may be spot treated just before flowering, when growth is most vigorous. Clopyralid, for example, (Defra 2007) is a selective herbicide that is translocated to the roots and inhibits growth in the following year; however, great care must be taken to avoid damage to semi-natural swards. Larger areas can be weed wiped following grazing to reduce the height of the sward. For more information on thistle control, see *Sustainable Management Strategies for Creeping Thistle* (Defra 2006b).

Lenient *grazing* by cattle or sheep has been recommended by Pywell *et al.* (2010) for the control of thistles, following a long-term study contrasting herbicides with more environmentally sustainable weed control strategies on two semi-improved grassland sites. These studies showed that the control of Creeping Thistle could be achieved without herbicides, if sites are leniently grazed in spring and autumn, as this can decrease thistle populations to low levels. This would be particularly desirable on more species-rich sites, where herbicides could damage non-target species.

3.3.5.2 Docks

Broad-leaved and Curled Dock are perennials with deep taproots. Broad-leaved Dock in particular can be a problem on more fertile grassland. Prevention is the primary means of control, by avoiding nutrient enrichment and excessive areas of bare ground that might facilitate colonisation while maintaining balanced grazing.

Hand control can be effective, topping before flowering and seeding or pulling, but docks will produce new shoots from root fragments left behind in the soil. Repeated cutting will weaken plants.

Chemical control can be achieved by spot treatment early in the growing season or weed wiping, but seedlings under the large leaves of mature plants may be missed. One selective herbicide, fluroxypyr, and a number of broad-spectrum herbicides, such as Grazon 90 (clopyralid + triclopyr) may be used (Defra 2007). Asulam was formerly recommended, but it has now been banned.

All Defra publications relating to weed control are available on Defra's website (https://www.gov.uk/government/policies/biodiversity-and-ecosystems).

Further information can be found in *The Scrub Management Handbook: Guidance on the Management of Scrub on Nature Conservation Sites* (Day *et al.* 2003), *The Upland Grassland*

Management Handbook (English Nature 2003) and *The Lowland Grassland Management Handbook* (Crofts and Jefferson 1999).

3.4 Scrub management

Scrub typically consists of low-growing communities of woody plants, such as shrubs or bushes, or underscrub species such as Bramble, usually less than 5 m tall (Section 1.6). It is an important component of many semi-natural dry grassland habitats, but it is often viewed as invasive where the conservation of grassland communities is the primary management aim. In some cases, scrub encroachment threatens such communities, but in others, a balance of scrub and open grassland is recognised as being important for wildlife conservation.

Traditional grassland management practices, such as grazing and cutting, generally keep scrub under control, but relaxation or abandonment of management often leads to scrub encroachment, overtopping grassland plants. If left unchecked, scrub may develop into woodland, a process called secondary succession. For example, in a study carried out on the South Downs between 1971 and 1991, aerial photographs showed that 4,729 ha of semi-natural dry grassland were lost. Although the majority was lost to agriculture, development of scrub and woodland accounted for a significant proportion (Burnside *et al.* 2003). In the wider British context, some experts consider inadequate grazing of semi-natural grassland, leading to rank grassland, scrub and trees to have been a moderate driver of change since the 1940s, continuing as a major driver at the present time and into the foreseeable future (Bullock *et al.* 2011). Scrub control is often one of the primary concerns of people managing semi-natural dry grassland, including sites with archaeological features, but it is important to remember that scrub can also have a high conservation value. Management should be based on the conservation value of both the scrub itself, and the grassland on which it occurs. Prevention of the encroachment of scrub of higher conservation value can be problematic, particularly when management plans demand a mosaic of species-rich open grassland and scrub.

Scrub management techniques are well understood, yet encroachment is still a significant problem, and its removal is a priority to avoid the loss of grassland species. Intimate mosaics of grassland and scrub can benefit wildlife, but are difficult to manage and it is easy to lose control of scrub. A safer policy is to have larger areas of grassland with smaller, more easily managed patches of scrub. For more information on the management of individual species, see *The Scrub Management Handbook: Guidance on the Management of Scrub on Nature Conservation Sites* (Day *et al.* 2003).

3.4.1 Assessment

The condition and conservation value of scrub should always be assessed before any management objectives are set. Scrub of high conservation value may be retained if it includes some of the following characteristics:

- supports a range of native woody species;
- is structurally diverse;
- forms a diverse mosaic of habitats with open grassland;
- supports fringe communities of plants; and
- supports fauna of conservation value, such as rare invertebrates.

On the other hand, management may be problematic if different species require differing management approaches. For example, scrub of low conservation value may comprise non-native species, such as cotoneasters, or it may be dominated by a single native shrub species, have poor structural diversity, form simple associations with grassland, and support few species of conservation concern.

Next, consider scrub condition and whether it is likely to be invasive. Preventative action is preferable to managing out-of-control scrub. Lowland scrub of lower conservation value should be prevented from encroachment into priority grassland habitat, and may also require reduction or eradication.

In general, the management of scrub of higher conservation value should focus on maintaining existing areas in favourable condition, i.e. preventing encroachment into priority grassland habitat; and in some circumstances enhancing areas of scrub in poor condition or under-represented. In the uplands, scrub has generally suffered from clearance and high grazing pressure from domestic and wild animals. Consequently, in many instances, maintenance or enhancement might be the preferred management option.

Scrub management is most likely to be successful when it is still open, and the sward is still present. Where scrub has been allowed to close canopy and become dense, its removal may not always produce the desired effect. Leaf litter may enrich the soil, and with the depletion of the seed bank, scrub removal might result in patches of invasive weeds and rank grassland. In many instances, such areas are unlikely to revert to semi-natural grassland communities without more robust restoration measures such as reseeding (Section 6.4). However, on steeper slopes and thinner soils, removal of dense scrub can result in colonisation by plant species from the adjacent sward. Consider the likely outcome before including scrub removal in a management strategy.

3.4.2 Management techniques

Options for scrub control include grazing/browsing (often the first option for site managers), mechanical and herbicides. Grazing/browsing can be used both to control scrub encroachment and to prevent regeneration following mechanical intervention such as mowing and flailing. However, palatability varies between animal species, and also between breeds. Mechanical control is most appropriate at the early stages of scrub invasion, when hand tools can be used. If scrub is neglected, heavier machinery may be the only viable option for either control or eradication. If scrub is to be removed completely, the effects of heavy machinery required for stump removal, grubbing out and disposal of arisings on fragile grassland ecosystems must be carefully considered. In all cases, conservation of scrub-dependent wildlife, accessibility, aftercare and cost must be considered.

3.4.2.1 Eradication/reduction

Where management plans call for the reduction or eradication of scrub, the main options are: cutting with hand-held brushcutters or heavy machinery; stump removal; grubbing out; weeding of young plants; and herbicide application. Cutting alone is not usually a viable option, as coppice regrowth may actually enhance areas of scrub. Cutting must be accompanied by follow-up treatments, such as glyphosate application to stumps, regular swiping of regrowth or stump removal. Hand tools, horses or mechanical diggers will remove stumps, depending on size and the likely impact of ground disturbance. For smaller areas, or open, patchy scrub hand tools such as brushcutters might be sufficient. For larger areas, heavy machinery such as tractor-mounted flails can be used, offering the potential for savings in time and cost. If heavy machinery is considered, slopes must not be too steep; the possibility of damage to fragile soils should also be considered. Gradual removal of scrub should be considered where cleared land is likely to be colonised by invasive weeds. In all cases, cuttings and litter should be collected and removed to avoid damage to the existing sward, and further enrichment of the cleared area.

Systemic herbicides can be applied to the stumps as they are cut, providing the necessary legislation and Code of Practice for using plant protection products is followed (Section 3.3.2). It may also be necessary to treat regrowth with herbicides.

3.4.2.2   Maintenance and improving quality

For maintaining or improving scrub of higher conservation value, the main options are: browsing/grazing; coppicing/thinning; mowing/flailing; burning; and management of edge zones and bare ground. Scrub may be cut (coppiced) on rotation in relatively small coups to create a mosaic of structural diversity and differently aged shrubs. The timing will depend to some extent on the growth rates of the species concerned. Scalloped edges are more natural and more diffuse, increasing the amount of scrub edge with varying microclimates and helping to diversify wildlife habitats. Following cutting, stump removal or treatment is not usually necessary, although species, like Dogwood and Blackthorn will respond by suckering, and may require further control to maintain a balance of species. Grazing animals will also exert some control of scrub, particularly along the edge. In some cases, it may be necessary to exclude animals on rotation, to protect this vulnerable habitat.

3.4.2.3   Enhancement

Scrub enhancement programmes are most likely to be required in upland areas, where scrub is under-represented or restricted to niche habitats, such as limestone pavement, or rocky ledges where it cannot be grazed or browsed. The main management options are natural regeneration or introducing shrubs through direct seeding or planting, particularly in upland areas where seed sources are scarce. Both options are likely to require protection from browsing, either by the removal of domestic stock and protection through fencing or tree guards. For more information on shrub planting, see *Woodland Creation for Wildlife and People in a Changing Climate* (Blakesley and Buckley 2010). The effects of excluding grazing animals need to be monitored closely, to avoid the development of tall grassland that would effectively compete with regenerating shrubs. Species such as Juniper, which are unable to compete in such circumstances, may require planting, or at least some management of the ground vegetation to support natural regeneration.

3.4.2.4   Grazing and browsing

In many situations, the choice of domestic animals primarily reflects the extent and quality of grassland grazing (Section 3.1). However, grazing and browsing can also have a significant impact, and can be used to reduce, maintain or enhance scrub (Table

Table 3.7 Summary of scrub browsing characteristics of domestic animals (after Day *et al.* 2003; © RSPB)

Livestock species	Suitability for conservation grazing
Cattle	Relatively unselective, browse a wide range of species such as Ash, Elder, Sycamore, oaks Bramble, thorny shrubs, birches are less favoured Create pathways through scrub by trampling Hardy breeds can cope with poor diet Primarily grazing animals, targeting more palatable grassland sward
Sheep	Selective browser, able to control sapling regrowth of palatable species such as Ash and Elder Alder and oaks are less favoured Browse Bramble in winter Wool can get caught up in thorny scrub, risking trapping and starvation of animal
Ponies	Selective; they browse evergreen shrubs and buds of deciduous species Strip bark from palatable species Palatable species include Blackthorn, Gorse and Holly Alder and Hawthorn are less favoured Some breeds are not hardy
Goats	Browse woody plants, often in preference to herbs and grasses Agile, can reach remote areas Strip bark on a wide range of shrubs and trees, which may include desirable species Require dry shelter

3.7). Using animals to control both the sward and scrubby areas is a more 'naturalistic' approach to vegetation management. Stocking densities need to be set with care, usually to maintain species diversity in semi-natural swards. The effects on scrub must then be carefully monitored, to ensure that the desired outcome is being achieved. In some cases, additional management will be needed if browsing pressure is too low, such as mechanical cutting of scrub.

For more information on the management of individual shrub species, and practical management by grazing/browsing, mechanical control or herbicide application, see Section 5.8.1 of *The Scrub Management Handbook: Guidance on the Management of Scrub on Nature Conservation Sites* (Day *et al.* 2003) (http://publications.naturalengland.org.uk/publication/72031).

3.5 Fertiliser application

It is well known that applications of high levels of inorganic fertiliser as N or N in combination with phosphorus (P) and potassium (K) to semi-natural dry grassland can cause rapid reductions in plant species diversity, and hence conservation value. This outcome is a result of the growth of more competitive grasses, which results in the decline and eventual loss of some slow-growing herbs and grasses. Inorganic fertilisers can also

Grassland on an ex-arable land adjacent to Hainault Forest in 2006 (left). Three years later, in the absence of grazing, scrub development is dramatic (right).

In an adjacent field, cattle grazing was introduced, resulting in a much more open grassland (top left), where cattle have had some control over willow growth (top right), but little impact on Hawthorn (bottom left); consequently, after 6 years of grazing/browsing, mechanical control is still required to remove Hawthorn scrub, and cut back some of the willows that have escaped the browse line (bottom right). Goats could be considered at sites prone to rapid natural regeneration.

adversely affect the soil microbial community. Consequently, Crofts and Jefferson (1999) recommended that P or K should not be applied to semi-natural grassland, and that any type of inorganic fertiliser would be unacceptable. For similar reasons, the use of animal slurry and sewage sludge should also be avoided. However, in a recent study on semi-natural hay meadows, Kirkham *et al.* (2008) suggested that more evidence is required, to determine whether intermittent applications of low amounts of inorganic fertilisers are more sustainable than farmyard manure (see Box 3.2).

Sheep and goats tend to have more impact on Bramble, but cattle may occasionally browse, as shown here along a woodland edge in Hainault Forest.

3.5.1   Farmyard manure

Light dressings of farmyard manure are widely viewed as more acceptable for certain types of semi-natural grassland, because nutrients are released much more slowly, over an extended period. Traditionally managed hay meadows, for example, have received occasional top dressings of well-rotted farmyard manure as well as applications of lime. These applications compensate for nutrient depletion by the removal of annual hay crops. Farmyard manure should only be applied to other semi-natural grassland where historically there is no evidence of damage to the flora and fauna. As a general rule, it should not be applied to semi-natural grassland managed as pasture (English Nature 2001).

In traditional systems, modest applications of 6–12 t/ha/yr add less than 75 kg N/ha/yr, of which only about 20% (< 15 kg N/ha) is immediately available to the sward and has relatively little impact on species diversity (Smith 2010). However, in upland hay meadows and similar neutral grasslands, applications equivalent to 18 kg N/ha/yr or more can cause significant reductions in floristic diversity. For farmyard manure, the same sources suggest that 12 t/ha/year can maintain current diversity in MG3 meadows with a past history of these levels of input, although increases in the cover of positive indicator species were associated with lower rates of only 6 t/ha/yr (Pinches *et al.* 2013). Many of the earlier agri-environmental schemes in the UK constrained annual fertiliser levels to zero or 'non-damaging' low applications, typically 25–50 kg N/ha, or farmyard manure applications averaging 12.5 t/ha/yr (Critchley *et al.* 2004). Annual hay yields from semi-natural grasslands range between 1.5 and 6 t/ha of dry matter, about 70% of the total annual production of 2–8 t/ha, compared with 10 t/ha or more for agriculturally improved and intensively managed grasslands (Tallowin and Jefferson 1999).

Manure is usually applied in the spring, or midsummer if ground-nesting birds are present. Recommended rates for lowland meadows suggest that these should not exceed

20 t/ha as a single dressing every 3–5 years (Simpson and Jefferson 1996), while annual or biennial applications of up to 12 t of well-rotted manure per hectare can be applied to upland hay meadows plant communities, providing they are known to be unaffected by this higher rate (Crofts and Jefferson 1999).

For more information on farmyard manure, see the *Use of Farmyard Manure on Semi-Natural (Meadow) Grassland* (Simpson and Jefferson 1996). Some concern has been expressed about the sustainability of regular farmyard manure applications for both existing semi-natural hay meadows and restoration of improved communities.

BOX 3.2 Studies on farmyard manure application

An important and pertinent study carried out over a period of 12 years looked at the effects of different rates and application times of farmyard manure, with or without lime, contrasted with equivalent applications of inorganic fertiliser on semi-natural species-rich meadows in upland Cumbria and lowland Monmouthshire (Kirkham *et al.* 2008; 2014). The highest rate of farmyard manure (24 t/ha annually) was detrimental to plant species diversity, favouring nutrient-demanding species. A similar result was found with 12 t/ha annually at the lowland site. At the upland site, which had a history of farmyard manure application, the lower rate was ecologically sustainable, although lower levels would be beneficial. At the lowland site, with no history of fertiliser application, modelling suggested that to maintain species richness, levels of ≤ 4 t/ha/year might be sustainable. But even these levels may be unsustainable in the long run. However, the overall conclusion of the work was that even relatively modest fertiliser inputs can reduce a site's ecological value if it has no recent history of fertiliser application and liming. The authors also concluded that inorganic fertilisers are no more damaging than farmyard manure if equivalent amounts of N, P and K are applied.

Liming alone (Section 3.5.2) was not found to have any detrimental effect on the vegetation at either site, but when applied with farmyard manure, it reduced vegetation quality at the lowland site only. The authors speculated that this might reflect past management at the site. Part way through the trial, liming was included as background management for all fertiliser treatments at both sites, as liming is widespread in hay meadows, and can be considered as traditional management.

Despite these interesting results, the authors emphasised that it would be unwise to extrapolate their results to meadows whose history is unknown. However, they suggested that low levels of fertiliser application to lowland meadows of similar botanical composition (see original paper) might be ecologically sustainable and gradually increase adaptation of the vegetation to slightly higher levels.

3.5.2 Lime

Liming of semi-natural grassland has traditionally been practised to raise the soil pH to around 6.0, to maintain productivity. It is commonly applied on northern hay meadows (MG3b types) where farmyard manure is also applied. The application of lime offsets losses by leaching and cropping, which otherwise may lead to increased soil acidity and it reduces the availability of aluminium (Al), iron and manganese; this can result in lower growth rates of grass in more acid soils. Soil acidity is complex and is influenced by many factors, including: soil type and the underlying geology; soil texture; rainfall; and management practices, such as the application of inorganic fertiliser. There is relatively little recent research on the effects of liming on semi-natural grassland, although its application, either alone or coupled with farmyard manure, has been reported to have a detrimental effect on species diversity on semi-natural hay meadows (Kirkham *et al.* 2008). In this case, the adversely affected meadow had no recent history of liming, and the authors cautioned against extrapolation of their results to meadows with a more recent history of liming or farmyard manure application.

Naturally acid grassland soils have a pH range of 3.5–5.5; neutral soils have a pH range of 4.9–6.5; and calcareous soils have a pH range of 6.5–8.0. Liming is generally viewed as unacceptable on semi-natural acidic grassland, where it would favour neutral or calcicole species, unless there is very clear evidence of a recent shift between communities. Specialist advice should be sought in such cases. Occasional applications (every 5–10 years) on semi-natural neutral dry grassland may be acceptable, to maintain pH between 5.5 and 6.0 if there is a risk of acidification at a site, and providing there is a long history of liming (Walsh *et al.* 2011). Special care is required for Crested Dog's-tail–Common Knapweed communities (MG5c), which may have a soil pH in the range of 4.9–5.4. Liming is acceptable in this case, providing that the grassland is sustainably managed, allowing for natural leaching and regular herbage removal (Walsh *et al.* 2011). Site managers must ensure that any species of conservation interest would not be adversely affected; if calcifuge species are present, which will not tolerate increased soil pH, then liming should not be undertaken. Lime is not normally applied to calcareous soils, and should not be applied to chalk heath or intimate mosaics of chalk and acid grassland (Crofts and Jefferson 1999).

Where liming is being contemplated, soil pH should be tested, directly in the field with a portable soil test kit or by sending samples away for analysis. Follow the sampling methodology outlined in *Soil Sampling for Habitat Recreation and Restoration* (Natural England 2008c) (http://publications.naturalengland.org.uk/publication/31015). Soil texture and organic matter are also important for determining the amount of lime to be applied. Products should declare both the fineness of grinding and their neutralising value (NV) (effectiveness in raising pH compared to calcium oxide), and application rates should be adjusted accordingly, as this affects the reaction speed in the soil. Lime is usually derived from chalk and limestone, comprising calcium carbonate and magnesium (Mg), but marine calcareous sand may also be used. Lime is normally applied in spring, although the nesting season should be avoided if the grassland supports ground-nesting birds or other species of conservation concern that may be vulnerable at this time. It should not be stored in piles on semi-natural grassland, to avoid scorching the sward. For more information on recommended rates and NVs, see *Fertilizer recommendations for agricultural and horticultural crops* (Defra 2010) available on Defra's website. Rates would typically be less than 7 t/ha, with rates towards the high end applied over 2 years, providing the meadow is not adversely impacted by a second application (Walsh *et al.* 2010). For more information on liming materials see *Agricultural lime – the natural solution* (http://www.aglime.org.uk/).

BOX 3.3 Grassland restoration and management on historic sites

The earthworks at the Iron Age hill fort known as The Trundle (north of Chichester, Sussex) support a species-rich calcareous grassland flora.

Many archaeological sites owe their survival to grassland management that has ensured they have avoided cultivation (English Heritage 2004). Earthworks often support species-rich semi-natural grasslands, thus benefiting both archaeology and wildlife. Most local authorities maintain Historic Environment Records (HERs) as part of the Historic Landscape Characterisation (Clark *et al.* 2004), overseen by Historic England (formerly English Heritage). Programmes of grassland restoration or creation should not be undertaken without reference to HERs and historic maps; this should establish past land use and help to inform future management. This should also ensure that the historic character of the landscape is protected, and not adversely affected by any land preparation. It is important that both above- and below-ground archaeological sites, many of which are unrecorded, are protected. Features that might be present on a site include:

- above-ground features, such as earthworks, prehistoric humps or hollows, historic field systems identified by historic extant boundaries, ridge and furrow, and related earthworks, settlements of all periods; industrial sites of all periods; wetland sites that preserve palaeoarchaeological remains, such as pollen, invertebrate macrofossils, and so on;
- below-ground features, such as archaeological sites identified by soil and crop marks on aerial photographs; settlements of all periods; industrial features of all periods; prehistoric field systems;
- historic routeways, such as sunken lanes and relict boundaries with remnant features, such as veteran trees, earthbanks or collapsed stone walls;
- historic buildings of varying uses;
- parkland and designed landscapes.

Care has to be taken to avoid damage to archaeological sites during a grassland restoration or recreation; the following activities can result in damage, even in the course of managing existing grassland on an archaeological site:

- Poaching or erosion is a serious risk if the site is grazed with cattle or horses, particularly in places where animals congregate to lay up or drink, or move from one area to another. Care should be taken in siting feeders and water troughs (away from the archaeology). Temporary fencing can be used to protect damaged areas or developing erosion scars, and stocking densities should be adjusted if damage is noted (see Section 3.1). Animals should not be used on wet or boggy archaeological sites.
- Damage can be caused by heavy farm machinery, particularly on water-logged ground.
- Burrowing animals, such as Rabbits and Badgers, can be particularly problematic on earthworks and may require some intervention, as illustrated in the case study on Seven Barrows (see Box 5.1).
- Scrub encroachment can cause major damage to buried archaeology and may require control by hand tools and herbicides to avoid further damage to the archaeology (see Box 5.1).
- Bracken rhizomes can also damage buried archaeology; since the withdrawal of Asulam, mechanical options should be considered (see Section 3.3.4).
- Tree and shrub planting to enhance the structural diversity of a grassland restoration site is not generally recommended, and may require planning permission (see below).

In Scotland, environmental assessment and consultation processes ensure that habitat creation respects archaeological features and historic and designed landscapes. If planning permission were required for habitat creation in England, applications would be viewed by a county archaeologist and checked against the county HER; in Wales, one of the four regional HERs would be consulted. For any field-sized habitat creation scheme, an archaeological assessment should be undertaken; accredited experts may be engaged to undertake the necessary

Four thousand years of archaeology are on show at Jarlshof in Shetland, from late Neolithic and Bronze Age houses to a Norse settlement and Medieval farmstead. In the past the grassland was extensively managed, including the use of fertilisers and selective biocides, although these are no longer used across much of the site. Management of the outer area (where there is no visible archaeology) has been changed from a 75–100-mm cut to a 'cut and scatter' three times a year, based on a considerably higher cut. However, the grassland around the archaeology receives a 37-mm cut, and periodic treatments with a broadleaf herbicide. No heavy machinery is used on the site.

Species-rich chalk grassland flora (left) on the banks of the Iron Age hill fort (centre) and Bronze Age burial mounds (right) at Old Winchester Hill (Hampshire) (Tone Blakesley).

desk study and field survey. A surveyor should follow the methodology outlined in Royal Commission on Historical Monuments's *Recording Archaeological Field Monuments: A Descriptive Specification*. Historic England's guidelines are also summarised in *Understanding the Archaeology of Landscapes: A Guide to Good Recording Practice* (English Heritage, 2007).

Both the National Trust and Historic England's guidelines include different levels of survey that depend on the nature of the site and the resources and time available. In many cases involving habitat creation, a non-intensive survey would be appropriate, combining a desk-based assessment of known records together with field visits, equivalent to Level 2 of the Historic England guidelines. Remote sensing using LiDAR might be used to detect archaeological evidence in open landscapes such as grassland.

Initially, a desk study may be undertaken using sources of information such as:

- HER (formerly the Sites and Monuments Record);
- county Historic Landscape Characterisation map and documents;
- historical maps; pre-Ordnance Survey, such as tithe maps, estate maps and enclosure awards;
- aerial photographs; sources include the National Monuments Record and the Unit for Landscape Modelling, county councils and LiDAR (Forest Research and Environment Agency data);
- County Records Offices;
- archive material to trace past land use, such as manorial records, terriers and deeds.

The resulting report should present an historical perspective of the landscape; an outline of the historic or archaeological character of the site, including a summary of previous studies in the area; a site inventory and maps of the area; and a comprehensive list of features. It should also include management recommendations for the conservation of any archaeological features found, which can be combined with information from the landscape evaluation to determine the likely overall impact of habitat creation.

4. Grassland restoration: threats and challenges

4.1 Threats to semi-natural dry grassland communities

4.1.1 Agricultural improvement and land conversion

Although grassland improvement started well before the twentieth century, most fields were unaffected before the start of the Second World War. During wartime, much of our semi-natural grassland was ploughed and cultivated in response to food shortages; repeated ploughing and reseeding with cultivated grass varieties and clovers also improved productivity. Improvements continued for many decades after the war (Table 4.1), so that by the mid-1980s, 97% of pre-war enclosed semi-natural grassland in lowland England and Wales had been destroyed through ploughing, reseeding, fertilising or development (Fuller 1987), while similar losses probably occurred in lowland Scotland (Bullock *et al*. 2011). It was recognised at this time that semi-natural grassland communities should be priority habitats for conservation (UK Biodiversity group 1998a; 1998b). Other non-agricultural changes, such as conversion of grassland to forestry plantations, and the development of infrastructure, such as roads and housing, were more moderate by comparison, and these have now largely ceased (Bullock *et al*. 2011). After such massive losses, however, the remaining sites tend to be small and highly fragmented: the 2009 Natural England Inventory showed that across England, 70–80% of semi-natural grassland sites are less than 5 ha in size (Bullock *et al*. 2011). The Lowland Grassland Survey of Wales carried out extensive surveys of different NVC communities, recording a maximum mean patch size of just 0.45 ha for the most widespread meadow type, the Meadow Vetchling subcommunity of Crested Dog's-tail–Common Knapweed grassland (MG5a). The implications of fragmentation are serious, particularly in a changing climate (Section 4.2).

4.1.2 Lowering soil fertility

Intensive agriculture leaves behind lime and fertiliser, a residual fertility that encourages the dominance of competitive and weedy species, particularly coarse grasses and tall herbs, that tend to swamp other species. Liming also diminishes the prospect of recreating acid grassland, where pH values below 5 might be desirable. Low levels of nutrient

Table 4.1 Summary of past and present threats to semi-natural dry grasslands, and their impacts (based on Bullock *et al.* (2011) and reproduced with kind permission (© United Nations Environment Programme-World Conservation Monitoring Centre))

Threats	Semi-natural grassland type	Impact on semi-natural grassland	Role since 1940	Present role
Agricultural improvement	Priority habitats	Domination by fast-growing plants; biodiversity loss; soil processes compromised	Major	Reduced
Conversion to arable	Priority habitats	Loss of habitat	Major	Minor
Conversion to forestry	All	Loss of habitat	Moderate	Minor
Infrastructure development	All	Habitat destruction	Moderate	Minor
Nitrogen deposition and transfer	All	Increased soil fertility favouring fast-growing plants; loss of species diversity; acidifying effects; eutrophication	Major	Moderate
On-farm nutrient transfer and supplementary feeding	All	Increased soil fertility	Major	Moderate
Inappropriate cutting dates	Species-rich hay meadows	Reduction in species diversity	Moderate	Moderate
Undergrazing	Priority habitats	Undergrazing leading to rank grassland and scrub encroachment on many grassland types	Moderate	Major
Overgrazing	Upland calcareous	Overstocking or deer allows coarse grasses to invade; loss of species diversity	Major	Moderate
Habitat fragmentation	Priority habitats	Many sites are isolated and small, risking loss of species and invasions	Moderate	Moderate
Invasion by non-native plants	All	Possible reduction in species diversity	Minor	Minor
Climate change	All	Increasing risk of species loss, colonisation of novel species; increased openness	Minor	Becoming major

Semi-natural acid grassland in Cumbria with species such as Betony, Tormentil, Heath Bedstraw and Pignut; threatened by trees planted to create new native woodland.

availability, particularly of P, are recognised as critical for long-term species coexistence in grasslands (Gough and Marrs 1990; Marrs 1993; Janssens *et al.* 1998; Critchley *et al.* 2002a; 2002b). The opposite applies in agriculture and horticulture, where soil analysis has long been used to determine the quantity of nutrient available for uptake by the crop, and to make fertiliser recommendations. Typically, macronutrients are extracted from soils in solutions of sodium carbonate (the extract for Olsen P) or ammonium nitrate (K and Mg) in protocols originally developed by the Ministry of Agriculture, Fisheries and Food (Ministry of Agriculture, Fisheries and Food (MAFF) 1986). These are then calculated as milligrams per litre of soil and expressed in a scale of Indices, ranging from 0 (nutrient deficient) to 9 (Table 4.2). Most field soils have element Indices between

Table 4.2 Levels of 'available' nutrients in soils recovered using extract solutions (MAFF 1986; © Defra)

Index	Macronutrient (mg/L)		
	P	K	Mg
0	0–9	0–60	0–25
1	10–15	61–120	26–50
2	16–25	121–240	51–100
3	26–45	241–400	101–175
4	46–70	401–600	176–250
5	71–100	601–900	251–350
6	101–140	901–1,500	351–600
7	141–200	1,501–2,400	601–1,000
8	201–280	2,401–3,600	1,001–1,500
9	>280	>3,600	>1,500

Deficiency levels highlighted.

1 and 4, but unimproved grasslands typically return very low levels of extractable P and exchangeable K of 4–11 mg/L and 76–210 mg/L respectively (Walker *et al.* 2004b), equivalent to fertility Indices of 0–1 and 1–2 respectively.

The huge influence of nutrients in determining grassland community composition is well illustrated by long-term experiments. The most famous of these in England are the Park Grass Experiment at Rothamsted, begun in 1856 (Dodd *et al.* 1994a; Silvertown *et al.* 2006) and the Palace Leas Meadow Hay Trial in Northumberland (Shiel and Batten 1988), which started 40 years later but included grazing as well as hay cuts. The first fully replicated fertiliser trial, the Rengen Grassland Experiment in the Eifel Mountains (south-west Germany), was set up on acid grassland in 1941 (Chytrý *et al.* 2009). All of these experiments have shown how long-term fertiliser applications, using different levels and combinations of macronutrient treatments, can create distinctly different plant communities on plots only a few metres apart. Changes in species composition were relatively rapid in the first few decades, with major implications also for sward productivity, soil chemical properties and soil microbial populations. Some of these changes are still ongoing.

After analysing soil, productivity and plant species data from 132 semi-natural grasslands in different parts of Europe, Ceulemans *et al.* (2013) concluded that soil N enrichment, together with atmospheric N deposition, did contribute to species losses, largely through soil acidification and the sensitivity of some forbs to increased soil N (especially ammonium ion (NH_4^+). However, these effects were still subsidiary to the major influence of P enrichment, which was strongly negatively correlated with species richness. These negative effects of both N and P inputs appeared to operate independently, however, so that the management or limitation of both elements was critical in securing the highest levels of biodiversity.

In practice, evidence from upland hay meadows in Britain suggests that whichever macronutrient is in shortest supply will be instrumental in limiting vegetation growth and promoting diversity (Pinches *et al.* 2013).

A view from the westerly end of the Rothamsted Park Grass Experiment with subplots (left to right) receiving increasing amounts of lime. The main plot in the foreground is the farmyard manure plot illustrating a sharp contrast in botanical composition compared to the adjacent plot that receives no fertiliser inputs (© Rothamsted Research).

Aerial view of the Rothamsted Park Grass Experiment. The 3-ha field is divided into 20 main plots (top to bottom) with contrasting combinations and amounts of nutrients applied since 1856. Each main plot is then divided into four subplots (a–d; left to right), with varying amounts of lime applied to achieve a target pH of 7, 6 and 5; subplot d is left un-limed (© Rothamsted Research).

Plot 17 on the Park Grass Experiment receives 48 kg N/ha as sodium nitrate and no other nutrients; the relatively low soil fertility supports a diverse flora with over 30 species including the iconic species, Fritillary, which can be seen in the photograph (© Rothamsted Research).

Aerial view of the Rengen Grassland Experiment in the Eifel Mountains, south-west Germany (© Michal Hejcman).

Critchley *et al.* (2002a; 2002b) examined the influence of soil properties on lowland grassland communities in a wide-ranging sample of sites in Environmentally Sensitive Areas (ESAs) in England. The surveys covered mainly mesotrophic grasslands, many of which were semi-improved or improved (MG6 and MG7 types, respectively), with some calcareous and a few acidic communities. The survey showed that, as expected, unimproved MG3 and MG5 communities were associated with lower levels of nutrients than their equivalent in improved and semi-improved MG6 and MG7 swards, especially in terms of extractable P and K. Soil properties of MG1 types exhibited a very wide variation, reflecting their origin as different grassland types that had become progressively uniform and species-poor through undergrazing. Unimproved calcareous grasslands were generally associated with high pH levels and very low extractable P.

Overall, pH had the strongest influence on vegetation composition in the ESA samples, followed by total N and organic matter, both of which were correlated with available soil water content. Species richness showed a positive but weak linear relationship with pH and also tended to be correlated with low levels of low extractable P, but less so with exchangeable K and Mg (Figure 4.1). In fact, soil properties accounted for only a small proportion of the variation in the vegetation, indicating that other influences were overriding, such as management (cutting and grazing intensity and frequency; nutrient addition, and so on), hydrological and topographical variation, and access to colonisation sources.

Critchley *et al.* (2002a) also found typical pH ranges of 4.9–6.1, 6.0–6.4 and 6.8–7.9 for acid, mesotrophic and calcareous grasslands, respectively, with a strong gradient in species richness and pH mainly defined by extremes between heathland and calcareous grassland types, respectively (Figure 4.2). The main vegetation communities overlapped within the 5–15 mg/L range of extractable P, with a similar degree of separation in species richness. The authors concluded that the most promising subjects for restoration

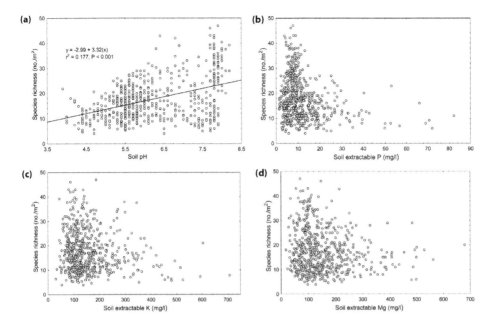

Figure 4.1 Results of soil and botanical samples from 571 locations in 14 ESAs in England, showing the relationship between species richness and (a) soil pH, and (b) soil extractable phosphorus, (c) potassium and (d) magnesium (reprinted from Critchley *et al.* 2002a, with permission from John Wiley and Sons).

were MG6 types, which with appropriate management could be encouraged to revert towards MG3 or MG5 community targets, whereas MG7 swards were often too heavily compromised by previous agricultural enrichment to be able to restore them easily. Where soil P levels were in excess of 10 mg/L, typical of brown earth soils, these were likely to require significant intervention measures.

Plant 'available' measures of nutrients are entirely dependent on the efficiency of chemicals used to extract them from soil samples. Using the Olsen method to investigate

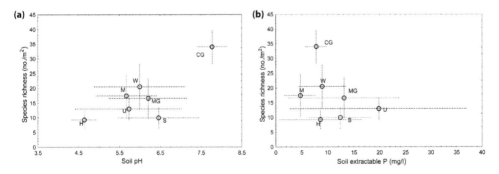

Figure 4.2 Relationship between species richness and soil pH (a) and soil extractable phosphorus (b) determined from the same ESA sample survey as Figure 4.1. Data are means for each vegetation type, with dotted standard deviation bars (reprinted from Critchley *et al.* 2002a, with permission from John Wiley and Sons). CG, calcareous grasslands; MG, mesotrophic grasslands; U, acidic grasslands; H, heaths; M, mires; W, underscrub.

P availability in soil samples from a range of different mesotrophic grasslands (MG1–10, 13 and 14), Gilbert *et al.* (2009) confirmed that the most species-rich communities (e.g. northern (MG3), flood (MG4), and old hay meadows (MG5)) tended to occur at the lowest P levels, contrasting with species-poor types associated with inundated soils, while intermediate P levels were associated with medium species richness.

To restore or recreate semi-natural grassland communities, the first priority on ex-arable or improved grassland sites is usually to reduce soil fertility. Marrs (1993; 2002) suggested three ways in which this could be achieved: by removing or diluting the nutrient pool; manipulating the stores and fluxes of nutrients to limit supplies; and increasing the removal of nutrients. In practical terms, techniques likely to achieve this are:

- ceasing fertiliser applications;
- fallowing, allowing natural leaching, or accelerating leaching using surface cultivation;
- arable cropping, cutting, burning and grazing;
- chemically manipulating soils to reduce nutrient availability, e.g. by adjusting pH;
- dilution with inert materials, or by deep cultivation;
- soil removal or stripping;
- reducing the productivity of competitive species, especially grasses, through long-term conservation management regimes.

4.1.3   Habitat deterioration

In Britain, our remaining semi-natural dry grassland is still threatened by a wide range of other factors, most of which cannot be accurately quantified, but their qualitative role has recently been summarised (Table 4.1). These include: atmospheric N deposition and run-off from adjacent agricultural fields; inadequate management through under- or overgrazing; and the timing of agricultural operations such as cutting. The cessation of traditional practices has increased on semi-natural grassland sites in recent decades (Bullock *et al.* 2011): for example, their species diversity may decline if they are inappropriately overgrazed in early summer, or if efforts are made to improve productivity through draining, liming, fertilising or herbicide use. Such grassland may still be classed as semi-improved if some of the original species composition remains, but improved if sown with Perennial Rye-grass cultivars after ploughing and fertilising, resulting in the loss of the native flora.

Threats are particularly acute on semi-natural dry grassland sites which do not have a statutory designation; for example, Hewins *et al.* (2005) found just 21% of a random selection of lowland dry acid grassland sites and 7% of upland hay meadow sites to be in 'favourable' or 'unfavourable recovering' condition, the remainder being poorly managed. The seriousness of the situation was further emphasised in a JNCC report published in 2006 on the condition of SSSIs, which focused on the special features for which those sites were designated (Williams 2006). The percentage of semi-natural dry grassland sites in favourable condition varied from just 25% of upland calcareous grassland to 42% of hay meadows. Undergrazing and abandonment were frequently cited as reasons, but examples of overgrazing and nutrient enrichment were also found. Undergrazing was thought to be the result of current agricultural economics and policies. By contrast, some semi-natural upland calcareous grassland sites have been subjected to overgrazing.

4.2 Climate change and dry grassland

The world's climate is changing, and this is already having a direct affect on species and populations in Britain. Climate change is a result of human activities, such as greenhouse gas emissions from burning fossil fuels and the continued destruction of tropical forests. Droughts in tropical rainforests may also reverse their sequestration of carbon, thus exacerbating the problem (Lewis *et al.* 2011). It is clear that human interference is affecting the world's climate, which poses risks for human and natural systems (Field *et al.* 2014).

4.2.1 Climate change in Britain

Climate projections for Britain have been produced in the form of maps and graphs, on a national and regional basis, using methodology designed by the Met Office (Murphy *et al.* 2009). Climate change has been projected to the end of the century, based on different carbon dioxide (CO_2) emission scenarios (http://ukclimateprojections.metoffice.gov.uk/). Generally, British summers are projected to get warmer and winters milder, with fewer frost days in many areas. Winters are also projected to be wetter over the majority of Britain, as was the case in 2013–2014. In summer, a south to north gradient is expected, with very much drier conditions in the south to little change in Shetland in the far north, but overall the annual rainfall is likely to remain constant. Under the medium emission scenario, more detailed projections are also made: for example, mean summer temperatures by the 2080s are projected to be just over 4 °C higher in southern England, but only about 2.5 °C higher in parts of northern Scotland. These predictions are based on a 50% probability or the 'median', so there is still considerable variability.

Climate change is also likely to alter the length and duration of the seasons, affect wind speeds and increase the frequency of extreme weather events. Weather events such as the relatively cold winters of 2008–2009 and 2009–2010, and the heavy snowfall across Britain early in the winter of 2010–2011 do not alter the overall conclusion that global climate is warming; in this millennium, each of the 12 years (2001–2012) features as one of the 14 warmest years on record. Despite a wealth of climate information, the future is still uncertain, as the projections are just that – uncertain signposts towards future options for adapting to climate change. But the precautionary principle demands that restoration practitioners and managers of semi-natural habitats think seriously about how to respond, focusing on adapting to the warming climate, as the time for mitigation is probably past.

Biodiversity, ecosystems and climate change are inextricably linked, and as a result a number of changes have already occurred in Britain, including examples of:

- species range;
- habitat preference;
- phenology and timing of reproduction;
- length of the growing season;
- timing of migration;
- frequency of pests and diseases.

Some of these issues have a direct impact on semi-natural dry grassland ecosystems, and these are discussed in the following section.

4.2.2 Impacts on grassland habitats and wildlife

4.2.2.1 Species range (and habitat preference)

A climate envelope is the area in which a species is able to survive and reproduce. In reality, if the climate envelope shifts, then a plant or animal must either move, essentially tracking climate change, or evolve to keep pace with the changing climate, otherwise it may become extinct. There have already been many examples in Britain of species from the warmer south expanding northwards; and colder-adapted species with a more northerly distribution in Europe, at the southern edge of their range, moving north or to higher elevations (where these exist). For example, in a study of 329 vertebrates and invertebrates from 16 taxonomic groups at the northern edge of their range in Britain, Hickling *et al.* (2006) found that 275 had expanded their range northwards and to a higher elevation since 1960. New species are also colonising Britain from the warmer south. Species better equipped to do this include some birds, bats, flying insects, intertidal invertebrates and marine fish. Similar northward (and southward) movements across Europe have happened in the past, in response to glacial periods, and the cooling and subsequent warming of the climate, most recently in the early Holocene.

Butterflies provide an interesting example, as many species reach the climate-induced limits of their range in Britain. They also have short life cycles and respond quickly to environmental change, therefore acting as valuable indicators. Between 1970 and 1982, and between 2000 and 2004, a number of species demonstrated predominately northerly range extensions, in keeping with the warming climate (Fox *et al.* 2007). These tend to be 'southern' wider countryside species, with the largest proportional increases for the Essex Skipper, with others including the Comma, Brown Argus and Speckled Wood. Climate change has also had some benefits for a small number of habitat specialists, for example,

The Comma butterfly has benefitted from the warming climate by extending its range (Tone Blakesley).

Climate change has benefitted some habitat specialists such as the Silver-spotted Skipper.

the Silver-spotted Skipper (Fox *et al.* 2011). This thermophilic species, at the northern edge of its range on lowland calcareous grassland in southern Britain, suffered a severe decline during the twentieth century due to intensive agriculture, declines in grazing and myxomatosis. In response to the warming climate, it is now able to exploit chalk grassland on previously cooler 'suboptimal' aspects, such as north-facing slopes and taller, cooler grassland (Davies *et al.* 2006), and as a consequence its range is expanding once more. This demonstrates the importance of microclimatic variability on a site, and the potential for some species to adapt to climate change by moving relatively short distances, to occupy parts of a site where previously the microclimate was unsuitable.

However, most habitat specialists have declined because they have very limited dispersal capability and the loss of breeding habitat prevents even very gradual range expansions. Three of the four northern species with southern range margins (Northern Brown Argus, Scotch Argus and Mountain Ringlet) have already exhibited range contractions to more northerly latitudes and higher altitudes (Franco *et al.* 2006). There have been no records of new colonists in Britain, but Red Admiral and Clouded Yellow, which are migrants, breeding in the summer months, are now able to overwinter, and may soon become residents.

From a wider European perspective, the *Climatic Risk Atlas of European Butterflies* (Settele *et al.* 2008) investigated the possible effects of various climate change scenarios on European butterflies to the year 2080. Distribution maps of European butterflies were used to predict future ranges of 293 of the approximately 450 European species under three climate scenarios. The results showed profound and largely adverse effects on the vast majority of species, with many potentially mobile species needing to move to avoid extinction, with immobile species facing a very uncertain future.

Although some wider countryside butterflies may be able to move in response to climate change, many other invertebrates and plants may not, due to their inherently poor dispersal mechanisms. Species that do have the capacity to move will find that the landscape today is very different to that in the past. Some amphibians and reptiles that used to migrate in the past, may not be able to navigate through a hostile landscape completely changed by man; habitat fragmentation is actually causing the range of such species to collapse southwards (Hickling *et al.* 2006). In the lowlands in particular, small fragments of suitable habitat may be separated by large areas of intensively farmed land, urban development and roads. Even species that are able to track the changing climate, may find that the environments into which they have moved are unable to support them, e.g. through a lack of food or nesting opportunities.

Although there are many examples of species facing enormous challenges with the changing climate, particularly those that are habitat specialists, many others are responding positively to management, such as the Cirl Bunting and the Marsh Fritillary butterfly. There are also similar examples among moths, grasshoppers and bush crickets, amphibians and reptiles (Maclean 2010). Some species are even colonising Britain, including well-documented breeding birds such as the Little Egret and Cetti's Warbler, recent breeding records of Purple Heron, Great White Egret and Spoonbill, and several new breeding species of dragonflies and damselflies in the South-East. The Long-tailed Blue butterfly raised broods at several south coast sites in 2013, and some species of continental moth may also be colonising Britain. There are also reports of tongue orchids and new records of bugs, bees, wasps and spiders (Maclean 2010).

Long-tailed Blues successfully bred along the south coast at several sites in 2013 (© Ross Newham).

It would be very useful to be able to predict how climate change will affect the range of British species, and potential colonisers of semi-natural grassland and other habitats over the coming decades. Scientists have already studied the impact of climate change on the distribution of selected species in Britain using computer simulations, based on the UKCIP02 climate models (which preceded UKCP09). These simulations allow future suitable climate space to be modelled for a variety of species (both 'winners' and 'losers') under varying climate scenarios (e.g. Walmsley *et al.* 2007). They begin to quantify the effects of climate change on wildlife, and inform nature conservation policy and wildlife issues in adaptation planning across the British Isles. More recently, Ausden *et al.* (2015) have considered how Britain's birdlife might change in the coming decades, based on the results of climate envelope modelling by Huntley *et al.* (2007), which assumes a 3 °C

The Willow Emerald damselfly is one of several members of the Odonata that have colonised South-East England in recent years (Tone Blakesley).

rise in global temperatures. However, while models provide a useful insight into the potentially dramatic impact of climate change on biodiversity, they cannot forecast with any precision the actual effects on communities, species or the competitive interaction between species. Each is subject to a range of caveats and assumptions, any of which could significantly change the projected outcomes (Smithers *et al.* 2007).

4.2.2.2 Phenology and the timing of reproduction

Climate change is already advancing some spring and summer events while shortening the winter season. Grassland ecosystems are complex, and changes in synchrony and competitive advantage between species could have profound consequences for the community as a whole, especially as different organisms are responding to climate change at different rates.

Already there are clearly documented changes in seasonal events in Britain, for example:

- earlier leafing of a range of trees, such as oaks, Ash, Lime and Hawthorn (Collinson and Sparks 2008);
- advancement of the first flowering date of 385 plants in central England; this had advanced by an average of 4.5 days in the 1990s compared to the previous four decades, with the greatest advances recorded for White Dead-nettle (55 days) and Ivy-leaved Toadflax (35 days) (Fitter and Fitter 2002). Insect-pollinated plants may react more strongly than wind-pollinated species, and species flowering earlier in the season appear to be more sensitive; this is probably correlated with temperatures in the month before flowering (Menzel *et al.* 2006);
- earlier flying of most British butterflies and many moths; they are flying earlier in the year now than in the 1970s (Roy and Sparks 2000; Burton and Sparks 2002);
- advanced egg-laying by birds;

- earlier arrival dates of summer migrant birds, particularly long-distance migrants (Cotton 2003);
- advanced amphibian spawning over a 16-year period (Beebee 1994).

Furthermore, by studying 542 plant and 19 animal species in 21 European countries (from 1971 to 2000), Menzel *et al.* (2006) showed that changes in phenological responses were not localised, but Europe-wide. Seventy-eight per cent of all leafing, flowering and fruiting events were advanced (30%, significantly). A serious concern arising from these changes is that different species respond differently, potentially disrupting crucial links in the food chain, such as flushes of caterpillars coinciding with the hatching of young insectivorous birds. In this case, birds may respond to photoperiod while their insect prey may respond to temperature, altering the previous balance. Information on how populations of plants and pollinators might be affected by climate warming is limited, although there is some evidence that there may be temporal mismatches in specialist pollinators, such as bees and hoverflies, and the plants they pollinate (Hegland *et al.* 2009).

4.2.2.3 Grassland communities

While many species have the capacity to move in response to climate change, habitats of course are fixed, and must therefore adapt. It seems inevitable then that as the climate changes, the structure and functioning of dry grassland plant communities will change to some extent, although far less is known about the response of whole communities than those of individual species.

A study on hayfields in northern Britain found that rising accumulated temperatures from February to June 2009–2012 accounted for 93% of the associated increase in sward height (University of Newcastle upon Tyne 2013). This suggests that swards will be taller in warmer, wetter springs and that this will greatly influence the time when a meadow is 'shut up' for hay (after an initial spring grazing). As accumulated temperatures also affect later nesting and fledging of some birds like the Yellow Wagtail, there is a risk that earlier hay harvests, stimulated by warmer years, could result in increased mortality (Pinches *et al.* 2013).

One of the most likely causes of disturbance to dry grassland ecosystems in the future is summer drought. Severe droughts may cause the mortality of established species and open up gaps for possible colonisation (Stampfli and Zeiter 2004), thereby changing the vegetation dynamics, or allowing in new species. Warmer, drier summers in the south, for example, would increase moisture deficits, particularly on thin calcareous soils of south-facing slopes, and well-draining sandy acid soils. Fire risks might increase on heathland and dry acid grassland. Where soil moisture deficits are less problematic, warmer temperatures, longer growing seasons and elevated CO_2 might stimulate more luxuriant growth, which may affect species composition. This will have implications for management, possibly requiring grazing earlier in the year, or in the case of meadows that are cut, this may need to be carried out earlier or possibly more frequently.

In one of the rare studies on dry grassland ecosystems, Grime *et al.* (2008) found that semi-natural calcareous grassland vegetation in northern England was highly resistant to simulated climate change treatments. They concluded that in this case, the impacts of nutrient enrichment and changes in management practices are a greater threat to plant communities than climate change at this time. In contrast, similar manipulations

The climate change experiment in northern England at Buxton Climate, situated in calcareous grassland in Derbyshire, has been the subject of continuous climate manipulation since 1993, including simulated summer drought imposed with the automated rain shelters seen here (© Jason Fridley).

of calcareous grassland in an early successional stage in Oxfordshire caused substantial vegetation changes.

Some indications of how habitats might be affected have been developed based on the UKCIP02 climate models. No similar assessments have yet been made based on the more recent UKCP09 projections. Seven of the 32 former UK BAP priority habitats were assessed to be at high risk, with a further 14 assessed to be at medium risk (Mitchell *et al.* 2007). Most semi-natural dry grassland habitat was assessed at low risk of direct impact of climate change, with impacts predominately influenced by other factors. Lowland heathland was assessed at medium risk, partly due to potential shifts in the competitive balance between heather and acid grassland. However, conclusions on the UK BAP priority habitats at risk were based on moderate or poor evidence (Mitchell *et al.* 2007). Catchpole (2011) set out a methodology for evaluating the vulnerability of priority habitats identified as being at risk from climate change; the key aim was to advise on climate change adaptation measures. Natural England subsequently developed a National Biodiversity Climate Change Vulnerability Model to allow non-specialist biologists to identify areas most at risk from climate change (Taylor *et al.* 2014). What is clear from all the evidence of recorded changes in plants and animals, and the climate change projections, is that maintaining habitats and ecosystems as they are today will be impossible (Lawton 2010).

4.3   Challenges in dry grassland restoration

Traditionally, grasslands were maintained either as pastures (using grazing animals exclusively) or meadows (removing a summer hay crop to produce winter-feed for livestock, usually accompanied by grazing periods in the spring, autumn or winter). With the massive destruction of semi-natural grassland during the twentieth century (Section 4.1.1), the least productive farm areas – on steep slopes that could not be ploughed – or wetland, were often abandoned altogether and allowed to scrub over.

Cumulative annual fertiliser applications of up to 400 kg N/ha, plus phosphate and potash, are now routinely spread on winter cereals and productive grassland. Grass cutting for silage is more economic than hay crops and starts much earlier in the year (May to mid-June), followed by one or two further cuts during the growing season. With added fertiliser, this increases dry matter production threefold compared with unfertilised, semi-natural grasslands, but the early cutting prevents the flowering and seeding of many forbs and so favours a closed canopy of grasses. Similarly, on fertilised pastures, very heavy stocking rates are possible, sometimes up to three times greater than could be supported on semi-natural grassland, but this can damage sites by contributing to soil compaction and nutrient run-off.

A further source of increasing fertility in grasslands is atmospheric pollution, or deposition of N and sulphur (S) from fossil fuel burning, the former including ammonia production from intensive dairy, poultry and pig farming enterprises. In parts of north-west Europe, deposition rates of up to 30–55 kg N/ha/yr may actually exceed 'critical loads' likely to cause changes in the function and composition of semi-natural grasslands, which normally are of the order of 15–30 kg N/ha/yr (Bobbink *et al.* 1998). In the UK, atmospheric deposition of N on acid grasslands varies from 5 to 35 kg N/ha/yr, the middle range of which is associated with a reduction of species numbers of at least 20% (Stevens *et al.* 2004). Higher deposition will in turn tend to increase N mineralisation, switching the balance of the community in favour of tall grasses, such as Tufted Hair-grass, Tor-grass, False Oat-grass and Wood Small-reed, depending on the soil type and region. At the same time, high deposition rates may cause a lowering of soil pH on base-poor substrates, a factor that may be at least as important for plant species richness as the increased fertility (Maskell *et al.* 2010). In general, however, eutrophication through atmospheric deposition is consistent with long-term vegetation changes like those observed by Bennie *et al.* (2006) in calcareous grassland, which became progressively more mesotrophic over 50 years. Countryside Surveys between 1998 and 2007 in England, Scotland and Wales have also detected decreasing plant species richness in neutral and acid grasslands, corresponding with an increase in nutrient-demanding and competitive species (Countryside Survey 2009; Norton *et al.* 2009; Smart *et al.* 2009). Other factors contributing to these trends include decreased grazing pressure and habitat fragmentation, the latter reducing the size of viable plant populations.

In addition to the impact of eutrophication on species diversity, there are formidable biotic limitations to be overcome. Soil seed banks and soil microbial communities of the original habitat will have become impoverished and altered following agricultural intensification. Key plant species will also be lacking; this is compounded by the limited opportunities for propagule dispersal between the remaining patches of species-rich grasslands, which have become separated from each other in the landscape (Bakker and Berendse 1999). Typically, in ancient grasslands the majority of species were long-lived perennials, with perhaps only half of the community members present in the seed bank.

Some of these are relatively transient and will quickly die out if management practices alter, often leaving behind the persistent seed banks of undesirable weedy annuals and biennials that are not strictly part of the target vegetation community, but which always have the potential to dominate following disturbance.

Taken together, both abiotic and biotic constraints can severely challenge restoration efforts, whether to recreate species-rich grassland communities on fertile, former arable land, or to diversify already agriculturally improved, species-poor pastures. There are also a large number of abandoned former grassland sites that have undergone or are undergoing succession as they become dominated by tall grasses, shrubs and, eventually, trees. To address these problems, restoration tasks must include a range of measures (Walker *et al.* 2004b). They must:

- reverse successional processes on abandoned or undergrazed pastures;
- reduce high residual soil fertility resulting from previous agricultural practices and eutrophication from other sources;
- address topographical and hydrological constraints;
- restore appropriate management to encourage sward diversification, often by reinstating traditional practices, by reducing competition for space and resources with less desirable species, and providing microsites for the establishment of target species;
- provide sources of propagules, where lacking, to rebuild the desired target community;
- reconnect and reintegrate restoration sites with remaining areas of species-rich grassland sites in the wider landscape.

When deciding which of these measures are most appropriate for sites requiring restoration, it is important to consider:

- the nature conservation value of the site and whether restoration would harm any species of conservation importance;
- the management history resulting in the present condition of the site, including soil characteristics;
- whether to reinstate traditional management, or use enrichment planting or habitat creation;
- the proximity of semi-natural grassland;
- the grassland community most suited to the site in its current condition (consider reference sites);
- the chances of success and the likely cost, including the impact of activities on adjacent land.

The appropriate restoration tasks can then be implemented, together with a programme to monitor their effectiveness (Section 7.6). In some cases, the first step will be choosing a suitable site, particularly when creating a new grassland habitat. Making the correct choice could be the difference between success and failure.

BOX 4.1 Case study: Winterbourne Downs arable reversion, part of the RSPB's Wiltshire Chalk Country Project

Topic: Recreating landscape-scale, species-rich chalk grassland
Location: Winterbourne Downs, on the Wiltshire Downs
Area: 296 ha
Designation of site: Arable farmland linking SSSI chalk grassland sites
Management and community use: RSPB-owned, but managed as a nature reserve and working farm with public access in some areas
Condition of site: Arable farmland
Website: http://www.rspb.org.uk/discoverandenjoynature/seenature/reserves/guide/w/winterbournedowns

Stone-curlews require short vegetation or bare ground for breeding, close to insect-rich feeding areas such as chalk grassland (Andy Hay, http://www.rspb-images.com/).

Background

There have been substantial losses of calcareous grassland in Britain over the past 50 years. A significant proportion of the remaining unimproved grassland is found on chalk in the Wiltshire Downs, supporting a diverse array of invertebrates, including many Red List species and chalk grassland specialists. The farmland landscape also supports strong populations of red- and amber-listed farmland birds, including Stone-curlew, Lapwing, Grey Partridge, Yellow Wagtail, Linnet, Yellowhammer and Corn Bunting. Of particular concern to the RSPB is the Stone-curlew, which is closely associated with the existing areas of chalk grassland on Salisbury Plain and Porton Down. Stone-curlews also breed on Marlborough Downs to the north, with a few pairs on the northern edge of the West Wiltshire Downs to the south-west. By the 1980s, just 100 pairs were estimated to breed nationally, which included 30 pairs in Wessex (Wiltshire, Hampshire, Berkshire and southern Oxfordshire). Numbers have now increased to over 130 pairs in Wessex, out of a national total of approximately 350 pairs. This increase in Wessex reflects co-operation between farmers and conservation bodies such as the RSPB and Natural England. Stone-curlews require short vegetation or bare ground for breeding, close to insect-rich feeding areas, such as species-rich chalk grassland. Creating unsown fallow plots in crops, or tightly grazed chalk grassland can provide breeding sites.

The aim of the RSPB's Wiltshire Chalk Country project is to create a landscape-scale area of species-rich chalk grassland, by expanding grassland areas between Salisbury Plain and Porton Down. Corridors and stepping stones of chalk grassland habitat are being created to support the area's mixed farming, which is rich in wildlife, and reverse the declines in farmland birds, including the provision of permanent safe havens for nesting Stone-curlew in the heart of their Wessex range. Several reserves have already been acquired in the area, including the purchase of Manor Farm by the RSPB in 2005, which covers 296 ha of Winterbourne Downs.

Arable reversion, 5 years after seeding.

Chalk grassland creation methods at Winterbourne Downs

Machine sowing of brush-harvested seed is the method of choice for the creation of species-rich chalk grassland on the ex-arable land of Manor Farm. This reflects the large areas involved (160 ha have been sown in recent years), the cost and the option to mix seed from different sources. A sowing mix is typically made up of seed from three or four sources, to maximise the diversity of species in the developing sward. Seeds have been sourced primarily from the Salisbury Plain Training Area and peripheral sites such as Tytherington Down, as well as seeds harvested from the Marlborough Downs and from a limestone meadow near Bath. Sowing is normally undertaken in early to mid-September. This is too late for green hay, which is only used locally for enrichment of existing semi-improved grassland.

Following harvesting of an arable crop, a sterile seedbed is created by spraying glyphosate to eradicate arable weeds such as thistles, Common Couch, Barren Brome and Black-grass, which are first allowed to germinate. Spraying is undertaken at least three weeks prior to sowing. Cultivation is then undertaken with a spring tine cultivator to provide a good tilth. Wildflower and grass seed are mixed with kiln-dried sand to prevent bridging in the pneumatic fertiliser spreader which blows out the seed from regularly spaced tubes (1-m apart) along an 18-m boom width. Seed is sown with 3–4 complete passes. The seedbed is then rolled with a Cambridge roller to give good contact with the soil. A sowing rate of 20 kg/ha typically costs about £800 per hectare. For the large-scale arable reversion undertaken at Winterbourne Downs, it is estimated that spraying, cultivation, seeding and subsequent rolling takes about 2 hr/ha. Smaller areas up to 6 ha have been sown by hand, using seed mixed with wood shavings to facilitate broadcasting. This method is as effective as machine sowing, but more expensive due to labour costs, unless volunteers can be employed.

In the first year sites can look very patchy, although many chalk grassland species can be found. To prevent arable plants from persisting in the sward, sites are topped once in early May, again in June or July and occasionally later in the year if required. (The first topping is timed to catch the majority of the Black-grass before setting seed, and to avoid topping more than 25% of the Yellow-rattle in flower.) Grazing with sheep in autumn/winter creates a short, open sward. Hay crops from second and third year reversions allow both nutrient stripping and the flowering and seeding of the developing chalk grassland flora. The P index can fall from 2 to 0 in 3 years using this technique but subsequent analyses have shown how the P locked up by the chalk can be made available at higher levels in later years, although the overall trend is downwards.

Both seed sowing with a pneumatic fertiliser spreader and by broadcasting by hand have been employed at Winterbourne Downs Nature Reserve (both images © Patrick Cashman, RSPB).

Left: Sites can look very patchy in the first growing season, although species such as Salad Burnet, Small Scabious and Yellow-rattle may already be evident by midsummer. Right: Yellow-rattle has been topped in spring, but the lateral shoots survive to flower later in the summer.

Monitoring

To assess the suitability of arable fields for reversion, soil analysis for P, K and pH is carried out in February, when the soil nutrient status is stable. This is repeated every 5 years for chalk grassland creation fields. Following seeding, vegetation is monitored annually for the first 3 years, and then every third year for the first 10 years of establishment. Typically, 10 random 1-m quadrats are surveyed in fields < 10 ha, and 20 quadrats in fields > 10 ha. An assessment of frequency is derived from the number of quadrats in which each species is found. To provide a measure of success, positive and negative indicator species are assessed, together with estimates of sward height and bare ground. The target for success is a minimum of five positive indicator species (two frequent, two occasional and one rare) within 10 years. For this attribute target to be met in botanically enhanced or newly created swards, at least three of the indicator species should be characteristic of nutrient-poor conditions. The list of positive botanical indicators for lowland calcareous grassland was taken from the HLS Farm Environment Plan (FEP) Manual: *Technical Guidance on the Completion of the FEP and Identification, Condition Assessment and Recording of HLS FEP Features. Third Edition* (Natural England 2010a). Some positive indicator species may be classed as negative if their frequency increases above 10%. Positive indicator species recorded in five-year-old grassland include Quaking-grass, Sheep's-fescue, Kidney Vetch, Clustered

Arable reversion at 1 year (left), 3 years (middle) and 5 years (right) after sowing, illustrating increasing diversity and consolidation.

Bellflower, Oxeye Daisy, Cowslip and Common Knapweed. Chalk grassland species that have colonised naturally include Pyramidal Orchid, Viper's-bugloss and Common Broomrape. Butterflies found in reversion chalk grassland include Small Blue and Dark Green Fritillary; other invertebrates include large spiders (*Araneus quadratus*) and Wasp Spider.

Information provided by Patrick Cashman (RSPB).

4.4   Limits to natural colonisation

Many different techniques have been used to create or diversify grasslands, depending on the starting point, whether industrial wasteland, former arable land or improved pastures. Each situation requires its own individual approach, usually involving some initial ground preparation, accompanied by seeding or planting of the desired species, and followed up by appropriate management. Although technical approaches tend to be the norm, in specific circumstances spontaneous succession may, without significant intervention, also deliver reasonable results, especially if the site conditions are not very extreme. An example would be a small, low-productivity site in close proximity to seed sources of the target vegetation type (Prach and Hobbs 2008). In Eastern and Central Europe, especially in the drier regions, grassland restoration through natural succession can be very successful, as seen in parts of Germany, the Czech Republic, Romania and Hungary, where remnant semi-natural grassland is still relatively abundant in the landscape. Here, species-rich perennial grasslands may develop on abandoned fields within 10 years and resist shrub invasion for many decades, whereas in higher rainfall areas the succession to forest is much more rapid. In South Moravia (Czech Republic), Sojneková and Chytrý (2015) found that perennial grassland established in less than 4 years after abandonment, and that native dry grassland species, including some Red List species, gradually increased after 40 years. However several 'target' species of ancient grasslands were still absent or rare after more than 50 years.

Another example of this is the spontaneous increase in abundance of species characteristic of steppe grasslands in older, abandoned agricultural fields in lower Transylvania (Ruprecht 2006). Most of these fields previously had favourable management, being cut for hay or summer-grazed, while potential seed sources were relatively abundant, leading to a repopulation of many target species within 14–20 years. Similarly, in the White Carpathians Protected Landscape Area in the south-east of the Czech Republic, where ancient, species-rich grassland sources are still relatively abundant, their proximity contributed to the ingress of unsown target species into restored, ex-arable fields restored over a decade (Mitchley *et al.* 2012; Prach *et al.* 2013). In the same region, Johanidesová *et al.* (2014) examined the ability of 11 target species to invade arable land previously restored by seeding regional seed mixtures. Although these mixtures did not contain original reference species, their ability to colonise was significantly influenced by their distribution in the surrounding landscape at distances of up to 500 m from the restored sites, the minimum distance to the nearest population, and the time since the start of the restoration. The most successful were the wind-dispersed composites Queen Anne's Thistle and Irish Fleabane, while the least successful was the ant-dispersed Leafy Spurge. Again, as much original grassland remained in the region, the potential for spontaneous colonisation was considered high compared with other more fragmented landscapes. Even on older lignite extraction sites within Central Germany, which were restored without topsoil, spontaneous vegetation cover converged towards that of ancient grassland within a decade, although more slowly than when actively restored with green hay or regional seed mixtures (Tischew *et al.* 2014). The probability of establishment increased when a species was abundant in the nearby surroundings, especially those with lighter seeds and slower terminal velocities. Natural colonisation, supported by traditional agricultural management, might therefore allow gradual diversification to occur on its own, and we consider this laissez-faire approach first.

Spontaneous colonisation, however, can be a very long drawn-out process, peculiarly unsuited to grassland communities because of the likely eventual succession to woodland. It will normally be limited by the lack of desirable plant propagules and the absence of appropriate management. Even when there is species-rich grassland nearby, the process of recovery may take several decades. In their study of areas disturbed by the manoeuvres of tracked army vehicles on Salisbury Plain, Hirst *et al.* (2005) found that although sward closure in disturbed calcareous (CG3) or mesotrophic (MG1) patches was rapid, taking less than 20 years, a close similarity in species composition to adjacent, undisturbed areas was only achieved after 35–50 years. Early colonisers tended to be annual and biennial forbs rather than grasses, but older patches had a higher abundance of perennial forbs than control areas. Other studies on the development of calcareous grassland on ex-arable sites have concluded that the process could take 100 years or more, and then only with the benefit of dispersal from nearby seed sources (Wells *et al.* 1976; Gibson and Brown 1991; 1992; Redland *et al.* 2014). Distance from the source is a key factor: on chalk soils in southern England, Fagan *et al.* (2008) found that the vegetation on arable restoration sites became progressively dissimilar with increasing distance from the nearest ancient (> 200 year) grassland sites with a similar slope and aspect. Although these restoration sites tended to move closer in composition to the reference sites over time, there was little overlap even after 60 years.

The same is likely to be true for other grassland types. Woodcock *et al.* (2011) estimated that it could take more than 150 years for an ex-arable site, sown with a seed mixture from a donor floodplain meadow, to return a similar species composition; although even then not in the same relative frequencies as those species found in the target community. Management can help to disperse the desirable species, although again this could be very long term. In a study in south-eastern Sweden, Öster *et al.* (2009) found that the abundance of colonising target species decreased rapidly across interfaces between semi-natural, mesotrophic grasslands and the adjacent ex-arable fields. Species richness and the similarity to semi-natural grassland communities declined over 12 m into the ex-arable sites, even though the sites were conjoined and grazed in common by sheep or cattle. On sites that had been conjoined for longer, species richness did increase further in the ex-arable sections after intervals of > 50 years, but was still significantly lower than in the semi-natural grassland. The authors concluded that establishment limitation, combined with dispersal limitation, were the main mechanisms restricting colonisation. Another study, also in Sweden, found more plant species in young (1–13 years) restored pastures adjacent to semi-natural grassland than in those adjoining crop fields, again highlighting the influence of adjacent habitat sources (Winsa *et al.* 2015). Even within this short time frame, species richness increased with the time since restoration and specialist species also tended to decrease with increasing distance from the edge.

Changes in management, even when seed sources are abundant, may also show delayed responses in the vegetation. In the classic Park Grass Experiment at Rothamsted, plots acidified by repeated low applications of ammonium sulphate fertiliser changed from an unfertilised, species-rich, MG5-type meadow towards a species-poor, Perennial Rye-grass–Meadow Foxtail (MG7d) community over 47 years (1856–1903), but then reverted back towards the original sward composition of the control plots during the next 70–90 years when regular liming treatments were applied (Dodd *et al.* 1994a). Similarly, Smith *et al.* (2000) calculated that to convert an improved, mesotrophic grassland to a northern hay meadow type (MG3) could take 20 years, assuming (although by no means certain) a linear rate of increase in species numbers.

Where semi-natural grasslands have been totally abandoned, the window for recovery is relatively short. On chalk grassland, for example, tall grasses often begin to outcompete the finer grasses and forbs after 10–15 years, shrubs in 15–20 years and trees after 35 years (Willems 2001). After a few decades the situation can therefore become gradually irreversible, making it impossible (without intervention) to reintroduce species that have disappeared. This is exacerbated by factors such as the paucity of target species with persistent seed banks, the susceptibility of the sward components to shade, and the build-up of nutrients due both to litter accumulation and atmospheric deposition of N.

4.4.1 Impoverished seed banks

Ancient grasslands managed as hayfields generally have low seed longevity (Bekker *et al.* 1997, 2000) and only those with persistent seed banks will therefore survive periods of neglect or abandonment. Grasslands subject to a high degree of natural disturbance, such as flooding in alluvial meadows (e.g. Hölzel and Otte 2004) tend to have more species with seed that can persist in the soil, some of them rare target species. However, in relatively undisturbed acid, mesotrophic and calcareous types, the soil seed bank seems to be of minor significance in determining sward composition (e.g. Öster *et al.* 2009). As a result, the seed bank composition of semi-natural dry grassland like a northern hay meadow (MG3) may be little different from its species-poor, fertilised equivalent (MG6) (Smith *et al.* 2002).

In arable situations, repeated soil inversion and aeration by ploughing tends to impoverish the seed banks of late successional species while favouring persistent ruderal species, many of them common weeds. The most dominant perennial grasses and forbs have only transient seed banks and therefore cannot survive cultivation and improvement. Even in the absence of cultivation, colonisation from adjacent, species-rich swards remains limited by the inability of key species either to persist as buried seed or to disperse any distance. On an ex-arable site that had been resown with grass 10 years previously, Hutchings and Booth (1996a) noted that the surrounding ancient chalk grassland flora of Castle Hill National Nature Reserve (NNR) in East Sussex, England penetrated only short distances into the improved sward. Across a narrow field of approximately 70 m, surrounded by chalk grassland on both sides, less than 20% of the total forb seed bank and less than 1% of the grass seed bank was characteristic of chalk grassland, with even fewer species recorded in the seed rain. Graham and Hutchings (1988) concluded that there was little potential for re-establishing chalk grassland vegetation either from the remnant seed bank or the adjacent seed rain, so that former arable sites on chalk soils were likely to become rapidly dominated by weeds of arable cultivation. As the proportion of target species with persistent seed banks in calcareous and many other grassland types is relatively low, often of the order of 25–30%, many species cannot be recovered from seed banks and would therefore need to be introduced.

This analysis – that the assembly of species-rich communities without intervention is ultimately seed-limited, both by a lack of target propagules in seed banks and their scarcity in the surrounding landscape – agrees with that of other researchers (Poschlod *et al.* 1998; Bossuyt and Honnay 2008). Seed banks of semi-natural vegetation can, however, be recovered in improved grasslands after appropriate restoration treatments. In the case of former alluvial meadows, Schmiede *et al.* (2009) recorded the slow build-up of target species that had been restored by introducing plant material from donor sites. After 5–6 years, weedy species such as Toad Rush still dominated and the transferred species tended to decline with depth, indicating that recovery was a relatively slow process.

Restoration is also less likely to succeed where abandoned grassland has begun to disappear under scrub and woodland. Davies and Waite (1998), for example, found that seed banks of calcareous grassland species decreased progressively with distance into scrub layers, and also with the age of the scrub. Similar findings were made in the Swabian Jura of south-west Germany, on sites where limestone grassland had been abandoned or afforested for up to 30 years (Poschlod *et al.* 1998; von Blanckenhagen and Poschlod 2005). Numbers of grassland species in the vegetation rapidly declined under the developing scrub and plantations, as did the seed bank, but at a slower rate. Even after cutting back scrub and forest and 9–10 consecutive years of grassland management, there were still major differences between the restored areas and remaining patches of open grassland. The cleared sites contained higher proportions of species with good dispersal mechanisms, copious seed production and transient seed banks, as well as more species with long-term and persistent seed (Figure 4.3).

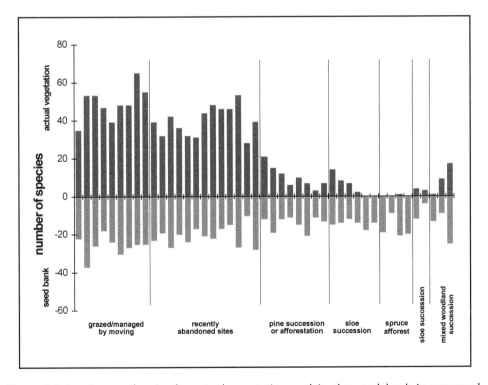

Figure 4.3 Species number in the actual vegetation and in the seed bank in managed, abandoned and afforested calcareous grassland sites (reprinted from Poschlod *et al.* 1998, with permission from John Wiley and Sons).

Nevertheless, reversion to species-rich grassland may be feasible if the succession is caught in time. A study contrasting open limestone (alvar) grassland in Estonia with areas overgrown by young pine forest aged 30–40 years found significantly lower species richness in seed banks under tree canopies, but an absolute loss of only three of the 69 species present in the open (Kalamees and Zobel 1997). Again, clearing a partial (< 50%) scrub cover of Juniper, Rose and Broom from limestone grassland plots in the French Prealps, followed by a grazing regime, caused almost a doubling of species richness of open grassland species after 4 years (Barbaro et al. 2001). In this particular case there were adjacent seed sources readily available.

4.4.2 Limited dispersal opportunities

Potential dispersal agents for seeds and propagules are wind, water, mammals, birds and human activities. Prior to agricultural intensification, the dispersal and exchange of propagules between areas of semi-natural grassland was effectively achieved by wandering herds of grazing animals or by hay harvesting machinery. In many parts of Europe these original grasslands exist only as small, isolated fragments of declining quality, vulnerable to neglect and to edge effects from the surrounding landscape. Small plant populations become vulnerable to inbreeding depression, increasing the probability of extinction. Dispersal of pollen by insects beyond 100 m, and certainly 1 km, is a relatively rare event. At the same time these fragments represent important food and nectar resources for insects whose wider dispersal declines with increasing distance between the fragments (Öckinger and Smith 2007). As the pollinators decline, this also threatens the survival of the plants that they pollinate (Biesmeijer et al. 2006).

Similarly, seed and propagules may disperse only a matter of metres unless they have specialist adaptations, such as plumes or wings for wind dispersal, or hooks or burs that attach to animals. Nevertheless, occasional long-distance dispersal events may occur because of the many chance processes involved in transporting seeds (Higgins et al. 2003). Dispersal distances of as little as 0.3–3.5 m have been found for some short-lived chalk grassland forbs (Verkaar et al. 1983), while wind transport of seed of many meadow species rarely exceeds 10 m (van Dorp et al. 1996). Colonising capacity is determined by a number of factors, such as the amount of seed produced by individual plants; seed weight, shape and inflorescence height relative to the rest of the vegetation; and finally germination success. In their study of isolated populations of wind-dispersed grassland forbs in the Netherlands, Soons and Heil (2002) showed that colonising capacity decreased with decreasing plant population size, which they attributed partly to lower seed production and less germination, but also to possible inbreeding depression and reduced pollination incidence. Furthermore, on sites with greater biomass production, the larger plants produced heavier seeds with better germination but poorer long-distance dispersal, suggesting that on isolated, relatively fertile sites with higher productivity – the likely situation for many restoration projects – the risks of extinction of small populations would be compounded. Dispersal limitation has often been reported in restoration projects and long-term sowing experiments, where some species remain confined to their original stations: for example, in a long-term study on ex-arable land using diverse mixtures, heavy-seeded species such as Meadow Vetchling were virtually absent from the plots where it was not sown, whereas light-seeded species such as Yellow Oat-grass colonised freely between plots (Lepŝ et al. 2007).

Grazing animals and other forms of human-mediated dispersal offer a possible transport mechanism for seeds and other propagules (Auffret 2011). Prior to recent agricultural intensification is it likely that livestock movement between fields, as well as unaided dispersal from local sources, brought about effective recolonisation. Fischer *et al.* (1996) found that a single sheep could transport more than 8,500 propagules, comprising 85 species of vascular plants, on its fleece, as well as some insects such as grasshoppers. They calculated that at least half the plant species present in calcareous, semi-natural pastures were capable of being transferred to other fields on the fleece of a single animal over a season, independent of seed or fruit morphology. Couvreur *et al.* (2004) found that a third of the seeds adhering to the coats of Galloway cattle and horses on Flemish nature reserves had awns, spiny calyx teeth or burs; 29% had wind dispersal appendages such as plumes or woolly hairs, while 27% had no obvious dispersal traits. They considered that isolated nature reserves (and by extension grassland restoration sites) could be effectively connected through seasonal grazing and the reciprocal transfer of seeds, possibly influencing long-term survival of plant populations. Research studies suggest that sheep retain more propagules than cattle, and cattle more than horses, and that different animals and breeds may disperse different sets of plant species.

Seeds are also carried in mud on animal hooves and coats, while some can survive ingestion by grazing stock and deposition in the dung, where they may be able to germinate and establish. Cattle, ponies and deer all disperse large quantities of seeds in their dung (Bakker and Olff 2003; Cosyns *et al.* 2005; Mouissie *et al.* 2005) and several investigations have carried out germination tests on dung samples collected from the field, under glasshouse conditions. In one such study, the dung samples of Konik or Shetland ponies and Highland cattle grazed on sand dune nature reserves along the Belgian coast, were found to contain 27% of all (117) plant species present on the sites. Some species, mainly ruderals, were present in large quantities (Cosyns *et al.* 2005). There were differences in seed composition between ponies and cows reflecting their feeding preferences, with more frequent dunging occurring in their preferred feeding habitats. However, while germination tests can be very successful, establishment in the field may be much less certain, depending on the availability of microsites for regeneration (Pakeman and Small 2009).

Many seeds do not survive ingestion by grazing animals. Edwards and Younger (2006) showed that when hay harvested from MG3 upland meadows was fed to cattle, the dung contained heavily reduced seed numbers and far fewer species. Some species, such as Rough Meadow-grass, Meadowsweet and Great Burnet, were relatively resistant to damage, including ephemeral weeds such as Toad Rush. Nevertheless, though there may be low levels of seed survival in the gut, large herbivores have the potential to disperse seeds in dung over their considerable ranges (Pakeman 2001). Retention time in the digestive system varies from approximately 40 h in ponies to more than 70 h in cattle, allowing for long-distance transport of seed, particularly if animals are moved from species-rich grassland reserves to species-poor sites requiring restoration, mimicking historical shepherding practices of transhumance. The spreading of farmyard manure, containing hayseeds that have been trampled in by animals at the feeding stalls, is a further seed source that may introduce a few species to new sites. The viability of seeds in manure heaps is relatively short term, however, lasting little more than 6–12 months (Edwards and Younger 2006), and is only likely to be of significance where traditional agriculture is practised.

Finally, mowing machinery and hay-gathering activities are other major sources of seed dispersal (Strykstra and Verweij 1997), which are made positive use of in grassland restoration. The practice of transferring hay from donor to receptor sites (Section 6.4) to disperse seeds can also be effective in passively dispersing insects such as grasshoppers (Wagner 2004), although their survival at new restoration sites may then be prejudiced by poor habitat quality (Kiehl and Wagner 2006). Repeated hay transfers at later intervals, once an effective vegetation structure has been established, may allow more efficient transfers of invertebrates, as well as a wider complement of plant species.

4.4.3 Missing trophic levels

Convergence towards a target community needs to be judged not only from the plant species present, but should also consider the full range of species at other trophic levels: microbes, fungi, invertebrates and vertebrates. Restoration practitioners tend to focus on creating the right conditions for the plant community, often neglecting other trophic levels that strongly interact with, and influence these same sought-after target communities. Soil microorganisms are a key trophic level, determining the competitive balance between plant species, for example, by the activity of parasites, pathogens, root herbivores, mycorrhizal fungi and N-fixing bacteria. Decomposers also influence plant communities through the release of nutrients from soil organic matter and litter. In ex-arable soils, as in much improved grassland, many of the soil organisms found in semi-natural grasslands will initially be missing and may be slow to colonise. One approach might be to try to speed up the process by synchronising the introduction of plant and soil organisms at different successional stages (Kardol 2008), for example, through soil or turf inoculation, although the efficacy of such methods has not yet been widely demonstrated. The alternative is to allow time for colonisation to occur, and to consider restoration a multistage process.

As many invertebrates of grassland habitats, like the plants themselves, are also dispersal-limited, they are even less likely to arrive at sites undergoing restoration. Insect species make good indicators of restoration success as their short life cycles, relative mobility, and responses to small-scale environmental conditions make them more sensitive subjects than plants. Phytophagous beetles, for example, are good subjects as they show considerable functional diversity, and are a major component of invertebrates in species-rich grasslands. The most diverse insect assemblages usually occur in grasslands with high structural diversity, incorporating variable sward heights and a range of microclimatic conditions associated with topography and aspect. Some of this can be achieved by applying appropriate cutting and grazing regimes and manipulating soil fertility, but key plant hosts, vital for some specialist insect feeders, may still be missing.

The ability of insects to locate their host plants will be weaker in diverse plant species mixtures but particularly when the host patches are small or isolated. Specialist butterflies that are dependent on species-rich grasslands often have maximum colonisation distances of 0.5–2.5 km (Thomas *et al.* 1992), although in situ habitat quality, determined by factors such as turf height, shelter and successional stage of the food plant, can also be important in the case of some colony-forming grassland species (Thomas *et al.* 2001). Reviewing evidence of the effects of grassland fragmentation on insects, Steffan-Dewenter and Tscharntke (2002) concluded that butterfly species richness, especially monophagous feeders, was generally greater in larger fragments (Figure 4.4). Habitat fragmentation also disrupted some plant-pollinator and predator-prey interactions, but where there was good connectivity, inter-patch movement and population densities of insects tended to increase. Declines observed in bumblebee populations have been attributed to losses of semi-natural grassland and forage availability, especially the pea family (Fabaceae) (Goulson *et al.* 2005; Carvell *et al.* 2006). Pollinators may not be able to locate their food plants or are limited in their ability to disperse: the range of bumblebee and honeybee species is usually restricted to, at most, a few kilometres (Woodcock *et al.* 2010). They may, however, be encouraged to forage more widely when pollen sources are provided. In one study of field margins on arable farmland in northern England, those sown with a grass and wildflower mixture attracted comparatively greater bumblebee populations than unsown margins (Carvell *et al.* 2004).

A recent study in eastern Austria highlighted the importance of considering a wide range of taxa, not just plants, when considering the best conservation prospects for grassland (Zulka *et al.* 2014). In the context of a landscape dominated by intensive agriculture, they studied two plant taxa (vascular plants, bryophytes) and 11 invertebrate taxa (gastropods, spiders, springtails, grasshoppers, true bugs, leafhoppers and plant hoppers, ground beetles, rove beetles, butterflies and burnet moths, ants and wild bees) in small patches of surviving dry semi-natural grassland. When considering all these taxa together, species richness of the dry grassland specialists was influenced by the historical grassland patch size (prior to when most fragmentation took place in the 1950s), rather than the current patch size, suggesting that with increasing fragmentation some species would eventually succumb to extinction debt. Patch area per se was therefore important, while species richness tended to decline with increasing distance from the largest grassland fragment > 15 ha, especially in the case of mobile animal taxa. There were, however, spillover effects of species present on arable land. The authors concluded that a network of high-quality patches, not too far separated from each other, may be the best conservation strategy where grasslands are highly fragmented.

The siting of newly created grassland areas in close proximity to source populations is a key principle in restoration strategy. On an experimental restoration site some distance from semi-natural grassland, Mortimer *et al.* (2002) found that sowing seed mixes of different diversity did not accelerate colonisation by beetle assemblages typically associated with target grassland communities, although plant and beetle diversity was generally positively correlated. Similarly, arable fields near Munich in southern Germany, previously restored with green hay additions, had little influence on grasshopper communities (Kiehl and Wagner 2006); this could be attributed to the polyphagous feeding habit of this group. However, removing topsoil layers and thus increasing the cover of bare soil and reducing vegetation height, created a sparser and more variable vegetation structure that benefited and increased grasshopper diversity, including some species which colonised from ancient grasslands located nearby.

In contrast, in their study of restored grassland areas in the Hortobágy National Park (eastern Hungary), Mérő *et al.* (2015) found that populations of mice, voles and shrews were positively influenced by the proportion of natural and restored grasslands at a landscape scale. In only 3–6 years, these grasslands became important refuges for small mammals, which survived better in early-cut swards (with taller vegetation later in the growing season, thus avoiding the risk of predation by birds and mammals). The authors concluded that management was more important than restoration per se, and that a mosaic of restored and appropriately managed grasslands with tall vegetation would most benefit small mammals.

Like small mammals, birds are far less dispersal-limited than invertebrates, but are equally affected by agricultural intensification. The main impacts are the destruction of nests by frequent, and especially early mowing for silage, and trampling by high

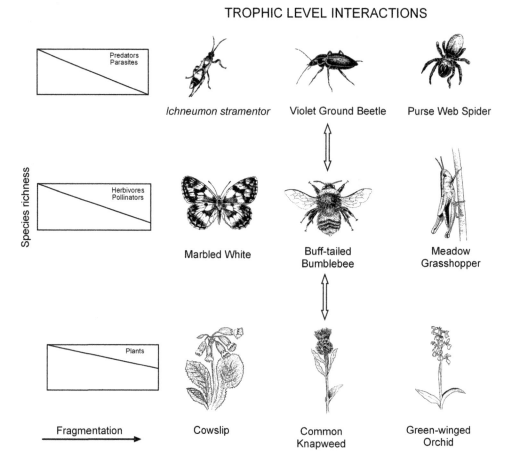

TROPHIC LEVEL INTERACTIONS

Figure 4.4 The impact of increasing habitat fragmentation in grassland on species richness and community interactions at three trophic levels. Predators and parasites are most likely to show the highest sensitivity (a steeper decline in species richness), followed by herbivores/pollinators and plants (illustrated by Tharada Blakesley).

levels of grazing stock. The uniformity of improved, fertilised grassland also reduces the availability of both invertebrate food, and limits flowering and seed taken by birds (Vickery *et al.* 2001). To improve habitat quality for bird populations, restoration measures require lower-input, traditional systems of management, some of which are currently supported by agri-environment schemes.

4.5 Assessing hydrological and topographic constraints

Before beginning any restoration work, it is essential to take full account of the site's physical constraints. In rare cases there may be opportunities to use heavy earthmoving machinery for topographic modelling work, perhaps after quarrying, or where the construction of new infrastructure allows scope for major landscaping works. Building in topographic variation by varying slope and aspect and altering topsoil and subsoil specifications can provide a wider range of environmental conditions that may benefit

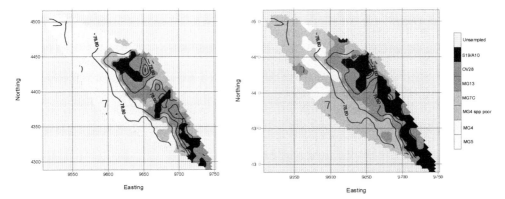

Figure 4.5 Maps of plant community distribution within the drainage basin at Cricklade North Meadow National Nature Reserve in 1998 (left) and 2001 (right), after exceptional weather during 1999–2001, the wettest 36-month period on record. Waterlogging resulted in the retreat of relatively 'dry' mesotrophic MG5 and MG4 communities, with a concomitant increase in swamp and aquatic S19/A10 vegetation, illustrating the sensitivity and rapid responses of these communities to inundation. Black lines represent contours at 0.2-m spacing (Gowing *et al.* 2002, reproduced with kind permission; © Defra)

and give scope to a number of different restored communities and species. Steeply sloping ground with southerly or south-westerly aspects are more likely to be susceptible to summer drought, and, if the soil profiles are also thinner, less nutrient rich. In calcareous grassland these situations are considered less susceptible to invasion by competitive species and more likely to support stress-tolerant vegetation (Bennie *et al.* 2006).

The more usual interventions in agriculture are to provide field drainage to increase crop production, or its converse in agri-environmental schemes, raising water levels through bund construction (or removal to allow flooding), and creating ponds and scrapes for conservation management. Although we are primarily concerned here with dry grasslands, ground water levels and periodic flooding can have a major influence on plant community composition. Species-rich, mesotrophic grasslands such as MG3–5 and MG8 in particular, are highly sensitive to waterlogging (Gowing *et al.* 2002) and can change rapidly when the hydrological regime is altered (Figure 4.5). To reconstruct the hydrological regime of a particular site requires detailed topographic surveying and the collection of water table depth data over a number of years using dipwells and piezometers, from which threshold depths of waterlogging or surface drying can be calculated. When related to botanical composition, it is possible to show how the moisture regime is subtly correlated with particular community types. At one site at Castle Meadows in the Thames Valley, Oxfordshire, Duranel *et al.* (2007) came to the conclusion that the duration of winter flood events and the depth of the water table in the summer made it unlikely that the proposed MG4 and MG8 meadow types could be reinstated, recommending instead MG5 and MG13 vegetation as more realistic targets that took account of the local topography.

Where detailed hydrological data are lacking, matching community types to prevailing conditions at the restoration site can also be done more crudely, especially where some semi-natural vegetation persists, by noting the species present and relating this to their environmental preferences. As an individual species' presence and abundance is strongly related to major gradients such as light, fertility and moisture, species indicator values have been proposed by Ellenberg (1988) for European vegetation, transcribed to the British situation by Hill *et al.* (1999), where each species is allocated a score according to its expected response to each environmental variable. Averaging the scores of all the species present in an indigenous community therefore gives a value, or approximation of the importance of that particular parameter. In the earlier example, the degree of waterlogging in mesotrophic grassland communities was well correlated with the mean Ellenberg moisture values (F values) of species present in sample quadrats (Gowing *et al.* 2002). A similar approach can be used to make a preliminary assessment of other important gradients such as site fertility (Ellenberg N values), or soil pH (Ellenberg R values).

5. Opportunities in grassland restoration

5.1 Conservation of semi-natural dry grassland habitats

Designated semi-natural dry grasslands may be afforded protection at various levels. The total area of land under the five former UK BAP priority dry grassland habitats in Britain was estimated to be 133,747 ha in 2006 (UK BAP Targets Review 2006). In England, an area of 70,921 ha of semi-natural dry grassland was designated as SSSIs (approximately 70%). Semi-natural dry grassland may also be designated as Special Areas of Conservation (SACs), Special Protection Areas (SPAs), NNRs and Local Nature Reserves (LNRs). Sites within National Parks and AONBs are also afforded some planning protection. Some priority grassland is also protected within SSSIs and National Parks in Scotland and Wales.

Agri-environment schemes are designed to reward farmers and landowners for environmentally sensitive land management. They aim to protect landscapes and wildlife without compromising farmers' ability to produce food. In England, the Environmental Stewardship Higher Level Scheme in operation in 2013 included incentives to maintain semi-natural grassland in good condition, and also to restore and create species-rich grassland. A more competitive scheme operated in Scotland, where the Scotland Rural Development Programme funded similar aims. The Welsh Countryside Management Scheme (Tir Gofal) administered by the Countryside Council for Wales was replaced in 2012 by one scheme, Glastir. This is a whole farm sustainable land management scheme aimed at enhancing biodiversity and mitigating climate change.

While designating semi-natural grassland sites is a step forward, a constant effort is needed to manage these (usually small) sites effectively, including those already in favourable condition. Sites in unfavourable condition must first be restored to more favourable condition. However, the ability of small, isolated sites to maintain metapopulations of key species can no longer be guaranteed. The landscape surrounding these fragments is likely to be hostile to wildlife, threatening the viability of populations of many invertebrates and plants. With predicted climate change, more patches of suitable habitat will be essential to support these populations and, critically, to facilitate the dispersal of wildlife through the landscape. This was emphasised in a report for Defra, *Making Space for Nature*, chaired by Professor Sir John Lawton (2010), which identified the need for 'large-scale habitat restoration and recreation, under-pinned by the re-establishment of ecological processes and ecosystem services, for the benefits of both people and wildlife'. The report called for the creation of ecological networks, through habitat creation and restoration, and the expansion of existing protected areas. These actions are widely viewed as the most important mechanisms for minimising the effects of climate change on all semi-natural habitats and wildlife in Britain (e.g. Hopkins 2007; Wildlife Trusts 2007). In addition, the UK Biodiversity Partnership's guiding principles to help practitioners take account of climate change also highlight the development of ecologically resilient and varied landscapes, and the reduction of threats not linked to climate, such as the intensification of farming practices and the abandonment of traditional management (Hopkins *et al.* 2007).

Cuttings along the 15-km A27 Brighton and Shoreham bypasses act as a wildlife corridor, linking fragments of semi-natural calcareous grassland in the adjacent landscape; chalk grassland plants, such as Common Spotted-orchids (top right), and butterflies, such as the Adonis Blue (bottom left) and Small Blue (bottom right), can be found.

Targets for restoration of priority habitats in England are currently being set as part of Biodiversity 2020 (Defra 2011). These follow earlier targets for restoration to the year 2020, set under the Biodiversity Action Reporting System, which are no longer being used. In England, an increase in the overall extent of priority habitats by at least 200,000 ha is included in the 2020 strategy.

5.2 Opportunities for dry grassland restoration

Public interest in habitat creation and restoration dates back to the late 1960s and early 1970s, when derelict land and urban wasteland, a legacy from Britain's industrial past, began to feature in government-led clearance and landscaping programmes. At the same time landscape architects were advocating an 'ecological' approach to greening expanding motorway networks and public open space in New Towns, much inspired by Dutch nature parks. This urban conservation movement found an echo in concerns for losses of semi-natural habitats forced by intensive agriculture and, somewhat ironically, urban development itself.

An early review of amenity grassland highlighted the scope for habitat creation and enhancement of species-poor grassland in parks, golf courses, roadside verges, country parks and the wider countryside (Natural Environment Research Council 1977). Much of this was considered unattractive and expensive to maintain, but with considerable potential to improve its aesthetic and wildlife value. Resowing was one possibility, using commercial wildflower seed mixes that became widely available in the 1970s, although these then contained agricultural varieties and some non-native species, mostly imported from Europe (Brown 1989). By the early 1980s, research by the Institute of Terrestrial Ecology into the feasibility of producing native grassland seed mixes led to the development of more sophisticated ranges, designed to emulate alluvial, clay and limestone grassland communities, described in landmark publications such as *Creating Attractive Grasslands Using Native Plant Species* (Wells *et al.* 1981; 1989). These 'native' mixes were taken up by commercial seed houses and became freely available to the landscaping industry. Further developments in seed mix design were specified for motorway and trunk road verges in *The Wildflower Handbook* (Department of Transport 1993) and for urban and derelict land in *Flowers in the Grass* (Ash *et al.* 1992). At the same time, the potential use of hay bales and green hay as seed sources, harvested from premier grassland sites, was also being investigated, while commercially mass-produced plugs and pot-grown 'wild' plants were also becoming available for grassland enhancement projects.

The importance of using plant materials of local provenance for grassland restoration has long been stressed by the conservation movement and more recently by the charity Flora Locale. In 2012 they launched an online Restoration Library of resources for ecological restoration. Other initiatives have included the opening in 2000 of the National Wildflower Centre near Liverpool, founded by Landlife, with a mission to educate the public in all matters concerning the creation and management of wildflower habitats. In 2011 the Millennium Seed Bank of the Royal Botanic Gardens, Kew launched the UK Native Seed Hub to support restoration efforts, initially of lowland meadows, by working to develop high-quality seed sources to be made available to commercial seed companies and conservation organisations for large-scale restoration projects.

The practice of creating and restoring species-rich grassland habitats has now expanded well beyond its urban and landscaping origins to embrace the wider countryside, encouraged by agri-environment scheme options and targets set for the recovery of semi-natural grassland priority habitats under the UK BAP (Defra 2008). The first schemes in the UK were introduced in designated Environmentally Sensitive Areas (ESAs) in 1987, recognised for their landscape, wildlife or historic value, where farmers could receive incentive payments if they volunteered to reduce agricultural inputs and outputs to maximise environmental benefits. In England these voluntary agreements were later expanded to include all agricultural land outside ESAs under the Countryside Stewardship Scheme (Defra 2001), until the Environmental Stewardship superseded both schemes in 2005 (Defra and Natural England 2008). Parallel schemes operate in Wales and Scotland. In the UK, agri-environmental schemes involving semi-natural grasslands currently cover over 440,000 ha (Bullock *et al.* 2011). The majority of these support the maintenance and restoration of existing, species-rich sites, but a significant minority, more than 10% of the total area, consists of grassland creation and expansion schemes.

Merely adopting traditional, extensive management in lower-tier agri-environment schemes does not guarantee sward diversification, mainly because many of the key species are missing or because the soil nutrient status is too high (Kleijn and Sutherland 2003). Interventionist restoration (e.g. ground preparation, seeding additional species, manipulating soil fertility, etc.) will usually be more effective and quicker, but can be expensive to implement. Nevertheless, techniques such as establishing a low-intensity grazing regime during the growing season, less frequent mowing or litter removal can be successful preliminaries to re-establishing sward diversity. Reducing or ceasing fertiliser applications altogether may also form part of this strategy (Section 5.4.3). In traditional management, the intensity, frequency and seasonality of interventions such as grazing, mowing or burning will ultimately determine sward composition and structure. To achieve restoration success, it follows that management regimes should closely match those supporting existing target, semi-natural vegetation.

One hectare of semi-improved grassland in Cumbria sown with species characteristic of northern hay meadows, including Wood Crane's-bill, Great Burnet, Oxeye Daisy and Primrose.

5.3 Reinstating traditional management

5.3.1 Grazing effects

Grazing, like cutting, acts to suppress dominant species in the sward and reduce competitive exclusion, while at the same time creating disturbances that provide regeneration niches for seed germination and establishment. Grazing is particularly effective as it opens gaps in the sward and creates bare patches, through defoliation, trampling and dunging. In a single field the opportunities for disturbance and colonisation will arise through many different agencies besides livestock trampling and dung, for example, in the bare patches created by burrowing Rabbits (Bakker and Olff 2003). Sheep and cattle husbandry routinely allows grazing of swards down to a height of 5 cm, sometimes with hard sheep grazing to less than 3 cm. Short, disturbed swards have increased susceptibility to invasion and diversification, but this can also facilitate the ingress of undesirable invasive species, such as agricultural weeds and exotics (Burke and Grime 1996; Davis *et al.* 2000).

Grazing intensity and grazing regime – rotational, continuous, seasonal and irregular – and to some extent the characteristics of the different grazing animals themselves, all influence grassland community structure (Section 3.1). The less palatable species and those with grazing-resistant traits, such as rosette-forming forbs and grasses, tend to suffer less selective pressure. Vegetation complexity, in terms of species richness, sward height and heterogeneity, strongly affects invertebrate species richness. Heavy grazing, for example, reduces sward height and increases uniformity, particularly when potential flowering and seeding niches for insects are removed. In an investigation of cattle grazing on pastures in northern Germany, samples of insect taxa, including beetles, bugs, plant hoppers and parasitic wasps, became increasingly diverse along a gradient ranging from intensively grazed swards to those that had remained ungrazed for at least 5 years (Kruess and Tscharntke 2002).

Grazing impacts tend to be more subject to variation in stocking and seasonality and less predictable in timing than, say, hay cuts undertaken regularly over many growing seasons. In a 12-year sheep grazing experiment on a species-poor, mesotrophic grassland, Bullock *et al.* (2001) found that species richness increased under heavy spring grazing, but decreased with heavy summer grazing and was unaffected by winter grazing. Plant responses to heavy summer grazing were linked to their ability to colonise gaps, while responses to lower spring and winter grazing were positively related to grazing selectivity. Warren *et al.* (2002) also compared the effects of summer grazing, in this case with sheep or cattle, with hay cutting and aftermath grazing on plots that had been sown with a neutral meadow (MG5) mixture, or left unsown. After 6 years, the sown plots had retained most of the original species under the summer grazing regime with cattle, whereas with summer sheep grazing the number of forb species in particular fell rapidly, probably due to selective grazing, illustrating the importance of husbandry in determining the trajectory of vegetation succession.

Other studies have found little effect of grazing season on plant species richness per se (e.g. Gibson and Brown 1991; Smith *et al.* 1996), probably because species gains as a result of changes in management regime tend to be balanced by losses. Most conservation managers advocate low to intermediate levels of stocking, aimed at preventing scrub succession while avoiding losses of stress-tolerant species through overgrazing, often with a heavy emphasis on the merits of particular livestock types or breeds (Section 3.1). However, a systematic review by Stewart and Pullin (2008) found little conclusive evidence that species diversity was directly affected by stock type in mesotrophic pastures, although they conceded that heavy sheep grazing was probably deleterious. Much more critical was the level of grazing intensity and in particular the need to strike a balance between maintaining openness for bryophyte species without prejudicing the abundance of forbs (Figure 5.1). A similar conclusion was reached by Dumont *et al.* (2011) in their study of intensively managed, upland grassland in the French Massif Central,

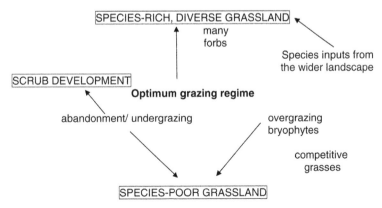

Figure 5.1 Grazing management regimes determining species-rich or species-poor grassland.

but here a detailed comparison of different stocking rates of cattle and sheep, grazed rotationally, showed clear selective differences over time. Both cows and sheep preferred legumes and forbs and selected short vegetation rather than tall grass areas, but at the equivalent stocking rate to cattle, sheep selected legumes and forbs more strongly, resulting in greater grass dominance, while the greater patchiness produced by cattle grazing gave the sward greater structural diversity.

Vegetation responses to changes in grazing regime may initially be rapid, particularly in terms of overall sward structure, but plant species composition may take many years to resolve. Humphrey and Patterson (2000) found that low-intensity cattle grazing in late summer maintained the species richness of acid Sheep's-fescue–Common Bent–Heath Bedstraw grassland (U4b), but ungrazed plots became less diverse over 9 years, principally through the build-up of litter and the exclusion of bryophyte species. In the study by Dumont et al. (2011), the differences between sheep and cattle grazing on legumes and forbs was not evident for 6 years, while in Welsh hill pastures, Hill et al. (1992) noted the rapid decline of low-growing plants in the first 7 years after excluding sheep, but only after 30 years did some plots eventually become dominated by a dwarf shrub cover of Heather and Bell Heather. Similarly, on limestone grassland in Germany, Kahmen et al. (2002) noted major changes over 25 years in species composition in ungrazed, compared with sheep-grazed plots, but in this case the major change took place only after about 20 years.

5.3.2 Cutting effects

Cutting reduces the abundance of tall vegetation, reducing shading, altering the competitive balance and allowing more stress-tolerant species, especially low-growing forbs, to survive (Parr and Way 1988). Unlike grazing, cutting is unselective and produces a relatively uniform sward structure, removing biomass and returning ripe seeds – but only if the vegetation is left long enough to flower; early silage cuts mean reduced numbers of species regenerating from seed, and smaller quantities of seed shed. Furthermore, there will usually be much less bare and disturbed ground caused by trampling by animals, little redistribution of nutrients in dung patches, and a greater build-up of litter and thatch at the soil surface, particularly where the cuttings are not collected. In the Hortobágy National Park in eastern Hungary, plots sown with alkali or loess grassland mixtures on former cropland, but left unmown, accumulated litter over 6 years, typically at > 300 g/m^2 (Kelemen et al. 2014). This decreased species diversity, total vegetation and sown grass cover relative to mown sites, but increased undesirable perennial species such as Common Couch and Creeping Thistle. Hay cutting, conversely, may also increase the dominance of coarse grasses (Hayes and Sackville Hamilton 2001) and favour more palatable, tall forbs. Cutting at the same point in the growing season favours a suite of species producing ripe seed at that time, while increasing the frequency of cutting may prevent flowering of some species altogether.

Up to a certain point, carefully timed removal of biomass by mowing can control the tendency for competitive species to dominate, particularly on relatively fertile arable reversion sites (Section 3.2). Under these conditions Prach et al. (2013) found that in sites sown with diverse regional seed mixes for up to 12 years, forbs tended to benefit from early rather than late mowing because this more effectively reduced grass biomass and cover. In an arable reversion experiment sown with 18 mesotrophic grassland species, more frequently cut plots (in July and again in September) also developed greater species richness and forb cover than plots cut only once in midsummer, or the uncut controls (Lawson *et al.* 2004). Smith *et al.* (2010) also found that cutting twice per year retained significantly more of the original components of a wildflower mix sown on arable field margins compared with similar, but uncut plots. After 13 years, however, the cutting treatments had little overall effect on the species richness of sown and naturally regenerated field margins, respectively, but plots in which the cut material was left lying on naturally regenerated plots had consistently fewer species.

Several long-running experiments have further investigated the impact of single or multiple cuts on semi-natural and improved grass swards. Compared with cutting in alternate years or complete abandonment, annual hay cutting was found to be the most effective treatment promoting species richness in a former Mat-grass–Heath Bedstraw grassland over a 25 year period, once fertiliser applications were withdrawn and regular hay cutting resumed (Bakker *et al.* 2002). In the Eifel Mountains of Germany, Pavlů *et al.* (2011) noted a shift over 20 years in the composition of a Rye-grass–Crested Dogs-tail sward cut twice or four times a year compared with controls. Although species numbers were not affected, Yorkshire-fog and Meadow Foxtail gradually replaced dominants such as Perennial Rye-grass and Common Couch in the twice-yearly cutting treatments. Cutting four times a year resulted in an increase of well-adapted species, such as the rosette forbs of dandelions and hawkbits and tolerant grasses including Red Fescue and Creeping Bent. Antonsen and Olsson (2005) also found that low-growing species such as White Clover, Thyme-leaved Speedwell and dandelions increased in a Norwegian boreal grassland under a twice-yearly mowing regime. This was again the case in a ten-year experiment in mesic, Mediterranean grasslands along the Californian coast, where cutting up to six times per year favoured forbs and shorter-statured species rather than grasses, although not necessarily benefiting native species; several non-natives were equally well adapted to such frequent cutting (Hayes and Holl 2011). Even after considerable periods of time, the community composition of the optimum treatments in most of these examples still deviated considerably from the local reference sites and retained several species characteristic of more fertile conditions.

Askrigg Bottoms hay meadow in the Yorkshire Dales (© Robert Goodison).

Hay time at Worton Bottoms in the Yorkshire Dales (© Roger Smith).

Although such experiments may determine an optimum cutting regime that delivers the greatest species richness, practical management issues often intervene, and the best results may even be at odds with agri-environmental scheme prescriptions. For example, Leng *et al.* (2011), by applying different treatments to ditch banks in the South Holland province, showed that cutting twice, on 1 July and 1 September, resulted in the maximum seed set of species, whereas the local agri-environmental scheme allowed farmers flexibility to mow at the beginning of June and August, respectively, thus reducing the likelihood of seed set and dispersal. Smith and Jones (1991) also pointed out that meadow management prescriptions for the former Pennine Dales ESA allowed haymaking to begin before many species had begun to set ripe seed. They recommended setting back the cutting period to later in the year, or allowing late cuts for two years in five. Crofts and Jefferson (1999) also suggested occasional late mowing in September, to benefit late-flowering species and maintain greater species diversity than in a standard hay meadow.

Burning in winter or spring is another option that has been used to prevent succession, control coarse grasses and remove litter, particularly in undergrazed or abandoned vegetation. In a short-term experiment on commercially sown hayfield vegetation in eastern Norway, spring burning had little impact on species composition and the field remained similar to uncut control plots, compared with mowing twice a year where there was a marked increase in species diversity (Antonsen and Olsson 2005). However, a much longer-term, 25-year investigation on limestone grassland in the Jura Mountains of south-west Germany showed an increase in rhizomatous species, such as Tor-grass, in plots subjected to annual winter burning. This was accompanied by a corresponding loss of low-growing species, compared with plots grazed by sheep at low intensity, thereby defeating the object of management (Kahmen *et al.* 2002). These results tend to confirm that mowing in midseason is more effective than burning in promoting coexistence of different sward components. Burning is also likely to have a deleterious effect on overwintering invertebrates.

5.3.3 Influences of restoration management on sward diversity

Many experiments have been designed to determine the best cutting, grazing or combined management regimes to deliver particular types of species-rich, semi-natural grassland. The approach is well illustrated by a study on chalk grassland vegetation by Jacquemyn *et al.* (2003), who contrasted cutting and grazing treatments. Three basic treatments were applied over four growing seasons: mowing at the end of June or beginning of July; intermittent grazing by cows (stocked at 15 animals/ha, from late May to the end of August); and unmanaged controls. The vegetation was recorded in mid-May each year, prior to management (Figure 5.2). As expected, sward heights were much greater in

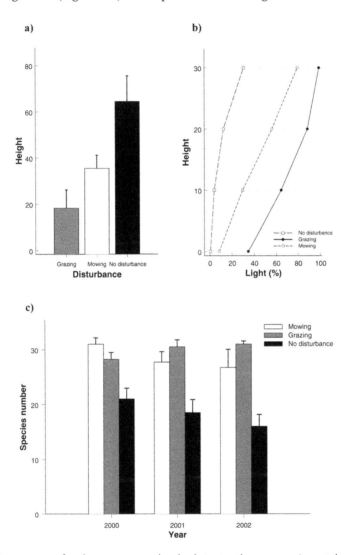

Figure 5.2 Responses of calcareous grassland plots to three experimental management regimes: intermittent summer grazing, mowing and unmanaged controls; (a) vegetation height and (b) light penetration profiles are shown in late May at the end of the experiment (2002) and (c) species density in 1-m² plots (reprinted from Jacquemyn *et al.* 2003, with permission from Elsevier).

unmanaged plots, where species richness declined by 30% within 2 years as tall grasses began to shade out low-growing forbs and fewer gaps became available for regeneration. Overall defoliation intensity in grazed plots was greater than in mown plots, with the result that sward heights were lower and more light was available. Notably, mowing once a year was insufficient to maintain high species diversity in this case.

Some authors have speculated that using well-designed combinations of cutting or grazing treatments, together with ceasing fertiliser applications, can override the tendency towards competitive exclusion and declines in species richness, even on sites where there is strong residual fertility. For example, when applying a range of cutting and grazing treatments to an improved MG6 pasture over 8 years, Hayes and Sackville Hamilton (2001) found increased species diversity in the most extensive treatments as the soil fertility and cover of dominant grasses declined. The authors recommended a cutting frequency of twice a year with aftermath sheep grazing as an initial treatment for the first 3–4 years to increase the regeneration niche for forbs and to speed up reversion. Wagner *et al.* (2014) found that 3 years after reseeding arable land with a mixture containing ten species of calcareous grassland, spring grazing by sheep gave a less pronounced increase in similarity to reference calcareous grasslands, compared with conventional hay cuts and aftermath grazing by cattle. In the earlier example (Figure 5.2), Jacquemyn *et al.* (2003) applied different levels of N fertiliser (up to 30 kg/ha/yr) to their cut or grazed plots. Increasing N supply caused reduced light penetration in the sward, with fertilised plots receiving an average of > 20% less light than unfertilised plots. Although grazing was again more effective than mowing in countering the effects of increased N, neither could prevent an increase in dominance of competitive species, especially grasses. Thus, while management is important, if nutrient levels differ markedly from those of the target community, there is likely to be a negative shift in community diversity over time. This is already occurring in semi-natural grasslands in industrialised areas where there are substantially increasing levels of atmospheric N deposition. On the other hand, withdrawing fertiliser treatment may, with appropriate management, allow species richness to gradually increase.

Semi-natural grasslands that have been abandoned can sometimes be revitalised by management treatments that effectively break up the accumulation of dense litter layers over time. Thick litter cover has been shown to lower seedling emergence in experiments on floodplain grassland (Donath and Eckstein 2012). Its mechanical and chemical effects can reduce the availability of microsites for seedling regeneration and establishment by creating a physical barrier and reducing light levels, while the leaching of secondary compounds may inhibit seedling growth. Ruprecht *et al.* (2010) tested the effectiveness of litter removal from dry, steppe-like grasslands in lowland Transylvania, which had been abandoned for about 40 years. Raking, or cutting and raking, increased cumulative seedling counts over 2 years and resulted in the re-emergence of some rare species, while cutting and litter removal further increased seedling survival. As an initial restorative treatment, therefore, light scarification or harrowing is likely to be beneficial, but following up with traditional management, such as cutting and hay baling, with the additional trampling of grazing animals, will also encourage litter break-up.

Sowing the target species, the success of which will in turn be strongly influenced by management practices, can also accelerate the diversification of species-poor grasslands. On an ex-arable site, cutting, producing a short and open structure, improved the germination success of four sown chalk grassland species (Hutchings and Booth 1996b). This treatment was not only superior to uncut plots, and also to others where the surrounding vegetation was cleared; the bare ground created a less favourable microclimate for the introduced seedlings, with the first two weeks after germination proving to be critical for establishment. Again, Jones and Hayes (1999) sowed five common forb species (Common Knapweed, Ribwort Plantain, Yarrow, Selfheal and Hedge Woundwort) in an improved, Perennial Rye-grass-dominated pasture under different management regimes. Two years after autumn sowing, the unfertilised treatments, cut twice in summer and followed by grazing, showed the greatest seedling establishment. Summer cutting alone gave the poorest establishment, even less than in continuously grazed plots receiving annual fertiliser inputs.

The crucial role of grazing management has been well illustrated on restoration sites. On a degraded tall grass prairie restoration site in Iowa, USA, low-intensity grazing by bison and deer was critical in promoting germination of seed supplements of the 'missing' forbs and grasses. Grazed plots had better light and water availability than ungrazed plots where establishment was poor, while grazing without sown seed supplements failed to increase diversity, due to local seed limitation (Martin and Wilsey 2006). Similarly, a long-running restoration of a floodplain site on the river Thames, formerly managed as intensive grassland, benefited markedly from aftermath grazing (McDonald 2001). The site was cultivated and sown with a diverse Meadow Foxtail–Great Burnet (MG4) seed mixture from a donor site, after which three different management treatments were applied: July hay cutting, or hay cutting with aftermath grazing by cattle or sheep. Over the 22-year recording period, the plots with aftermath grazing diverged from the cut-only plots, becoming progressively more similar to the species complement of nearby target MG4 floodmeadow communities, while the cut-only plots remained relatively species-poor False Oat-grass MG1 communities. However, this increasing similarity to MG4 grassland was expressed as species presence, rather than relative species abundance in the target community. This failure to achieve closer similarity over two decades could have several explanations, such as uneven distribution of species in the original sowing mix, less than ideal environmental conditions at the restoration site (such as hydrology and fertility), or management regimes differing subtly from the previous, historic regimes from which MG4 communities developed (Woodcock *et al.* 2011). On the other hand, the functional traits of plants comprising the restored community (i.e. assessed as their environmental requirements, life forms, regenerative strategies and seed biology) tended to converge towards the target community at a more rapid rate. Assuming a linear progression, convergence in species similarity to the target was projected to take more than 150 years, but in terms of the functional traits the time frame was considerably shorter (> 70 years).

5.3.4   Impacts of restoration management on invertebrates

Apart from its influence on the vegetation, restoration management affects other species groups, including invertebrates, both directly and indirectly. The greater structural diversity of tall, extensively managed grassland usually provides more suitable breeding and foraging niches than short, more uniform and intensively managed vegetation, where the regularity of disturbance may be particularly detrimental to the smaller populations of invertebrates at higher trophic levels. Furthermore, periods of abandonment or neglect allows more time for invertebrates with poor powers of dispersal to colonise, up to a point where succession of scrub and woodland begins to dominate the sward. Phytophagous and predatory insects are particularly dependent on sward structure and species diversity. In a multisite experiment on lowland sites in Somerset and Devon, improved grasslands were subjected to seven different levels of management intensification for 3 years (Woodcock *et al.* 2007). Greater species richness, and particularly species richness of beetles, were found in the extensive treatments involving a single annual cut, no grazing, no fertiliser or no management, compared with the more intensive treatments of cutting twice per year, N/P/K fertilisation and aftermath grazing. In more extensive regimes there was a greater proportion of seed- and flower-feeding species present and the more complex grass sward architecture was positively correlated with beetle abundance, with tussock grasses in particular providing increased diversity of niches for ground beetles.

Similar observations were made by Pöyry *et al.* (2005) who compared the abundance of butterflies and moths in old, cattle-grazed mesotrophic pastures in an area of south-west Finland with those that had been abandoned for 10 or more years. A greater abundance and species-richness of insects was found at the abandoned sites, whereas, in contrast, the vegetation of the continuously grazed sites was more species-rich (Pykälä 2003). Mean vegetation height and the presence of nectar-feeding species contributed most to insect species abundance, although some butterflies and moths with more specialised requirements were more restricted to shorter, managed vegetation containing a greater variety of plant species. Where restoration grazing had been resumed on previously abandoned sites for periods of 3–8 years, the species characteristic of continuously grazed grasslands did not immediately respond, probably because missing larval host plants were hindering them, or insect species were slow to colonise from suitable patches in the surrounding area. Such studies illustrate the need to balance different intensities of grazing to achieve a variety of sward patchiness and structure, both locally (perhaps through rotational grazing) and at a landscape level, where patches of extensive, intensive and even temporarily abandoned sites are juxtaposed.

BOX 5.1 Case study: Seven Barrows (Historic England)

Topic: Management of unimproved chalk grassland for archaeology and Marsh Fritillary
Location: Berkshire Downs
Area: 3.7 ha
Designation of site: Scheduled Ancient Monument and SSSI
Management and community use: privately owned, but managed as a nature reserve with public access by the Berkshire, Buckinghamshire and Oxfordshire Wildlife Trust
Condition of site: Unimproved chalk grassland in unfavourable, recovering condition
Website: http://www.bbowt.org.uk/reserves/seven-barrows

Background

Seven Barrows is a round barrow cemetery designated as a Scheduled Ancient Monument and SSSI for its unimproved chalk grassland. Round barrow cemeteries date to the Bronze Age (approximately 2000–700 BCE). They comprise closely spaced groups of up to 30 round barrows – rubble or earthen mounds covering single or multiple burials. Most cemeteries developed over a considerable period of time, often many centuries, and in some cases acted as a focus for burials as

Seven Barrows SSSI.

late as the early medieval period. They exhibit considerable diversity of burial rite, plan and form, frequently including several different types of round barrow, occasionally associated with earlier long barrows. Where large-scale investigation has been undertaken around them, contemporary or later 'flat' burials between the barrow mounds have often been revealed. Round barrow cemeteries occur across most of lowland Britain, with a marked concentration in Wessex. In some cases, they are clustered around other important contemporary monuments, such as henges. Often occupying prominent locations, they are a major historic element in the modern landscape, while their diversity and their longevity as a monument type provide important information on the variety of beliefs and social organisation among early prehistoric communities. They are particularly representative of their period and a substantial proportion of surviving or partly surviving examples are considered worthy of protection.

Seven Barrows is particularly important as it has survived well and, despite partial excavation of some of the barrow mounds, has considerable potential for the recovery of environmental and additional archaeological remains. It exhibits a diversity of barrow types and is therefore an outstanding example of its class. There are actually ten barrows arranged in parallel lines, forming the core of a widely scattered barrow cemetery on the floor of a dry valley in the Berkshire chalk downland. Eight of the mounds are bowl barrows, with one additional disc barrow and one saucer barrow. Antiquarians in the late nineteenth century explored many of them and recorded numerous finds, including cremation burials and animal bones.

The presence of the barrows has protected the site from the plough, and consequently it supports a species-rich calcareous grassland flora and a diverse butterfly community. Coarse grasses, such as Upright Brome and Tor-grass, and finer grasses such as Sheep's-fescue, Crested Hair-grass, Downy Oat-grass and Quaking-grass, dominate the sward. Abundant herbs include Greater Knapweed, Common Rock-rose, Cowslip and Harebell. Other herbs include Chalk Milkwort, Clustered Bellflower, Dropwort and Saw-wort. The site is also rich in invertebrates, including typical chalk grassland butterflies such as Small Blue, Brown Argus and Chalkhill Blue. The site has also recently been colonised by Marsh Fritillary, which feeds on Devil's-bit Scabious. This Red List butterfly has declined severely in recent years, and the population at Seven Barrows is an outlier of the south-western population on chalk.

Left to right: Clustered Bellflower, Dropwort and Six-spot Burnet on Small Scabious.

Management to protect the barrows

Management must accommodate both the archaeology and the features for which the site was designated as a SSSI. One of the main risks to the barrows in recent years has been succession, leading to scrub and tree cover, which can damage archaeology, especially if trees with large root systems fall over. Rabbit burrows can also cause significant damage to buried archaeology. Stock feeders and water troughs must also be located away from sensitive areas to avoid trampling.

Scrub control

Scrub is regularly checked, and if necessary, removed by hand using bow saws. Herbicide is applied to prevent regeneration from the cut stumps, usually late in the year when grazing stocks have been removed. One barrow has a 'feature' Beech tree growing on it; there are plans to carry out a crown reduction to reduce the weight and 'windsail' without losing the tree. However, if the tree shows signs of major weakness it will be removed as an uncontrolled failure, because a collapse could cause significant damage to the barrow beneath. Away from the actual barrows, a small amount of scrub provides some structural diversity, but this is not allowed to occupy more than 5% of the site.

Rabbit control
The site was formerly unfenced, and heavily grazed by Rabbits, to the detriment of both the archaeology and biodiversity. Rabbit infestation had become extremely severe by the winter of 2005, forcing Historic England (then known as English Heritage), English Nature and West Berkshire Council Archaeological Service, to agree to jointly fund Rabbit clearance, Rabbit fencing around the area of the SSSI (which includes the Scheduled Ancient Monument), and a long-term grazing plan.

A Rabbit fence was constructed in September 2006; by July 2008 the Rabbits had been almost entirely removed. The change in both the appearance of the site, and the potential damage to the archaeology was striking. Dexter cattle and ponies were overwintered on the site in 2008–2009, grazing the excessive and rank long grass down to a short sward. By March 2009, the site was well grazed and the areas of former Rabbit burrowing were largely settled and covered by grass.

Protected through the presence of the barrows, the site supports a rich chalk grassland flora.

Management for biodiversity

Grazing
Management of the chalk grassland is not typical, due to the requirements of the Marsh Fritillary, which prefers a taller sward, and the late flowering of its food plant, Devil's-bit Scabious. Grazing has to be extensive, which compromises the chalk grassland flora a little. The sward is grazed by cattle to support a wildflower/species-rich open grassland, while allowing enough tall and tussocky vegetation to remain to provide the habitat for the Marsh Fritillary to complete its life cycle. Dexter cattle are preferred, as they are smaller and therefore lighter on their feet than some other breeds, thereby minimising the risk of erosion on the sensitive monuments. Dexters are also relatively hardy, meaning they can cope better with poor-quality forage, and generally require less husbandry. Ponies are grazed from time to time on the areas of uniform rank grasses.

The timing of grazing on the site is also affected by the Marsh Fritillary. In their absence, grazing would be routinely undertaken in autumn and spring. However, to avoid direct damage to larval webs and caterpillars, autumn grazing is held back until mid-October and continued as long as possible into December. Spring grazing begins in March and continues until early May. Spring grazing is viewed as restoration grazing, and autumn grazing as maintenance grazing. Consequently, if the diversity of the sward improves on the eastern portion of the site, spring grazing to tackle the thick stands of Tor-grass and False Oat-grass may be reduced, or become unnecessary. Stock density depends on the quantity of growth present in any given year, but does not exceed ten Dexters (two LSU/ha). If ponies are also used, then the number of cattle is reduced.

Mowing

The eastern end of the site and parts of the northern boundary have come to be dominated by tall rank grasses, notably Tor-grass and False Oat-grass, as well as Creeping Thistle and Hogweed. To help with the control of these species, targeted cutting is undertaken in early April with mowers or brushcutters. When cattle are returned to the site, they should preferentially graze the aftermath growth in these areas, further helping to reduce the dominance of these species.

Rank grasses dominate parts of the site

Information provided by Daniel Bashford (Historic England) and Andy Coulson-Phillips (Berkshire, Buckinghamshire and Oxfordshire Wildlife Trust).

5.4   Site limitations and solutions

5.4.1   Fallowing

Withdrawing fertiliser applications does not automatically or immediately increase species diversity. High residual levels of nutrients may be sufficient to maintain productive, vigorous swards for many years, preventing the ingress and establishment of stress-tolerant species, even when these are present nearby or are supplied as seed. Such action is likely to be least effective on inherently fertile soils such as brown earths and clay-rich gleys, which can maintain relatively high levels of productivity for long periods. However, when combined with extensive management practices, such as hay cuts and aftermath grazing, the cessation of fertiliser treatments may help to gradually restore a degree of species diversity. An experiment where N applications of 25, 50, 100 and 200 kg/ha, respectively, were applied for 4 consecutive years to an unfertilised, wet mesotrophic meadow, but then withdrawn for 3 further years, resulted in a gradual decline in dry matter yields and less dominant grass cover, but also reduced levels of soil N and P (Mountford *et al.* 1996). This reversion towards nutrient levels in the untreated control plots was calculated to take 3–9 years, assuming that good colonising sources were retained. Similarly, Hejcman *et al.* (2010) found that 14 years after ceasing N/P/K fertiliser applications to a Perennial Rye-grass–Crested Dog's-tail pasture in the Eifel Mountains, south-west Germany, levels of plant-available P had fallen substantially and K levels slightly, and were accompanied by a decrease in biomass productivity. On thin, free-draining and sandy soils levels of nutrients may be lost more rapidly through leaching and cropping. Annual cropping rates of dry matter exceeding 5 t/ha/y are unlikely to leave much space allowing the less competitive components of the sward to survive (Oomes 1992), and production from fertilised swards can be more than double this figure.

Extractable P levels in arable soils are often three or four times greater than in semi-natural grasslands on similar parent material. Fertiliser use is also likely to inflate available soil K, although this is more rapidly leached, whereas under grassland the pool of potentially mineralisable N increases, due to the gradual build-up of organic matter. One possible means of reducing the nutrient load is therefore to fallow the land for a period to allow natural leaching to occur, or accelerating it by continuous cultivation to maintain bare soil. Marrs *et al.* (1991) tested this experimentally. They examined leaching losses from semi-natural grassland and arable soils respectively on clay, chalk and sandy substrates. During an eight-month overwinter period of fallowing, the greatest N losses occurred in the arable soils kept bare of vegetation and were almost an order of magnitude greater than in treatments with a vegetation cover. However, P losses were small, amounting to less than 1% of the extractable total, although these were greatest in the sandy soil. The authors concluded that leaching was only a minor pathway for nutrient losses in fertile soils, and potentially a very long-term process.

Snow *et al.* (1997) examined reductions in extractable P over time in a calcareous gley soil over chalky boulder clay. The study contrasted four adjacent fields: one managed for intensive arable production; another converted from arable production to a grass/clover ley, then managed as an unfertilised hay meadow with autumn grazing; and two much older meadows, under similar traditional management. In the converted meadow, extractable P levels fell from 19.6 µg/g to 3.2 µg/g over 9 years, a calculated loss rate of 1.82 µg/g/y. This rate of decline is supported by other studies and suggests that a target index of extractable P of 0–1, typical of unfertilised semi-natural grasslands, can be achieved over time. The length of the period will depend critically on soil type, the residual levels of P present at the start and the subsequent management regime. Snow *et al.* (1997) considered that initial reductions were relatively easy to achieve, due to rapid P adsorption on soil minerals (especially on lime-rich substrates), together with chelation by increasing organic matter, but that long-term losses by sequestration, leaching and erosion were likely to be counterbalanced by atmospheric inputs, organic mineralisation and the weathering of parent material.

5.4.2   Plant and animal offtake

An annual grass cut removes relatively little of the soil nutrient pool (e.g. 1% of total P and exchangeable K and 2.5% of total N annually (Bakker 1987)), and N inputs may balance removals where there is heavy rainfall or atmospheric deposition, although repeated cuts will remove more. Bakker *et al.* (2002) found that after 25 years, twice-yearly cuts in July and September had reduced the dry matter yield of the sward to a quarter of that measured in the original improved grassland, but N offtake was still considered insufficient to compensate for inputs through atmospheric deposition. Similarly, after 18 years of cutting and removing roadside vegetation at two sites in Cambridgeshire, only extractable soil K declined while most other soil nutrients, including total and available N, were unaffected (Parr and Way 1988). In a 15-year experiment to determine whether intensive cutting could reduce grassland productivity through offtake, Hejcman *et al.* (2010) found that although productivity did decrease, this occurred equally in control and cut plots, including those cut four times per year. The authors concluded that withdrawal of fertiliser use was the cause of the decline, and that restoring low productivity grassland could not be achieved by cutting alone.

Offtake by grazing animals (e.g. as wool or carcasses) may remove even fewer nutrients than cutting regimes, as the nutrients are returned to the soil in dung and urine, and are also replenished with rainfall. The historic grazing practice of moving animals from hill pastures to valley paddocks overnight may have once been effective by impoverishing hill land while manuring the valley fields. Ironically, grazing can actually cause the reverse, eutrophication, if stock are alternated between semi-natural grasslands and nutrient-rich improved pastures, or if supplementary feeding is given in winter. Kirkham (2006) calculated that nutrient loads imported as excreta from stock grazed on improved sites to semi-natural grasslands could be equivalent to farmyard manure applications when repeated over a number of years, whereas supplementary feeding was likely to be less of a problem. Dual grazing of sites (alternately of donor and receptor sites) had the potential to cause losses in botanical diversity on semi-natural grassland, through N and P enrichment, although these effects would tend to be more localised around supplementary feeding stations. These effects can be relatively easily avoided by (a) reducing the frequency of stock commuting between one habitat and the other and (b) restricting overwintering of stock on valuable semi-natural grasslands.

An extreme method of removing nutrients involves carrying out enhanced arable cropping with fertiliser, as, for example, when applying inorganic N and K fertilisers to boost cereal production, increasing the removal of residual P *en passant* (Marrs 1985; Tallowin *et al.* 2002). The N and K not taken up by the harvested crop rapidly leach away and do not persist, but soil P levels are gradually lowered through biomass removal. However, in moderately fertile soils, mineralisation from the organic fraction and weathering of the non-available pools present in the soil can rapidly replenish that removed by herbage. At Minsmere in the Suffolk Sandlings, former arable fields that had either been left unfertilised, or fertilised with ammonium sulphate, both failed to show any measurable decrease in key soil nutrients, even after cropping for 6 years (Marrs *et al.* 1998). Similarly, applying N and K fertilisers to a restored mesotrophic grassland for three consecutive seasons failed to show any reductions in available soil P, or increases in species richness, of a subsequently sown diverse seed mixture compared with unfertilised plots (Pywell *et al.* 2004a; 2005). The amount of P removed was small relative to the large residual pool in the soil. Assuming a linear response, the authors concluded that it would take 12.5 years of cropping to reduce total P to levels equivalent to unimproved, diverse grassland. However, the timescale can be even longer in some soils. In the long-term cropping experiment on Exhaustion Land at Rothamsted, P and K residues remaining in the soil from fertilisers and manures applied prior to 1902 gradually declined; however, in the case of available P, it did not fall to critical levels for almost 70 years, severely restricting yields of spring barley (Johnston and Poulton 1977).

5.4.2.1   Soil stripping

More immediate and drastic steps to reduce high nutrient loads can involve turf removal, soil profile stripping and profile inversion (Figure 5.3). After determining extractable levels of nutrients at different depths in the profile, reference levels of fertility, such as < 10 mg/L P, could theoretically be achieved by simply stripping surface layers to the appropriate depth. For example, Tallowin and Smith (2001) found it necessary to remove 15–20 cm of topsoil from an improved rush pasture, more than 80% of the organo-mineral

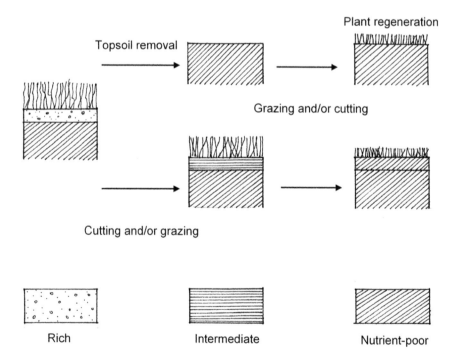

Figure 5.3 Topsoil removal on former agricultural land is by far and away the quickest way of reducing levels (illustrated by Tharada Blakesley).

layer, to match the lower nutrient status of local M24 fen meadows. This action lowered sward production from 10–15 t/ha to that of the target community (approximately 6 t/ha), as well as reducing plant uptake of N, P and K and significantly improving the survival and cover of transplanted M24 species. Soil removals of 20–40 cm, followed by seeding using hay transferred from a donor site, were similarly successful in restoring fen and floodmeadow communities, respectively, in Germany (Patzelt *et al.* 2001; Hölzel and Otte 2003). Removing a 20-cm layer of topsoil from an old orchard site in southern France was also effective in partially reinstating a type of reference steppe grassland, while reducing the non-target species in the seed bank. Soil transfer from a donor site was equally successful and strengthened target species dispersal, whereas sowing a three-species cover crop and hay transfer were less successful treatments (Jaunatre *et al.* 2014). In an analysis of wet meadow restoration sites in Western Europe, Klimkowska *et al.* (2007) found greater gains in species richness with progressively deeper (> 20 cm) removal of soil layers compared with shallower removals, possibly because the wetter conditions in the former case discouraged some common competitive species. In experiments in the Netherlands where topsoil was stripped from former agricultural land but left to naturally recolonise, some target species of low-production vegetation communities from the local species pool did increase over 9 years, but the process was very slow due to poor seed dispersal (Verhagen *et al.* 2001).

Soil stripping has a number of advantages and disadvantages. Apart from lowering fertility, removing even 5 cm of topsoil from ex-arable sites will drastically deplete the soil seed bank, a practical advantage if sowing late-successional species that would otherwise have to compete with high densities of weed seeds. Most of the soil seed bank is contained in the surface 0–10-cm layer. Hölzel and Otte (2003) found that stripping 30 cm of topsoil from a former arable field in northern Germany removed 60–80% of weed seeds and some common grasses; an even thicker 50-cm layer almost entirely eliminated the seed bank. However, soil stripping may also adversely affect some important soil biological properties by reducing bacterial numbers, fungal biomass and nematode abundance, which could have a positive interaction with resown vegetation, while deep excavation exposing the mineral surface risks creating a poor seedbed with low soil moisture, thus inhibiting germination (Kardol *et al.* 2008; 2009). Soil stripping is also liable to increase the soil pH by exposing the lower, mineral-rich horizons, a disadvantage if acidic vegetation types are the intended target. On acid soils, such as improved former heathlands, soil stripping can increase the availability of potentially toxic metals leached from the upper horizons. After removing 20 cm of topsoil, Diaz *et al.* (2011) found elevated Al concentrations compared with controls, despite pH levels remaining above 6. Because of these potential disadvantages, including high cost and the risk of damage to archaeological features, soil stripping needs to be weighed carefully against other alternatives.

Deep cultivation is another technique by which nutrient-rich surface layers are diluted or temporarily buried below rooting depth, along with undesirable weeds in the seed bank. Soil profiles can be inverted using trench excavators, drawn by powerful tractors, to depths of 40–80 cm. In experiments to restore species-rich vegetation on arable land, Pywell *et al.* (2002) also found that ploughing to 30–40 cm effectively reduced levels of P and K at the surface by inverting the profile and simultaneously diluting the upper fertile layers, a technique especially effective in shallow chalk soils. However, an attempt to create heathland on an improved grassland site in Kent, using deep ploughing, failed because this brought calcareous clay to the surface, altering the pH and nutrient levels (Allison and Ausden 2004). In a more successful treatment where topsoil was removed to a depth of 25 cm, this significantly lowered levels of total N and extractable P and K in the upper 10 cm of soil, after which heathland vegetation was established despite the high pH (5.9–6.8).

5.4.3   Chemical manipulation

Other ways of dealing with high nutrient levels use chemical manipulation, often by adjusting pH levels to reduce nutrient availability and by slowing mineralisation. Phosphorus availability, for example, can be fixed chemically by adjusting the soil pH to either high or low extremes. In a series of experiments carried out by the Groundwork Trust in Merseyside to reduce fertility in urban soils (Scott and Ash 1990; Ash et al. 1992), the treatments involved stripping 10-cm surface layers of soil, which were then substituted with waste mineral materials and mixed by rotovation to 20 cm and finally sown with forb-rich mixtures. Crushed concrete, colliery spoil, sand, brick rubble and subsoil were all effective in adjusting pH and controlling sward productivity, while at the same time promoting species diversity. Other techniques that avoided topsoil stripping by top-dressing the surface with thin (1–2 cm) surface layers of waste material, or inserting the material using a sand-slitter machine to 20 cm were ineffective, either because the amount of foreign material added was not enough, or because the disturbance caused enhanced mineralisation.

Liming with alkaline wastes such as pulverised fuel ash, Leblanc waste, chalk or ground limestone can effectively immobilise P, but limits the restoration targets to mesotrophic or calcareous grassland types for the foreseeable future. Acidification is another means to reduce nutrient availability when acid grassland is the intended result. Efforts to recreate heathland and acid grassland, using acidifying agents such as iron and aluminium sulphates, elemental S or acid peat can be effective in lowering pH, thereby reducing fertility (Dunsford et al. 1998; Owen et al. 1999). Bracken litter and pine bark chippings are also synergistic acidifying agents, providing sources of carbon for heterotrophic bacteria that are capable of oxidising S (Owen and Marrs 2001). At Minsmere in Suffolk, applying 2.6 t/ha of S to ex-arable land, a level assumed to reduce soil pH by one unit, and reseeding with a Bent/Red Fescue/Fine-leaved Sheep's-fescue mix, successfully lowered the soil pH after 2 years from an average pH of 7.1 to 4.9 (Ausden et al. 2010) and facilitated the development of acid grassland. Lower S additions of 1.7 t/ha plus a 2.5-cm layer of Bracken litter and Heather cuttings, resulted in a mixed acid grassland and heathland community (RSPB 2006). The increasing acidity was also effective in reducing the infestation of arable weeds from the second year after treatment.

Although the addition of S causes base elements such as calcium, K and Mg to be rapidly leached as the pH falls, P availability may actually increase as calcium phosphate residues in the soil are broken down (Walker et al. 2007; Diaz et al. 2011). This in turn may lead to vigorous growth of arable weeds or sown grasses on relatively productive, former arable sites, suggesting that acid grassland or grassy heath vegetation might be a more realistic target than pure heathland in such cases (Walker et al. 2004b). Another side effect of acidification is the increase, below pH 5, of soluble toxic metal ions, which through uptake in vegetation, can accumulate through the food chain in invertebrates and their predators. Diaz et al. (2011) showed that the shoots of Sheep's Sorrel and Common Bent, sown on a former improved pasture, accumulated significantly higher concentrations of Al and Zn with increasing applications of S, up to 3.6 t/ha. On another ex-arable site where S had been applied, they found that both low pH and high Al levels respectively

Following application of 2.6 t/ha of S to ex-arable land at Minsmere, pH was reduced from an average of 7.1 to 4.9. Subsequent seeding with acid grasses created a tight sward dominated by sown Sheep's-fescue, Fine-leaved Fescue and Common Bent. Few other plants are present, probably due to the low soil pH and lack of gaps between the perennial grasses (© Malcolm Ausden).

Following application of 3.3 t/ha of S with 2.5-cm deep Bracken litter, pH was reduced from an average of 6.9 to 5.1. This treatment, together with the addition of Heather clippings and litter created a sward dominated by Common Bent, and a mixed acid grassland and heathland community, including Heather, Bell Heather, Sheep's Sorrel, Brown Bent, Western Gorse and the mosses *Hypnum jutlandicum* (Heath Plait-moss) and *Polytrichum juniperinum* (Juniper Haircap) (© Malcolm Ausden).

decreased the abundance of spiders collected in pitfall traps, although ground beetles were unaffected.

5.4.4 Immobilisation

Immobilisation of nutrients using alkaline or acid waste materials can be successful, as the Groundwork Trust experiments demonstrate (Ash *et al.* 1992), by preventing dominant species responding to high fertility and encouraging grassland diversification. However, a common technique, much used in prairie grassland restoration in the USA, is to apply organic carbon sources such as straw, sawdust, wood shavings or peat, widening the carbon-nitrogen (C:N) ratio of the growing medium and thus immobilising N mineralisation by diverting plant uptake to the soil microbial population, rather than the vegetation. The quantities required to reduce N levels and vegetation productivity may be considerable, with up to 10–30 t/ha of carbon equivalent applications needed (Blumenthal *et al.* 2003; Averett *et al.* 2004; Wilson *et al.* 2004; Eschen *et al.* 2007).

Carbon treatments may be only temporary in their effect, especially when materials readily metabolised by microorganisms, such as sucrose, are used. Laboratory-based experiments have shown that combinations of sucrose, starch, cellulose and sawdust all effectively decreased available N, the sucrose providing the most rapid short-term response while the other materials maintained immobilisation of nutrients for longer (Török *et al.* 2000). In field experiments where straw or birchwood fragments were disced into the top 10 cm of soil on an ex-arable site, Kardol *et al.* (2008) failed to find any reduction in available N and P over 3 years. A glasshouse experiment showed some short-term immobilisation with straw, but not with wood fragments, probably because the high lignin content would take many years to break down. However, it could be argued that persistent immobilisation is not necessary if materials applied at modest rates can prevent domination by annual weeds and grasses in the initial stages of restoration, encouraging the late-seral species characteristic of species-rich grassland to establish (Eschen *et al.* 2006). With vegetation development and the gradual incorporation of organic matter into the soil, the C:N ratio would be expected to widen naturally. Baer *et al.* (2002) recorded a significant increase after 12 years in soil carbon and the C:N ratio during the restoration of tallgrass prairie from cultivated fields, showing that native grasses were effective in driving ecosystem processes towards the original system.

5.4.5 Reinstating soil communities

In general, improved grasslands retaining high residual fertility tend to have bacteria-dominated soil microbial systems, characterised by high nutrient availability through rapid recycling of organic matter. By contrast, soil communities of traditional, semi-natural grassland are often fungal-dominated (Bardgett and McAlister 1999) and are associated with less fertile soils, slower rates of nutrient cycling and availability, and greater plant species richness. Plants on these lower fertility soils also tend to be more highly colonised by mycorrhizal fungi (Eriksson 2001). Arbuscular mycorrhizal fungi (AMF) have been shown to exert a major influence on plant biodiversity, nutrient capture and biomass productivity, all of which increase with AMF-species richness (van der Heijden *et al.* 1998). The abundance of soil-active fungi and bacteria can be determined in soil assays that measure phospholipid and neutral lipid fatty acids (PLFAs and NLFAs), components derived from the cell membranes of these organisms. For example, additions of fertilisers and farmyard manure have been shown to reduce soil fungal: bacterial (F:B) PLFA ratios (Smith 2005; Smith *et al.* 2003; 2008), while fungal-rich soils appear to retain N more effectively (Gordon *et al.* 2008). Mowing has also been found to stimulate the NFLA signatures of AMF in a species-poor hayfield, probably in a feedback response to above-ground increases in species diversity promoted by this management (Antonsen and Olsson 2005).

Management can have a strong influence on soil organisms and hence their interaction with restored plant communities. In the long-running experiment to diversify an improved upland meadow at Colt Park in the Yorkshire Dales, F:B ratios generally increased over time in plots where native seeds had been added and in the absence of fertiliser, whereas reductions in soil F:B ratio were linked to intensive management with mineral fertiliser. In the absence of nutrients added in organic or inorganic form, there was a continuing trend towards the fungal dominance normally associated with traditional meadows, demonstrating that over time it was possible to reverse the trend from the bacteria-dominated soil communities of improved grasslands to fungal-dominated ones (Smith 2005; Smith *et al.* 2008). Increases in the F:B ratio were also associated with more conservative nutrient cycling, increased abundance of mycorrhizal fungi and enhanced plant species diversity. Fungal PFLA was linked to less competitive species such as Yellow-rattle, Sweet Vernal-grass and Meadow Buttercup, while lower F:B ratios were associated with dominant grasses such as Rough Meadow-grass, Meadow Foxtail and Cock's-foot.

The importance of plant–soil feedback – the influence of plants on soil biological and chemical properties, which in turn influence plant performance – is often underemphasised in restoration. Different plant species can bring about a specific conditioning of soil properties, depending on soil type, which then affect the performance of that species or others growing in that soil (Bezemer *et al.* 2006). Changes in soil microbial or plant communities may act directly or indirectly upon each other. Microbial populations may be driven by changes in plant species composition, which can be manipulated by seed additions of 'facilitator' plants; these in turn have properties or traits promoting fungal dominance, thus providing niches for mid- to late-successional species. These mechanisms are not fully understood, and there is also evidence that some plant species may have the opposite effect. However, it is well established that hemiparasitic plants such as Yellow-rattle are major drivers of both above- and below-ground properties, both through their litter inputs and root exudates as well as their selective parasitism of some plant species. Counter-intuitively, Yellow-rattle is known to promote N mineralisation (Bardgett *et al.* 2006), suggesting that it should encourage bacteria-dominated soil communities, but its influence may ultimately favour other species that act as fungal facilitators, resulting in higher F:B ratios, as was observed in the Colt Park trials.

Öster *et al.* (2009) found that when 16 forbs typical of mesotrophic grassland were seeded into either ex-arable sites or adjoining fields of semi-natural grassland, the latter invariably showed better recruitment after one year. However, since there was no appreciable difference in soil chemistry between the two grassland types, the authors concluded that aspects of soil quality, such as soil texture and microbial and soil faunal development, could explain the greater success in the established semi-natural grassland. This suggests that a way of accelerating the colonisation of soil microbial communities would be to inoculate the site to be restored with soil or turves from donor, species-rich grasslands. However, in a four-year study where shredded topsoil from a donor grassland was incorporated into an improved grassland and sown with a diverse seed mix, this had no impact on botanical diversity (Pywell *et al.* 2007a). The timing of inoculation in relation to the development of the plant community and the underlying soil nutrient status may be critical to the success of this technique.

Restoration of soil communities can also potentially be accelerated by incorporating carbon substrates, and by mycorrhizal inoculations. In a microcosm experiment, bacteria-feeding and root hair-feeding nematodes were more abundant in soil amended with straw and wood treatments than in the control, but not in the field, where these materials were disced 10 cm into the surface (Kardol *et al.* 2008). Similarly, transferring soil layers or turf transplants from a species-rich donor grassland to a topsoil-stripped ex-arable field had little effect on nematode community composition (Kardol *et al.* 2009), although this may have been due to differences in environmental conditions (e.g. alternate wetting and drying) between the donor and receptor site, and limited resources in the lower organic matter of the topsoil-stripped treatments.

Macroorganisms such as soil nematodes are convenient indicators of ecosystem functioning, being sensitive to disturbances and successional changes. Both nematodes and collembolans are relatively slow to increase over time in restored grassland sites (Kardol *et al.* 2005; Chauvat *et al.* 2007), and colonisation by macro-invertebrates is also likely to be slow. Investigating the effects of diversifying arable field margins using different vegetation covers and management treatments, Smith *et al.* (2008) detected no significant influence after 4 years on the soil macrofauna communities of three different seed mixtures, consisting of grasses only, tussock grasses with forbs or fine grasses with forbs, respectively. However, the macrofauna diversity of these field margins was much greater than in the arable crop, which contained fewer earthworm, woodlice and beetle species. Management treatments also had a strong effect, with annual scarification reducing the abundance of sensitive litter feeders such as woodlice and some predator beetles, although spring cutting and gramicide-treated plots had similar macrofaunal communities. The authors concluded that the build-up of substantial litter layers would encourage litter-dwelling invertebrates, although this might not be compatible with restoration schemes promoting high floristic and pollinator diversity. Gormsen *et al.* (2004) compared earthworm populations in experimental plots of restored grasslands in four European countries, sown with either low- (4 species) or high-diversity (15 species) mixtures, or in unsown control plots. Results varied between the four sites, but although the earthworm communities had changed in all treatments after 3 years from those present in the original agricultural field, they were still very different from long-established reference sites. In these cases it must be accepted that restoration of the soil community is likely to remain a lengthy process, unless active steps are taken to import soil from donor sites.

6. Plant material for dry grassland restoration

6.1 Surveying the restoration site

6.1.1 Soil analysis

Information on geology and soil type(s) will help to inform the choice of species mix for grassland restoration. Geological maps are valuable for predicting soil types, but only if they show drift deposits as well as solid geology, because whichever is at the surface forms the soil parent material. Geological information, including Countryside Survey topsoil maps, can be obtained from the 'Geology of Britain' viewer (http://www.bgs.ac.uk/discoveringGeology/geologyOfBritain/viewer.html), and the Natural Environment Research Council Soil Portal (http://www.ukso.org/). For England and Wales, the Land Information System (http://www.landis.org.uk/), hosted by the National Soil Resources Institute at Cranfield University provides a range of services, including site-specific soil information with maps, soil descriptions and the relationship between soils and the environment. A charge is made for the 1:250,000 scale digital National Soil Map and other services. The 1:250,000 maps show the boundaries of polygons of different soil associations, listing the main soil type or group to which it belongs, as well as brief details of soil characteristics and typical land usage. In addition to the Soil Survey of England and Wales, there is a separate classification for Scotland, the information for which is held at the James Hutton Institute. In England, the UK Government's web-based interactive map, MAGIC, includes a 'Soilscape' option at a scale of 1:250,000, but this is only indicative of generic soil types within a region and more detailed information would be required to inform grassland creation.

6.1.1.1 *Undertaking a soil survey*

Soil surveys are important, because they indicate the suitability of a site for grassland creation, and inform decisions on the target community. Digital national soil maps indicate general soil types and hydrological properties, but laboratory-based analyses are required to determine pH, P index and other measures of soil nutrients. A detailed soil survey involves traversing an area and sampling the nature of the soil at intervals using auger borings and digging occasional inspection pits. Depending on the complexity of the site, a density of five samples per hectare is usually sufficient for a very detailed survey at a scale of 1:2,500 and for areas up to 20 ha, which might take an experienced surveyor 2–3 days to complete. For larger areas and where less detailed information is wanted, this could be reduced to perhaps one sample per hectare. At each sample location a note is made of the colour, texture and stoniness of the soil, together with a record of the soil layers or horizons present, up to a depth of at least half a metre. Rooting density and depth is also noted, because in some soils the roots are killed off by a lack of oxygen if the lower profile is waterlogged. Such soils have gleyed horizons, where the iron compounds in the soil are reduced (deoxygenated), turning them grey or greenish in colour. If there is seasonal waterlogging, the upper parts of the profile become oxygenated as they dry out during the growing season and the iron compounds change to red and orange colours, causing a characteristic grey/orange mottling at the transition zone.

The type of humus at the soil surface is also a good indicator of fertility: on very acid, infertile soils the litter layer breaks down very slowly, and often forms a thick (> 5 cm) deposit in recognisable layers representing varying states of decay – known as mor humus. Mull humus, on the other hand, is much thinner and forms over more fertile, less acid soils when there is a rapid turnover of litter. Moder humus types are intermediate. Altogether, the descriptive methodology outlined here should enable the surveyor to classify and map soil units to a reference soil series.

The soil can be further tested quantitatively by removing soil samples with a screw, Dutch or tubular auger, usually inserted to 15-cm depth if a topsoil sample is required, and returning them to a laboratory for analysis. Many authorities suggest walking a W-shaped path across the sampling area, taking 5–6 samples per limb. Ideally at each sample point 3–5 separate cores should be taken and bulked. A simple physical 'finger' test can be used to assess the soil's textural composition of sand, silt and clay particles. It is based on the degree to which soils cohere when moistened and rolled into a ball, or into thin cylinders and finer threads, which indicates their textural status.

Simple chemical tests, measuring the level of soil acidity (pH) or the soil nutrient status, can be carried out using kits available from specialised suppliers and garden centres. Indicator papers or water testing kits can measure the whole, or part of the pH range (1–14), but for more accurate determinations a portable or laboratory pH meter with an electrode is used to test a solution of soil, made up with distilled water. Other routine laboratory tests of soil fertility usually measure quantities of the major plant nutrients – available phosphate, K and Mg levels, plus total N and S.

6.1.2 Vegetation

For arable land, sophisticated survey techniques may not always be necessary, unless rare or endangered plants such as cornfield annuals are present, or if remnants of unimproved grassland remain along field margins. In other areas, such as existing grassland, an NVC might be appropriate. Consult the *National Vegetation Classification: Users' handbook* (Rodwell 2006), which provides details of the methodology for sampling and describing vegetation communities. Refer to the keys given in Rodwell (1991 *et seq.*) to determine the most appropriate NVC community classification. Computer programs are also available to calculate the 'goodness of fit' of data collected from quadrats to the expected species composition of semi-natural woodland communities and subcommunities recognised by the NVC. One such program, 'Tablefit' (Hill 1996), is available freely from the Centre for Ecology & Hydrology (http://eidc.nerc.ac.uk/products/software/tablefit/download.asp).

Consider also the key to botanical enhancement of species-poor grassland illustrated in Figure 6.1.

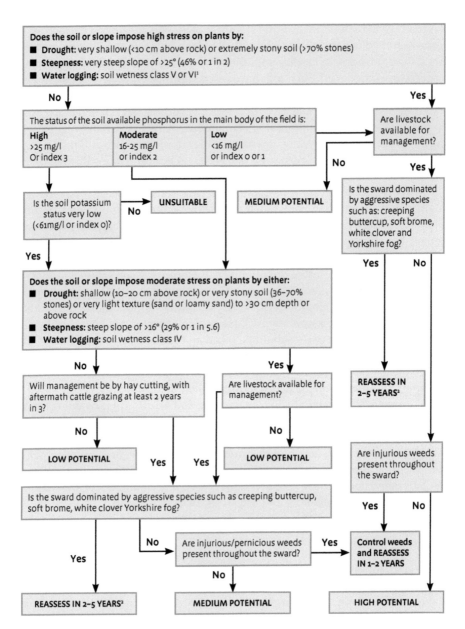

Figure 6.1 Key to the botanical enhancement potential of species-poor grassland (Farm Environment Plan Manual (Natural England 2010a) (© Natural England)). [1]Soils with wetness class V and VI are wet for long periods into the growing season, or permanently waterlogged near the surface. Soils with wetness class IV are waterlogged for long periods in winter. [2]Where these species dominate there is no existing botanical interest; it may be more effective to remove the sward (taking into account historical, bird and other interests on the site, and the risk of soil erosion).

6.2   Surveying reference sites

Species mixes for grassland restoration should be based on semi-natural grassland communities present on similar soil types in the local landscape, so-called 'reference sites'. This assumes that the geology and soil type of the restoration site is known and that it has not been radically altered by agricultural production, or that measures have been taken to lower soil fertility if necessary.

To locate reference sites in England, the UK Government's web-based interactive MAGIC map is a useful starting point for key environmental schemes and designations (statutory and non-statutory). It also includes some information on the main dry grassland types (former UK BAP priority habitats). The BAP priority habitats can also be found on Natural England's web-based interactive 'Nature on the Map'; this includes a comprehensive coverage of SSSIs, nature reserves and agri-environment schemes. Scottish SSSIs are documented on the Registers of Scotland website (https://www.ros.gov.uk/), and information on Welsh SSSIs can be found on the NRW website (https://naturalresources.wales/splash?orig=/).

In Wales, NRW is developing new maps for displaying environmental data, including maps of protected areas; these will supersede LANDMAP, which is no longer available. Scottish Natural Heritage provides information on sites with conservation designations, such as NNRs, LNRs, SSSIs, and SPAs. Sites can either be found by searching specific regions, or in the case of NNRs, SSSIs, SACs and SPAs, by using various maps on the Scottish Natural Heritage's information service.

Non-statutory wildlife sites should also be identified in the local landscape. Because these have evolved separately in different counties, they are known by many names, including County Wildlife Sites, Sites of Nature Conservation Interest/Importance (SNCIs), Sites of Interest/Importance for Nature Conservation (SINCs) and District Wildlife Sites. Selection of these sites in England and Wales is often undertaken by local Wildlife Trusts on behalf of Local Authorities, and details can be obtained from these bodies. In Scotland, 'Local Sites' identified for their high wildlife value by the Scottish Wildlife Trust are also officially recognised by local authorities through a non-statutory designation. Information on these sites can be obtained from the Scottish Wildlife Trust.

Surveying species-rich calcareous grassland along the margins of a wood.

Roadside Nature Reserves (RNRs) are included in the Welsh SINCs, but are listed separately in many English counties, where details should be obtained from the local Wildlife Trust. There is no Scottish equivalent of RNRs, although some local authorities do identify roadside areas that have some value to wildlife, and may 'mark' and manage them accordingly. The NBN Gateway (https://data.nbn.org.uk/) also includes an interactive map that allows designated sites such as SSSIs and LNRs to be located.

Semi-natural grassland communities – particularly in the lowlands – are highly fragmented and rare. In many cases, it will simply not be possible to find good reference sites in the local landscape. Reference to historical records or typical NVC communities will then be the method of choice.

If information on potential reference sites is not available or is not sufficiently detailed, a simple NVC survey can be undertaken, either by a professional ecologist, or anyone confident in identifying grassland plants. A simple assessment of the NVC plant community should suffice in most circumstances, for example as follows:

1. Identify a homogeneous area within the reference site for survey.
2. Set up the first of three 2 × 2 m quadrats and record the species present and their percentage cover. The abundance of each species can be recorded using percentage cover or the Domin scale; note any other species present in the environs.

Cover (%)	Domin scale
91–100	10
76–90	9
51–75	8
34–50	7
26–33	6
11–25	5
4–10	4
< 4 with many individuals	3
< 4 with several individuals	2
< 4 with few individuals	1

3. Repeat this process three times, with quadrats spread evenly around the area.
4. If different community types are present in other homogeneous areas, repeat the process for each.

Use the data collected to determine the most appropriate NVC community classification, referring to the keys given in Rodwell (1991 *et seq.*).

6.3 Selecting plant material

6.3.1 Use of ecological traits

Plant selections may be influenced by the desire to reproduce a reference grassland community, or to create a habitat that will support other species (such as rare butterflies) of conservation value. Although such abstractions may be tempting, the first priority is to make sure that species appropriate to the site are chosen; this is essential if they are to survive and thrive in the long term. Pragmatically, the generic ecological characteristics of species should be matched to the site in question, based on ecological traits or functional strategies, such as life form, competitiveness, phenology, regeneration, seed biology and known environmental associations. This process of prior 'ecological sieving' of the suitable candidates is more likely to meet with success than allowing 'environmental sieving' to naturally eliminate species in the field. The point was well illustrated by Gilbert *et al.* (2003) in an experiment to create inundation grassland communities (MG13 and MG8) for wintering wildfowl and breeding waders. They found that the precise composition of the seed mixture and the sowing rate was less important than achieving the correct hydrological conditions. Although the original composition of the seed mixtures influenced how the proportions of species were established initially, after 3 years the vegetation communities were determined primarily by the variability in aeration stress (i.e. inundation) operating over the site during the growing season. With hindsight, a better selection of species might have been made.

Many examples of disappointment and failure in restoration trials have been reported in the literature. In a review of seed sowing experiments, Turnbull *et al.* (2000) observed that seeds of new species introduced into existing vegetation often germinated successfully, but survival to adulthood was less certain; few such experiments have been monitored for long enough to determine the outcome. An early survey of 19 sites by Wells (1987) found that of the 108 species sown as species-rich grass mixtures, almost half failed to establish. Hodgson (1989), reanalysing Wells's results, showed that the most successful species were those associated with improved grasslands, i.e. competitive ruderals and competitors, while slow-growing, stress-tolerant species tended to disappear. There is also some evidence that large-seeded species tend to have broader, less restrictive regeneration niches, allowing more consistent recruitment, while small-seeded species tend to produce more seeds and have greater colonising ability, but require more specialised niches, such as lighter conditions and large vegetation gaps (Pakeman *et al.* 2002; Turnbull *et al.* 2005). In their experiment on chalk grassland creation, Pakeman *et al.* (2002) also found that species with poorer germination were less sensitive to light or dark, but tended to have greater long-term persistence once established.

A review by Pywell *et al.* (2003) investigated the performance of a range of different seed mixes, representing species-rich grasslands, that had been sown on formerly intensively managed agricultural land in lowland Britain (ex-arable and species-poor improved grasslands). In 25 restoration experiments, 58 species (13 grasses and 45 forbs) were present in at least 5 experiments. A performance index was calculated as the proportion of quadrats containing the species during the first four establishment years, while plant traits,[1] or functional groups, were also assigned to each species. These traits were then correlated with the performance indices calculated from each experiment. Successful species tended to have the following characteristics:

- good gap colonising ability (able to capitalise on short-lived space);
- strong competitive ability (becoming increasingly important over time);
- good vegetative regeneration capacity (particularly important in closed, competitive communities).

---

[1] These included competitive strategy (*sensu* Grime 1979); plant life form; Ellenberg environmental value (Ellenberg 1988); seed bank longevity; germination rate; seedling growth rate; mycorrhizal association; specificity to a NVC; and method of dispersal.

Most were generalists, associated with relatively fertile habitats, with grasses usually more successful than forbs. Species that established well in the first year, typically ruderals with autumn germination and a high percentage of germination, were liable to gain an advantage over spring-germinating species such as Cowslip, which requires a period of chilling. Examples of particularly successful species were grasses such as Red Fescue, Yellow Oat-grass, Crested Dog's-tail, and forbs such as Ribwort Plantain, Selfheal and Common Bird's-foot-trefoil. Stress-tolerant species, including Quaking-grass, Sheep's-fescue, Yellow-rattle, Fairy Flax and Clustered Bellflower, performed relatively badly. Recommendations following from this meta-analysis were:

- Species with proven and consistently good performance would increase the viability of restoration schemes, although this might result in uniformity of schemes.
- Projects should be aimed at lower-fertility soils, or fertility reduction treatments should be implemented and the abiotic environment otherwise modified to increase suitability for the target species.
- Introductions of less robust, stress-tolerant and late-successional species should be postponed until a later date, once the plant community has achieved greater stability and giving time for environmental conditions to improve.

Functional traits can be useful measures for interpreting community assembly processes in restoration and understanding the impacts of land use management. For example, measures of simple traits (using either Ellenberg or competitive strategy (CSR) scores) enabled Lewis *et al.* (2014) to identify potential causes of shifts over time in the composition of machair grassland in Scotland.

Other researchers have also argued that grassland restoration is best managed as a two-stage process, with early-successional species being sown first, only introducing late-successional species once the sward has established and plant–soil feedback mechanisms have begun to take effect. This has the advantage that a few dominant 'foundation' species may effectively suppress competing weeds, while 'subordinate' and dispersal-limited species can be established in subsequent sowings or in hay transfer (Coiffait-Gombault *et al.* 2012). However, under fertile conditions there is still a danger that such later sowings may be outcompeted. Warren (2000) noted a decline in the species richness of a sown mesotrophic grassland community as White Clover became dominant in grazed plots. He suggested that the N-fixing clover might be favoured on fertile ex-arable soils with low N, encouraging the spread of sown grasses, which in turn suppress the forb component. Best practice in these cases would be to reduce the proportion of such competitors in seed mixes, or to control their dominance using cutting and grazing, to maintain space for late-successional species. A default position might be to select for robust meadow communities of moderately fertile, non-calcareous, neutral soils in the first place, such as a modified Common Knapweed–Crested Dog's-tail (MG5) community that contains a high proportion of species with combined competitive, ruderal and stress-tolerant (CSR) strategies (Hodgson 1989). Vigorous agricultural cultivars of grasses and legumes should be avoided, and hemiparasitic species could be used to control dominant species (Pywell *et al.* 2004b; see Section 6.2.3).

6.3.2   Complex or simple mixtures?

Sowing complex communities may appeal on the grounds that these may be stable in the face of environmental change or perturbation, and because unwanted, unsown exotic species are less likely to invade a species-rich community where most of the niches are already filled. In a chalk grassland recreation experiment, Pakeman *et al.* (2002) found that plots sown with up to 37 species showed little flux initially, although later this increased as populations built up and spread between plots, whereas simpler mixtures and unsown controls showed a much higher turnover from the outset. Experiments also suggest that there is likely to be a natural limit to the total number of species surviving as the site's resources are increasingly exploited. In an investigation by Tilman (1997) where seeds of up to 54 species were sown into native grassland, fewer survived in plots that were more species-rich at the start as recruitment limitations began to operate.

Diverse seed mixtures, i.e. comprising 10–40 species, are commonly used in restoration projects, based on a particular reference grassland community (e.g. Lawson *et al.* 2004; Foster *et al.* 2007; Jongepierová *et al.* 2007; Kirmer *et al.* 2012; Pywell *et al.* 2002; Warren *et al.* 2002), but the drawback is cost, and some species may not be available commercially. An alternative is to sow lower-diversity mixtures, to be later supplemented either by natural colonisation or further species introductions. Robust, low-diversity mixes also have the advantage of suppressing early weed growth, particularly annuals on ex-arable sites, and may also eventually develop into relatively species-rich grasslands (Manchester *et al.* 1999; Mortimer *et al.* 2002a; Bullock *et al.* 2007; Lepš *et al.* 2007). Török *et al.* (2010) reported a large-scale restoration project in which only two or three target grass species were sown on previously arable fields in the Hortobágy National Park in eastern Hungary, an area formerly occupied by alkali grasslands over the loess substrate. The fields were first deep-ploughed to bury weed seeds and to invert the fertile topsoil, then harrowed and sown, followed by hay cuts taken after year 1. In the first year, sparse weed growth cover provided a beneficial microclimate for the germination and establishment of the sown grasses. By the second year, the sown grasses had practically replaced short-lived arable weeds, achieving a cover of > 50% after 3 years. Several other unsown species became established and there was a slow immigration of species characteristic of the reference grassland sites into the restored areas.

Conversely, sowing grasses and other productive vegetation as low-diversity seed mixes, particularly at higher seeding rates, may delay or even suppress the inflow of potential desirable colonists (Stevenson *et al.* 1995; van der Putten *et al.* 2000; Warren *et al.* 2002; Walker *et al.* 2004b; Lepš *et al.* 2007; Smith *et al.* 2010; Lengyel *et al.* 2012). Productivity may also be compromised where fewer species are used. Although this may not be a goal of conservation sowings, there is a broad consensus from many experimental studies that greater species diversity also tends also to increase biomass production in semi-natural grassland communities (e.g. Dodd *et al.* 1994b; Tilman *et al.* 1996; Hector *et al.* 1999; Loreau *et al.* 2001; Hooper *et al.* 2005; Stein *et al.* 2008). Theoretically, this is because the differences in physiology, morphology, resource requirements and life histories of a large number of individual species allow site resources to be used more fully than with fewer species. Kirmer *et al.* (2012) also found that a high-diversity seed mixture of 51 species was more productive than a low-diversity mix (consisting of 3 grass cultivars) sown on surface-mined land. For these harsh conditions, they recommended using high-density mixtures comprising 6–10 grasses and 15–20 forbs to accelerate vegetation development. In long-term arable reversion experiments in southern England, Bullock *et al.* (2007) found that a species-rich mixture comprising 11 grasses and 28 forbs yielded 43% more hay than the 7-species grass mixture prescribed by an agri-environmental scheme, although the latter did colonise with a number of forbs over time. Higher yields were strongly correlated with forb richness, although not with legume content per se (which might be presumed to increase productivity through N fixation). In such cases, management would play a key factor in providing the right conditions for colonisation, for example, by introducing sheep and cattle pastured on nearby species-rich grasslands, or importing hay from pristine sites.

6.3.3 Hemiparasites: Yellow-rattle

Hemiparasitic plants of the figwort family (Scrophulariaceae) appear to have much potential in reducing the productivity of competitive species in grassland swards. The performance of Yellow-rattle in particular has been examined in a number of restoration experiments. Bullock and Pywell (2005) listed its advantages as follows:

- Seed is relatively inexpensive and easily obtainable.
- Hemiparasites reduce the vigour of competitive species, facilitating the establishment and persistence of target species.
- Yellow-rattle can colonise rapidly and persist in fertile grasslands.
- Excessive populations can be controlled relatively easily, if necessary by close mowing before flowering and seed set.

Yellow-rattle occurs naturally in 14 main grassland communities, particularly in lowland and submontane, mesotrophic and calcicolous types, but also in acidic grasslands with a northern and upland distribution. Only one subspecies (*minor*) is common and

Yellow-rattle in a hay meadow at Askrigg Bottoms in the Yorkshire Dales (© Roger Smith).

widespread, while the ssp. *stenophyllus* is restricted to damp grasslands particularly in the north and west, ssp. *calcareus* to southern, calcareous soils, and ssp. *lintonii*, *monticola* and *borealis* to montane habitats (Stace 2010). Furthermore, it parasitises a wide range of hosts. Gibson and Watkinson (1989) reported at least 50 affected species from 18 plant families from Britain and Central Europe, with 22% in the Leguminosae and 30% in the Gramineae, the hosts varying with grassland type and site conditions. Several studies have shown that the parasite causes a general reduction in sward productivity by extracting water, nutrients and carbon compounds from its hosts, in particular grasses and legumes, which is not compensated for by its own biomass. There is less evidence that Yellow-rattle directly influences species diversity, especially on fertile sites, but non-leguminous dicotyledons appear to be the greatest beneficiaries where it is present (Ameloot *et al.* 2005). As an annual, Yellow-rattle dies down in late summer, creating gaps in the sward that are then colonised by other species.

On fertile sites with high levels of sward productivity (e.g. with a P index > 2), or more than 5 t dry weight/ha, (Mudrák *et al.* 2014)) it is unlikely that Yellow-rattle will establish satisfactorily without some nutrient stripping beforehand. Scarification, applying herbicide patches or hard grazing, preferably with some bare patches 10–30 cm in diameter, is also likely to improve establishment (Westbury *et al.* 2006). Before introducing other wildflower species, better results may be obtained by phasing the restoration work – i.e. sowing Yellow-rattle first, to reduce the vigour of the sward, and following up with native seed mixtures or spreading donor plant material (such as green hay) in subsequent growing seasons (Pywell *et al.* 2004b). When sown in conjunction with standard meadow seed mixtures, Yellow-rattle has been found to facilitate their establishment and development, even in agriculturally improved situations (Westbury *et al.* 2006). Compared with graminicide (fluazifop-P-butyl) applications as an alternative to reducing grass competition with introduced meadow species, introducing a Yellow-rattle treatment increased sown wildflower species numbers and cover, whereas the more specific chemical treatment promoted several unsown species (Westbury and Dunnett 2008).

Typical sowing rates for Yellow-rattle are 1–2.5 kg/ha, but lower rates of 0.5 kg/ha or less can be effective with precision slot seeding (Pywell *et al.* 2005). Scarification, or the removal of litter prior to sowing is important to aid seed establishment, both on new and in already established swards (Mudrák *et al.* 2014). A suitable establishment target is 100–200 plants/m², a density range that will normally reduce sward production significantly (Natural England 2009). The winged seed is also rapidly dispersed by agricultural machinery, spreading at rates of 4–5 m/y with the turning of the hay crop. Sowing should be done in the autumn, as the seed requires a period of prolonged chilling for successful germination. Smith *et al.* (2000) showed that early hay cuts (14 June) virtually eliminated the species by removing the seed heads before ripening, whereas it remained abundant in swards when cut later in summer (21 July). Heavy spring grazing and all-year-round pasture management will also lead to its decline. However, in the first year of an arable reversion project, it may be necessary to top the sward to discourage seeding of arable weeds such as thistles, docks, Common Ragwort and Black-grass. Provided the cutter height is set no lower than 10 cm, the lateral shoots of Yellow-rattle should develop to produce quantities of seed later in the year.

Apart from Yellow-rattle and some other related *Rhinanthus* species, few other hemiparasites are likely to be as universally effective for grassland creation and restoration. Red Bartsia, for example, is more usually associated with sparse and disturbed grassland; eyebright species are much less robust but one or two species are available from seed merchants, while many perennial parasites, such as the broomrapes, tend to be too host-specific for general use.

6.4 Sourcing plant materials

During the 1980s the demand for native wildflower seed rose from approximately 3 to 10 t/y (Wells, 1987; Brown 1989). In a survey of 14 commercial seed and plant companies supplying the UK wildflower market in 2004, this had further increased to at least 20–30 t/y of over 100 native species, in addition to just over 5 million plug plants (Walker *et al.* 2004a). Put in perspective, this represents only a fraction of 1% of the agricultural varieties of native species, mostly grasses, sown annually. Significant amounts of the seed used in agri-environmental schemes and conservation projects are now sourced directly from semi-natural grasslands: in 2010, under the HLS provision for maintenance, restoration or creation of species-rich grassland, 17% of agreements received a supplement for using native seed originating from semi-natural sites (Bullock *et al.* 2011).

Seed for creating new or enhancing existing grasslands can be sourced from:

- nursery-grown plants, harvested for seed or grown on as container plants or pre-grown turf;
- harvested plant material, such as green hay and litter from donor grasslands, imported and spread directly on to the restoration site;
- existing species-rich grasslands, using vacuum or brush harvesting of the sward in situ without mowing;
- hay bales or seed-rich chaff;
- translocation of turf and seed-containing soil.[2]

Apart from the traditional method of hay transfer, various hand-held, tractor-drawn and specialist machinery has been developed to harvest this material in bulk, or to dislodge and separate seed from the flower heads using seed strippers, combine harvester threshers or vacuum collectors. Typically, operations at the restoration site will include some form of ground preparation prior to receiving this seed. Collections from donor sites may be of hay, litter or soil. The seed itself can be gleaned from them directly, or bulked up in commercial nurseries and then broadcast, direct drilled or slot-seeded at the restoration site. Very occasionally seeds may be grown on, ex situ, as plug-grown transplants or turf for planting out. In a review of 38 published international studies on grassland restoration, Hedberg and Kotowski (2010) found that broadcast seeding accounted for more than half of these methods, while a quarter used hay spreading. Brush harvesting (Section 6.4.2), slot seeding (Section 6.7.1) and plug planting (Section 6.7.4) were all comparatively rare. The main sources of wildflower material are summarised in Table 6.1.

In the case of both nursery-grown and wild-harvested plant material, its genetic variation, fitness and suitability for the intended receptor site needs to be carefully considered. Commercial wildflower gene pools are likely to be very restricted because there are relatively few collection sites, with many of these biased towards particular regions and habitats. Local seed sources are more likely to be reliable. For example, Auestad *et al.* (2015) found that hay meadow seed derived from a distant source (85 km

---

[2] Topsoil and turf stripping are, of course, highly destructive to the donor site, and would only come into play when major development projects (e.g. for transport, housing, mining and industry) override conservation objectives, precipitating efforts to salvage or compensate for the loss of valuable grassland.

Table 6.1 Types of wildflower material used for habitat re-creation (after Walker *et al.* 2004a; © Defra)

Type	Description	Main uses
Native seed	Seed of native British species (mainly herbs), originally sourced from wild populations in the UK and then grown on as stocks from which seeds are harvested. The vast majority of seed is supplied as individual species either in bulk or in pre-prepared amounts by smaller specialist suppliers	Wildlife gardening Small-scale habitat recreation projects Experimental studies Inclusion within mixtures for large-scale sowing (e.g. AE[1] schemes)
Native seed mixtures	Pre-prepared mixtures of the above, but often including large amounts of agricultural grass and legume seed of non-native provenance. Often tailored to suit the soil type or desired target community (85:15% grass:herb ratio). Sold by both smaller specialist growers as well as large commercial seed houses	Large-scale habitat recreation (e.g. AE schemes) Civil engineering projects (e.g. new roads, flood defences, mine-workings, etc.) Amenity plantings (e.g. urban developments, landscaping, etc.)
Harvested mixtures	Mixtures cut as hay from species-rich grasslands (often conservation sites such as SSSIs). Often harvested under licence as contracts for specialist projects (e.g. High Weald). However, the number of donor sites is currently limited in number	Specialised habitat recreation/restoration projects (e.g. on SSSIs, etc.)
Native transplants	Usually pot-grown herbs, especially aquatic species, originally harvested (as seed) from wild populations in the UK. Usually supplied as plug plants, rhizomes, etc. by smaller specialist growers	Specialised habitat recreation/restoration (e.g. new wetlands, etc.) Civil engineering (e.g. flood defence, bioremediation, etc.)
Agricultural cultivars and varieties	A small number of forage grasses and legumes, many of which are certified in Europe. Sold in large quantities direct to farmers via large commercial seed houses, although also supplied by smaller specialist growers of native seed. Largely imported varieties bred in other European countries (e.g. Denmark, Holland)	Mainly used for reseeding pastures, but also used for 'bulking-up' wildflower mixtures for large-scale habitat recreation (e.g. AE schemes, etc.)
Amenity cultivars and varieties	A small number of 'turfgrass' varieties bred specifically for hard-wearing qualities. Often bought direct from large commercial seed houses, although also supplied by smaller specialist growers of native seed. The majority of seed is imported from Europe and New Zealand	Mainly used as 'turf' for sports pitches, lawns, parks, golf courses, etc. Landscaping of contaminated or waste ground 'Bulking-up' wildflower mixtures for large-scale habitat recreation (e.g. AE schemes)

[1]AE, agri-environment.

away) was less successful than local seed, illustrating the mismatch between donor and receiver sites that was probably climatically influenced. In a survey of wildflower seed companies actually carrying out their own harvesting of wild populations, Walker *et al.* (2004a) found that over half of these companies sourced their supplies from only ten or more donor sites. Those producing their own seed attempted to avoid artificial selection pressures arising through multiplication in the nursery by routinely supplementing their stocks with wild-collected seed every 1–5 years. Nevertheless, given the potential bottlenecks of limited wild sources and nursery multiplication, the possibility of founder effects influencing the performance of vegetation at the restoration site is real. Similar restrictions could also apply to wild-harvested seed or hay collected from a single field, but then sown at a site remote from the source and where the ecological conditions may be markedly different.

6.4.1 Nursery production

A number of commercial seed companies supply wildflower seeds from plants grown as single-species rows in standard nursery beds, which are then harvested, extracted and cleaned with specialised sorting machinery, and finally made up into appropriate mixtures. Typically these mixtures contain 20–25 wildflower species, made up to serve broad soil types, such as calcareous, sandy, acid, clay (neutral), alluvial or wetland. Occasionally the composition may be more refined, so as to mirror specific NVC communities, but the seed of most individual species can also be bought separately, allowing clients to construct or specify their own community specification. In all, about 60 common species account for more than three-quarters of the seed and plants used in restoration projects for a range of different habitat types (Walker *et al.* 2004a).

Commercial 'native' meadow and pasture seed mixes have improved in quality and authenticity over the years and, on the whole, now contain fewer highly productive agricultural cultivars of grasses such as Perennial Rye-grass, Cock's-foot, Tall Fescue and Timothy. Non-natives such as Highland Bent are also largely absent, and there is

Ripe seed of Ragged-Robin ready for collection (Tone Blakesley).

Propagation of Common Knapweed (left), Betony (centre) and Common Sorrell (right) in nurseries at Wakehurst Place (Royal Botanic Gardens Kew).

less emphasis on short-term annuals or species with highly specialised germination and edaphic requirements. The species included now tend be robust, often competitors or ruderals rather than stress-tolerators, these ecological traits being selected to maximise success. However, whereas grasses might account for only 60% cover in many semi-natural sward types, commercial wildflower seed mixes commonly contain 80–85% grasses and only 15–20% forbs by weight. In terms of seed numbers, the grass component may be even greater if small-seeded grasses such as bents and meadow-grasses are significant components and greater still when an additional grass nurse crop is advocated. As grasses are likely to become dominant on fertile sites, a much lower grass:forb ratio is advisable. Conrad and Tischew (2011) caution that highly competitive cultivars, particularly grasses such as Red Fescue, should never be used in commercial mixtures. Colonisation windows are quickly occupied by such competitive species, with lasting effects on the development of a plant community if the reference species are limited by their dispersal ability. Although much more expensive, regional, site-specific mixtures are likely to prove more successful, and even spontaneous regeneration should be preferred over standardised mixtures.

6.4.2 Wild harvesting

An alternative to sowing nursery-produced seed mixes is to harvest plant material directly from existing species-rich grasslands – often from sites that have already been designated for their conservation value, giving the product a marketing edge. Seed from these sources has the advantage of being of known provenance, free of improved cultivars, and often contains more species than are normally found in commercial wildflower mixes. Hay can either be dried in the field and baled, as in traditional agriculture, or freshly harvested 'green' for direct transfer to a receptor site. Green hay is cut at the peak of flowering, when the flower heads retain greater quantities of seed and a wider species composition than dried hay (Jones *et al*, 1995; Trueman and Millet 2003). Collected fresh, seed viability also remains high and losses are minimised during collection and transport. However, the high moisture content of green hay means that it cannot be stored, and ideally is transferred from the donor to the receptor site within 24 h to avoid fermentation. Even if left overnight in heaps or bales, the material can rapidly heat up, compromising seed viability.[3] Dry hay, on the other hand, can be stored, but is a comparatively depleted seed

---

[3] In a development of this method (the 'whole-crop method'), the green hay is rowed immediately after mowing, collected with a multi-chop forage harvester to reduce its bulk and transported in manure spreaders to the receptor site (see Box 6.1). The operation can be carried out in wet weather, but as for green hay generally, transport distances to the receptor site are necessarily limited (usually < 10 km) to avoid fermentation losses.

source because during the drying process the hay is turned, rowed and finally baled, so that as much seed may fall on to the soil surface during harvesting as that picked up and baled in the hay crop (Smith *et al.* 1996). Dry hay spreading has now been largely superseded by green hay and by seed stripping (brush harvesting) or combine-harvesting (in situ threshing) machinery.

Seed can also be extracted from dried hay bales, or by collecting the seed-rich chaff remaining after storing dried hay on the barn floor. Wells *et al.* (1986) found that the traditional, rectangular 21 kg bales (46 × 36 × 90 cm), harvested from the flower-rich alluvial North Meadow at Cricklade in Wiltshire, contained up to 450,000 individual seeds.[4] Seventeen grass species accounted for about 90% of the total seed, and 24 forbs were recorded. When this seed-bearing material was subsequently sown in experimental seedbeds, 44% of the species originally present in the hay were found to have established after two growing seasons.

Raking and scarifying the surface of a donor site to collect litter and seed is another potential harvesting method, though less frequently used: this has the additional advantage of transferring mosses and lichens that would not be present in a hay harvest. Other methods of collection use specialised seed stripping machinery to dislodge ripe seed, without mowing, thus reducing the bulk to 5% or less of whole hay harvests. Mechanical separation is done by rotating combs, nylon brushes and sieves (hence the term brush harvesting), or by vacuum collectors. The vegetation represented in such collections is often considerably less diverse than the original sward, both because of the timing of the harvest and the selective operation of the machinery. In chalk and limestone grassland, only approximately 50% of the species present were recovered in seed collected using vacuum harvesters (Stevenson *et al.* 1997; Riley *et al.* 2004). At Thrislington Plantation NNR in Durham, two main grasses – Fescue species and Quaking-grass – accounted for 74% of the seedlings germinating in seed trays from vacuum collections, with many other species contributing less than 1% of the total. Several species characteristic of the Magnesian Limestone vegetation were absent, probably due to the timing of the first harvest (mid-July) when some early flowering species would already have shed seed. Other possible reasons were: the presence of unripe seeds; the inability of the harvester to cope with low-growing species with small seeds; and the failure to germinate under glasshouse conditions (Riley *et al.* 2004).

Brush and, to some extent, vacuum harvesters tend to harvest selectively from the sward profile, so that species numbers and seed quantities collected will vary according to the machine height setting, the lower-growing species in particular often escaping. Working in meadows in the eastern Italian Alps, Scotton *et al.* (2009) found that only 20% of the available seed and half the number of species was recovered by a brush harvester in tall, False-oat grassland, compared with almost three-quarters of both the seed and species in shorter bent/fescue swards when the brush axle was set lower (Table 6.2). The direction of rotation of the leading edge of the brushes, upwards or downwards, can also affect collecting efficiency, as well as the speed of rotation and the ground speed of the pulling vehicle. A traditional hay harvesting treatment carried out at the same time, where the hay was turned and dried for 36 h, gave a better representation than brush

---

[4]   This was an overestimate of what would become established, because it did not take into account either seed germination or viability. Edwards and Younger (2006) found that seeds extracted from bales derived from MG3 hay meadows contained similar numbers of seeds to harvests by Wells *et al.* (1986), but germination in seed trays was less than 10 per cent of original numbers.

Table 6.2 Number and percentage of seed or species recovered from standing crops of (a) False oat-grass and (b) bent/fescue grassland by hay harvesting or by a brush harvester set at heights of 100 and 45 cm, respectively (reproduced from Scotton *et al.* 2009, with permission from the University of Wisconsin Press)

	False Oat-grass sward		Common Bent–Chewing's Fescue sward	
	Hay harvest	Brush harvest	Hay harvest	Brush harvest
Number of seeds/m²	7,033 (33%)	4,333 (20%)	3,547 (45%)	5,673 (73%)
Total number of species	27 (68%)	16 (50%)	24 (75%)	25 (78%)
Number of grasses	11 (100%)	9 (82%)	7 (54%)	11 (85%)
Number of legumes + forbs	16 (76%)	7 (33%)	17 (89%)	14 (74%)

harvesting of the species present. It also harvested legumes and forbs in the lower strata more efficiently, but in the shorter bent/fescue sward the brush harvester was generally the more effective overall.

More recently, modified combine-harvesters have been used for seed collections. These mow the crop, thresh and extract the seed simultaneously: both plot threshers (mini combine-harvesters) and larger, field-scale versions are available. Like brush harvesters, they have the advantage that bulk is reduced and the seed is then immediately available for broadcasting or drilling, although in both methods the seeds of species that remain attached to the flowering head tend to be under-collected, whereas they would be fully represented in hay. Germination tests on material collected from donor False Oat-grass meadows in Germany, using either on-site threshing or green hay collection, showed that both methods produced similar numbers of germinating seeds and species per unit area of the donor meadows harvested (Kirmer and Tischew 2010). Various advantages and limitations of hay cutting versus seed harvesting from donor sites are summarised in Table 6.3.

Hay transfer, green or dry, has the advantage of using tractor-mounted machinery available on most farms; it can also be relatively cheap compared with sowing high-diversity commercial seed mixes at standard rates, although the quantities of seed dispersed per unit area are likely to be significantly less (Donath *et al.* 2007). The material, if not spread too thickly, can also give surface protection from erosion and provide a favourable microclimate for seed germination. There is some evidence that strewing very thick layers or green or dried hay can adversely influence the outcome, suggesting that a balance needs to be struck between the quantity of seed supplied against the risk of poor germination due to fermentation or mulching, especially in wet weather. In one restoration experiment, Edwards *et al.* (2007) found that a somewhat more diverse sward developed after a higher strewing rate of 1:1 (donor to receptor) than a lower rate of 1:3, although it is common restoration practice to spread the material on areas at double that of the 'donor' site (2:1), spreading 5–10 t of fresh material per hectare in layers 5–15-cm thick (Donath *et al.* 2007; Kirmer *et al.* 2009; Kiehl *et al.* 2010). Several authors suggested even higher ratios of 3:1, although others recommended 1:2 or 1:3. For light applications, the original sward will be visible and cattle or sheep can be used to trample in the seeds. Heavy mulch thicknesses are best removed quickly, within three weeks to avoid inhibiting germination, and turned once to dislodge the seed (Trueman and Millett 2003).

Table 6.3 Advantages and limitations of restoring sites using seed collected by different harvesting methods from donor sites

	Hay harvests (green and dry)	Seed stripping and combine- harvesting
Advantages	Cheaper than seed stripping and combine-harvesting: no specialist machinery required, apart from normal farm machinery More complete recovery of seed is obtained, including species in the lower canopy (e.g. Selfheal and Common Bird's-foot-trefoil) and some orchid species	Harvested material can be stored and the material cleaned and sorted Collections are flexible and can be carried out at any time during the growing season, e.g. to recover seed of early- or late-flowering species Material can be transported or mixed easily
Limitations	Less flexible in timing: harvests are usually restricted to mid-summer, e.g. July/August, reducing the full complement of species Long-distance transport of green hay is usually unfeasible Baling and storage of dry hay results in loss of seeds Cut material must be spread immediately to optimise seed viability Hay may need to be collected after spreading to avoid mulching and facilitate establishment	Expensive compared with hay strewing; requires specialist machinery Removes only ripe seed from seed heads, within a narrower phenological window than hay harvests Seeds of some species are not easily dislodged from flower heads Recovery of seeds tends to be from the upper layers of the sward canopy with less from low-growing species Some species are more difficult to store and lose viability quickly

As part of an EC-funded programme promoting 'High Nature Value Farmland' in Central Europe, the recent SALVERE project (*Semi-Natural Grassland as a Source for Biodiversity Improvement*) has investigated practical ways in which semi-natural grasslands could be used to restore intensively used farmland (Kirmer and Tischew 2010; Krautzer *et al.* 2011; Scotten *et al.* 2012). Partners from six EU countries compared the effectiveness of different seed harvesting and storage techniques, including seed quantity, viability and quality, as well as the performance of sown material at restoration sites (see Table 6.4). Almost irrespective of the harvesting method, the high proportion of grass:forb seeds in the material was striking, despite the harvests being carried out at the same time of year. Relatively large quantities of pure seed were recovered from standing crops typically bearing 6,000–13,000 ripe seeds/m^2 (up to 200 kg/ha), sufficient to sow up to 20 times the original donor area. Green haymaking collected the most seed and took less time than dry haymaking, which for reasons already mentioned also gave the lowest yields.

Few other investigations have directly compared the quantities of seed collected by different harvesting methods from similar unit areas of the same donor grassland. In the two experiments described by Edwards *et al.* (2007) where seed was collected by brush harvesting or as green hay, the latter method delivered more species-rich swards at the receptor site. The authors attributed this to the greater operating height of the brush harvester, missing some low-growing species, while the green hay method was able

Table 6.4 Variation in the proportion of grasses and herbs, harvested seed yields and rate of working for different harvesting methods: examples from mesotrophic grassland and nutrient-poor floodplain and fen communities. Working rates do not include drying and cleaning the seed (from combined data of Krautzer *et al.* 2011 and Scotton *et al.* 2011, reproduced with permission from Bernhard Krautzer (Agricultural Research and Education Centre, Raumberg-Gumpenstein, Austria))

Harvesting method	Grass:herb ratio	Seeds harvested (kg/ha)	Percentage seed collection efficiency	Percentage seed in harvested material	Rate of working (h/ha)
Green hay	80:20	100–200	90–100	0.2–2	1–2
Dry hay	70:30	40	30–50		3–4[a]
Plot thresher	80:20	60–150	30–80	25–60	5–10[b]
Large thresher[c]	60:40	50–200	15–30 (dry hay)		1.5–3
Seed stripping	80:20	20–60	55–75[d]	30–45	1.5–3[b]

[a]Includes turning, rowing and baling.
[b]Depends on vegetation type.
[c]Large threshers can also be used to thresh green hay directly, when seed collection efficiency varies from 50–90%.
[d]Using downward leading-edge brush rotation.

to sample a wider phenological window because it removed whole seed heads rather than just the ripe seeds detached by the brush harvester. Some species, such as Upright Brome and Cock's-foot, also show marked seed retention and are resistant to stripping by brushing. In Table 6.4, threshing and brush harvesting gave intermediate collection efficiency, but the product was relatively impure and contained more than 50% of plant residues[5] (although comparing very favourably in this respect with bulk harvests of green or dry hay), making it likely that further seed cleaning and drying would be required. Smaller-scale versions of these machines, such as plot threshers and hand-held vacuum harvesters took longer to operate, but had the advantage of being more manoeuvrable on poorly accessible or steeply sloping sites. They are also useful for collecting from small sites, such as roadside verges (see Box 7.1) and churchyards. Further comparisons of efficiency are considered in Table 6.5.

Results from the SALVERE project suggest that while each technique showed differences in the quantity, quality and composition of the harvested seed, they all recovered similar *numbers* of the available species. All, however, incurred some species losses due to the timing of the harvest relative to flowering phenology, mechanical discrimination between species during harvesting and extraction, and a subsequent failure to establish at the restoration sites. Using brush harvesting or on-site threshing, Haslgrübler *et al.* (2014) extracted 9 grass species (80% by weight) and 19–20 forbs (20%) from a False Oat-grass meadow, but the composition and quality of seed material would be expected to vary greatly from year to year, depending on the time of harvest and the weather conditions at the time. With on-site threshing, 60–70% of seed germination was found: germination was slightly higher in the case of brush harvesting, perhaps due

---

[5] Stevenson *et al.* (1997) found that two-thirds of the material collected with a hand-operated vacuum harvester machine consisted of plant debris rather than seed.

Table 6.5 A comparison of different harvesting methods for recovering native seed and material from donor grassland sites (after Scotton *et al.* 2011, reproduced with permission of Bernhard Krautzer (Agricultural Research and Education Centre, Raumberg-Gumpenstein, Austria))

Operation requirements	Green hay	Dry hay	On-site threshing	Dry hay threshing	Brush harvesting	Vacuum harvesting
Efficiency of seed recovery	+++	++	+++	+	+/+++	+/++
Efficiency of species recovery	+++	++	++	++	++/+++	+/++
Availability of equipment	+++	+++	+/–	+/ –	+	+
Low harvesting costs	+++	++	+++	+	+++	++
Low transport costs	+	++	+++	+	+++	+++
Seed and mulch are produced together	+++	+++	–	–	+	+
Storage of seed and material	–	++	++	++	++	++
Good accessibility needed for machinery	++	++	++	++	++[a]	++[a]
Proximity of donor and receptor sites needed	+++	+	+	+	+	+

[a]Good accessibility is less of a requirement for more manoeuvrable, hand-held harvesters.

the greater likelihood of detaching ripe seed by the latter method. An earlier study by Kiehl *et al.* (2006) found that 69–89% of the species with viable seeds were successfully transferred in green hay from a species-rich nature reserve to different ex-arable receptor sites, but this had declined to 58–76% after 5 years. A possible way to increase the level of success is to carry out hand collections of minor species, but these are time-consuming and relatively inefficient (Stevenson *et al.* 1997). Another alternative is to supplement plant material with nursery-produced seed, allowing a more balanced community specification to be selected with probably higher germination rates, although some of the most desirable species will not be available commercially.

The composition and quality of harvested seed varies from season to season, depending on weather conditions and different flowering phenologies. These determine the composition of the seed mix, so that harvesting in midsummer will omit or under-represent both early- and especially late-flowering species, such as Harebell, Yarrow and Lady's Bedstraw. In central England almost twice as many species produce ripe seed in August as in July (Wells *et al.* 1986). An illustration of how varying harvest dates affect seed composition is Smith *et al.*'s (1996) study of a mesotrophic hay meadow in Upper Teesdale. Hand harvests of hay taken successively on 14 June, 21 July and 1 September accounted for 15, 34 and 51% of total seed production, respectively. The equivalent forb:grass ratios were 2.92, 0.82 and 0.15, respectively, indicating that many forbs had produced their ripe seed earlier, in June and July, in contrast to many grasses in August and September. However, Hölzel and Otte (2003) showed that late September harvests increased the share of forb seed production in species-rich floodplain meadows where Meadow Foxtail was the dominant grass. The practical implication is that the harvest date should be adjusted to match the management of the specific grassland type, or targeted so as to include particular species. One solution would be to supplement restoration sites with additional hay transfers taken at different times during the season (Kiehl and Wagner 2006) or to carry out successional sowings, especially from harvests taken later in the year (Pywell *et al.* 2003; Edwards *et al.* 2007; Natural England 2010a).

6.4.3   Adverse impacts of seed collection

A concern is that regular harvesting from donor grasslands could rapidly deplete seed stocks and their return to the soil, altering the balance of that plant community over time. Stevenson *et al.* (1997) suggested that only about 25% of a grassland nature reserve should be collected from in any one season, for example, in well-spaced strips (Figure 6.2). Restoration methods using economical seeding rates, such as seeding in localised patches or using slot seeding instead of broadcasting, could help to reduce pressure on the donor fields. Once established, the restored fields themselves might in turn be used as donors to enhance further sites.

All harvesting methods are likely to have some consequences for grassland fauna – ground-nesting birds, small mammals, amphibians and insects. Many studies have been concerned with the destruction of nests and chicks by mowing machinery or heavy grazing (e.g. Vickery *et al.* 2001) and how this may be avoided by carefully scheduling agricultural operations. Later summer cuts will avoid nesting birds, but many insect taxa – bugs (Hemiptera), butterflies, spiders and even bees – are sensitive to mowing at this time. Apart from direct mortality and injury, the conditions prevailing after mowing, such as the changes in temperature and humidity, the loss of food sources, or increased vulnerability to predation, can be deleterious to several species. Cutter type, mowing height, tedding, windrowing and baling all have their impact. In a review of the impact of meadow harvesting techniques on field fauna, Humbert *et al.* (2009) concluded that disc or drum rotary mowers were liable to cause twice the mortality of finger- or double-bladed cutter bar mowers, and three to four times more when a conditioner was attached to crush and dry the grass. Flail and suction flail mowers were the most potentially damaging. Low cutter heights set at 5–7 cm caused most mortality, but at 10 cm this was much reduced. In general, smaller-bodied insects were more likely to escape damage than larger ones.

Suction or brush harvesting machines also passively collect sedentary invertebrates, such as the forester or burnet moths found on flower heads in the upper canopy. A study by Waring (1990) showed that vacuum harvesting of wildflower seed in a hay meadow collected the equivalent of 1–3% of the insect fauna present, comprising large numbers of springtails, bugs and sawflies, together with 16 species of Lepidoptera. However, damage to invertebrate populations through mowing may be greater, taking account of both the cutting and collection processes. Turning and windrowing the cut material will tend to concentrate some species into these linear refuges, from where they are removed or killed by forage harvesters or baling machines. Woodcock *et al.* (2012) considered that transfers of green hay using manure spreaders could cause considerable mortality in phytophagous beetles. In one study, harvesting green hay from a restored donor site in southern Germany resulted in 42% mortality of the sedentary bush cricket (*Metrioptera*) population (Wagner 2004). On arrival at the receptor site only 5% of the original population was found to be present and capable of breeding, although this was thought sufficient to form new populations. Gardiner and Hill (2006), on farmland in southern England, also recorded considerable mortality in Meadow Grasshoppers after cutting in July with a rotary mower, set at a height of approximately 10 cm, but again observed that live individuals could be transported in the harvested material. Whether these survivors can then persist in their new location would seem unlikely in the absence of their host plants, unless suitable habitat was available nearby.

To avoid mortality of the fauna on potential donor sites, one solution might be to practise rotational harvesting in strips, for example, using a ratio of one in five, each uncut strip separated by four machine widths (Porter 1994; Figure 6.2). A study of Swiss hay meadows showed that in 10% uncut refuges, sited in the centre of 50-m diameter plots, grasshopper densities doubled following harvesting as additional insects moved in, while only a tenth of the number remained in the harvested surroundings (Humbert *et al.* 2010). The varied sward architecture provided by these refuges could potentially benefit other invertebrates by conserving food and nectar sources and providing shelter, while also serving to protect ground-nesting birds. Later hay cuts (e.g. in September) could also be considered, or harvesting in alternate years or at even longer intervals not only to protect fauna dependent on the grassland for breeding, but also to conserve seed stocks. Delaying the first mowing date from 15 June to 15 July and leaving uncut refuges of 10–20% of the meadow area both benefited significantly orthopteran abundance (Buri *et al.* 2013).

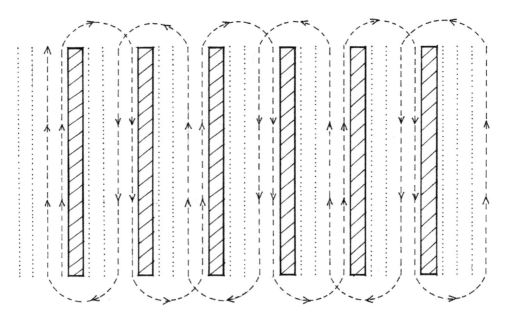

Figure 6.2 Mowing sequence, leaving one strip in five (cross hatched on diagram), maintaining seed stocks while also providing invertebrate refugia (after Porter 1994; © Natural England).

BOX 6.1 Case study: Whole-crop method, a type of green hay transfer, Beech Estate

Topic: Developing a novel green hay technique for meadow restoration
Location: Beech Farm Estate, Battle, East Sussex
Area: Donor sites: Ashburnham Meadows 2.85 ha; receptor sites: 8 ha
Designation of site: Donor site is a Site of Nature Conservation Interest
Management and community use: Beech Estate is a privately owned and managed as a series of working farms under Countryside Stewardship
Condition of site: Semi-natural grassland donor sites and semi-improved grassland receptor sites

Semi-natural grassland on the Beech Farm Estate.

Background

The Estate comprises a series of small farms and woodland, typical of the East Sussex High Weald, covering an area of 770 ha. For many years, the Beech Farm Estate operated in part as a medium-sized dairy farm with 60–100 cattle. In 1998 a decision was made to end the dairying operation, due to falling milk quotas and concerns over the viability of the business. Today, the Estate sells herbage, the grazing rights to fields, and timber as standing wood for local forestry businesses. Control of grassland management is retained by the Estate, and with support from Entry and Higher Level Stewardship grants, all improved pasture grazing land is being converted to species-rich grassland, with the whole estate certified as organic. Consequently, the Estate now practices traditional farming, while endeavouring to enhance habitats for wildlife. The farm is registered under the Countryside Stewardship and Environmental Stewardship schemes for Educational Access.

Whole-crop method at Beech Estate

The whole-crop method was originally developed because conventional hay strewing was found to be too slow to establish diversity, with poor results. Dry hay was also dismissed, essentially because of the loss of valuable seeds during baling, rowing and drying. The ethos of the Beech Farm Estate was to view semi-natural, species-rich grassland seed as a crop, and to use standard farm machinery to harvest, transport and sow harvested seed into receptor sites. In a development of the green hay transfer technique, known as the 'whole-crop method', hay is cut after 15 July (following Stewardship recommendations). Early work experimented with grass cutting using a mower-conditioner that crimps and crushes the hay, allowing some seed to be returned to the donor site; mowing without a conditioner was later adopted to reduce the impact of the operation on invertebrates, while harvesting in alternate strips or at longer intervals reduced the risk of depleting the seed rain in the donor meadows. After mowing the material was rowed with as little disturbance as possible and then immediately collected up with a multi-chop forage harvester, reducing its bulk, and blown into dung spreaders for transport directly to the receptor site, and finally spread. Receptor sites were previously prepared by hard grazing to reduce top growth, followed by power harrowing to approximately 8 cm to achieve up to 60% bare ground. Spreading of green hay was done at a ratio of 1:3, donor to receptor, giving a very thin covering that quickly disappeared. The transfer of material from donor to receptor site took place on the same day and could be carried out in wet weather; however, transport distances to the receptor site were necessarily limited to avoid fermentation losses (usually < 10 km, although greater distances of > 20 km were attempted). Cattle were grazed at the receptor site immediately afterwards, which may have helped to tread in the seed. The animals were removed from the site in the following February, after which a hay cutting regime was implemented in July with aftermath grazing.

Following a period of hard grazing, receptor sites are power harrowed to approximately 8 cm to achieve up to 60% bare ground (Keith Datchler).

Points to consider for the successful implementation of the whole-crop method include:

- weed-free, semi-natural grassland donor sites;
- survey of species and soil type at the donor sites;
- weed-free receptor sites;
- survey of species, soil type and fertility at the receptor sites;
- good access for both donor and receptor sites;
- identification of a best time after July 15 for maximising harvest of ripe seed;
- access to large scale farm machinery, or contractors, which can supply this;
- sites of low fertility (because they will always give better results when enhancing species richness).

The whole-crop method has several advantages over conventional green hay strewing, the most obvious being the reduction in bulk, so that the spread material does not need to be collected up, especially at a low ratio of 1:3. The operation uses standard farm machinery and, providing that there is good access at both the donor and receptor sites, it is much faster than conventional hay strewing; for example, transfers from a 3-ha donor site to an 8-ha receptor site could be readily accomplished in an afternoon. Theoretically, hay collected in this way also contains more seed, as there is relatively little disturbance to the flower heads apart from during mowing and rowing, and no baling is necessary. In an operation carried out in 2000, when green hay (with no turning or tedding), was transferred from a donor site at Coach Road Field (part of Ashburnham Meadows) to Rocks Farm, two rates of application, 1:1 and 1:3 were compared (equivalent to a fresh mass of 3.1–6.2 t/ha), with and without previous power harrowing of the receptor site.

Monitoring

Of 30 species recorded at the donor site, seed of 18 species were found in the brush harvested seed mix and 19 in the hay. Seed abundance varied considerably: > 50% of the brush-harvested mix was composed of Ribwort Plantain, with 10–20% grass seed (Figure 6.3). In contrast, Yellow-rattle was the most abundant seed in the green hay, which also included more low-growing species, with seed heads held low in the canopy. Mature seed attached to seed heads was transported to receptor sites in hay, but these would be lost during brush harvesting.

Botanical composition at the Rocks Farm receptor site was most enhanced by the higher level of hay, followed by the low level of hay and the brush harvested seed treatment. Most of the frequent species in the donor sites colonised, many at high frequency. These included Common Knapweed, Cat's-ear, Oxeye Daisy and Ribwort Plantain. Others, such as Common Bird's-foot-trefoil and Meadow Buttercup, became more frequent in successive years. Using a lower ratio of 1:3 might be a little slower, but will ultimately produce similar results, with less impact on the donor meadow. Species richness increased over a four-year evaluation period (Figure 6.4), with the power harrowing treatment at both application levels showing far greater similarity to the donor site community than no harrowing (Edwards *et al.* 2007). Without harrowing, plots treated at the lower rate of hay strewing were much less similar to the donor site than at the higher rate.

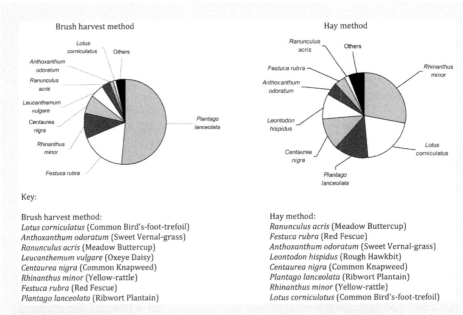

Key:

Brush harvest method:
Lotus corniculatus (Common Bird's-foot-trefoil)
Anthoxanthum odoratum (Sweet Vernal-grass)
Ranunculus acris (Meadow Buttercup)
Leucanthemum vulgare (Oxeye Daisy)
Centaurea nigra (Common Knapweed)
Rhinanthus minor (Yellow-rattle)
Festuca rubra (Red Fescue)
Plantago lanceolata (Ribwort Plantain)

Hay method:
Ranunculus acris (Meadow Buttercup)
Festuca rubra (Red Fescue)
Anthoxanthum odoratum (Sweet Vernal-grass)
Leontodon hispidus (Rough Hawkbit)
Centaurea nigra (Common Knapweed)
Plantago lanceolata (Ribwort Plantain)
Rhinanthus minor (Yellow-rattle)
Lotus corniculatus (Common Bird's-foot-trefoil)

Figure 6.3 Composition (percentage by weight) of the seed collected using whole-crop and brush harvesting.

Overall, results from the early trials showed that the whole-crop method at the higher rate of spread was superior to brush harvesting, with power harrowing increasing the similarity coefficients for all seed addition treatments (with the exception of the control receiving no seed from donor site). The cost of applying the whole-crop method to harvest and spread material on 10 ha was approximately £1,900, equivalent to approximately £190/ha. Power harrowing incurs an additional cost, but this was relatively inexpensive, at approximately £27/ha.

A further refinement of the system is whole-crop strip enhancement, in which six-metre enhancement strips are established by power harrowing and whole-crop

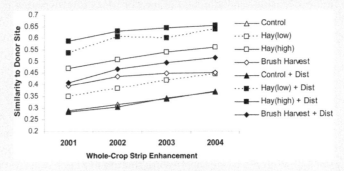

Figure 6.4 Similarity of plant community composition (Czekanowski coefficients) to the donor site for all whole-crop (hay) and brush harvesting treatments at the Rocks Farm experiment. Dist, power harrowing disturbance.

material is applied as bands running across the receptor site. When these enhanced strips are later harvested, the harvester can blow material sideways onto adjacent ground, previously prepared by power harrowing, thus gradually increasing the area until eventually the whole field is effectively enhanced over successive years. Although the conversion takes time, the advantage is that very little material is required and the donor meadows are less likely to be overworked.

Keith Datchler demonstrates a meadow enhanced by green hay from a donor site, sown using the 'whole-crop method' (Keith Datchler).

Betony flowering along the margins of a semi-natural meadow.

Information provided by Keith Datchler OBE (Beech Estate).

BOX 6.2 Case Study: UK Native Seed Hub, Royal Botanic Gardens, Kew

Topic: Methods for harvesting and handling seeds for wildflower meadow restoration
Location: Wakehurst Place, Royal Botanic Gardens, Kew

Introduction

The UK Native Seed Hub was launched in 2011 to mobilise the resources and expertise of the Royal Botanic Gardens, Kew and the Millennium Seed Bank in support of conservation and habitat restoration in the UK. The Millennium Seed Bank holds collections of more than 90% of the UK's native flora, including many multiple-origin collections and rare species. The Seed Hub makes these collections available to practitioners via an online seed list, with an option to bulk up small collections at its seed production site at Wakehurst Place, Sussex. New collections are added each year, with bespoke collecting and production providing difficult to source species or specific local origin seed for conservation or restoration projects.

The Seed Hub also works with a range of partners – native seed and plant producers, conservation organisations, companies, landowners and managers – to improve access to high-quality, native-origin seed and plants and overcome obstacles to the successful use of seed in the landscape. The Seed Hub offers technical advice and training on seed collection, processing and storage, for example, and are developing novel propagation and restoration protocols for problematic native

Trialling grassland enhancement techniques with staff and landscape maintenance contractors at Toyota Manufacturing UK, Derbyshire. Training and development activities are part of a broader collaboration between Toyota, Royal Botanic Gardens, Kew and local Wildlife Trusts aimed at reducing environmental impacts and increasing biodiversity potential at Toyota sites in the UK (Trustees of the Royal Botanic Gardens, Kew).

species. Information on the seed storage behaviour of thousands of species, as well as other seed biological trait data, can be found online on Kew's Seed Information Database (http://data.kew.org/sid/).

Seed collection/harvesting

When setting out to restore a wildflower meadow, we first identify a nearby donor site that supports our target plant community and has similar ecological conditions to the restoration site. We start by surveying the donor site to establish the location and density of target species and to plan harvest dates according to seed readiness. To capture a representative diversity of multiple target species, and to increase genetic diversity within individual species that have a long fruiting period, this will usually mean making an early/mid and a late summer harvest.

Hand harvesting individual species (Trustees of the Royal Botanic Gardens, Kew).

Hand harvesting is often needed for species that fruit particularly early, such as Wood-rushes and Common Sorrel, or in donor sites with difficult to reach areas, such as banks, ditches or uneven ground. It is also advantageous when there are low numbers of small-seeded species, such as orchids.

Brush harvesting provides quicker harvesting of large, mixed species collections and generally collects a greater percentage of mature material than the combine-harvesters, which remove everything regardless of readiness. Brush harvesting is also less damaging to plants and several passes of the same site may be performed during one season to capture species that mature at different times. If done correctly, it also allows for a hay crop to be taken at the end of the collecting season, which appeals to some landowners.

Green hay is another effective way of transferring plant material from one site to another, but is not generally used by the Seed Hub as the material has to be used quickly and cannot be stored.

Brush harvester towed by a Land Rover (Trustees of the Royal Botanic Gardens Kew).

Preliminary drying in the field immediately after harvesting (Trustees of the Royal Botanic Gardens, Kew).

Post-harvest handling

Following harvesting, the collected material (whether brush or combine-harvested) should be spread out on sheets or tarpaulins on site for 1–2 h to allow 'livestock'/invertebrates to escape back to the field and to start the drying process. 'Wet' green material will heat up during transit, so this preliminary drying reduces the problem. Following this step, material can then be bagged up loosely in breathable sacks ready for transport. The addition of a temperature/relative humidity (RH; %) hygrometer or data logger (we use Gemini data loggers) to one of the sacks is often used to monitor conditions during this period. An increase in temperature much above normal ambient conditions will have a detrimental effect on seed viability, even for a short time.

Freshly collected material needs to be spread out to dry as soon as possible (preferably at the end of that day) and an open barn is an ideal place, protected from sun and rain (preferably wind too) and again spread out on sheets or tarpaulins. This needs to be turned twice a day to start with and then daily until the material reaches equilibrium with the environment. Check the ambient RH of the air using a hygrometer and measure the equilibrium RH of the harvested material by placing the sensor in a small plastic bag with some of the material.

A data logger is used to monitor ambient temperature and RH of the seed-drying environment. To monitor the equilibrium RH of the seed material, the data logger is placed in a plastic bag with a quantity of seed (Trustees of the Royal Botanic Gardens, Kew).

Air-drying, brush harvested seed in a barn on tarpaulins placed over pallets (Royal Botanic Gardens, Kew; Trustees of the Royal Botanic Gardens, Kew).

Seed drying

One week to ten days will allow good drying and at the same time permit ripening of any immature material. Ripe mature seeds will often disperse under these conditions with minimal intervention. This process can be improved by spreading the harvested material on wire racks (25-mm chicken wire is useful). This assists drying and allows seed to fall through on to the sheet below where it is easier to collect. Daily turning is still required and helps to separate the seeds from the bulk harvest.

During this period when the harvested material is stored under barn conditions, drying will only proceed as far as ambient temperature/RH will allow. Warm sunny days will have RH levels below 50% and this is ideal for drying the seed for storage. Cool, rainy days have higher RH levels and drying will be minimal. This situation will also occur as autumn/winter approaches and also at night, when falling temperatures result in high RH.

Plant material drying on a frame constructed with 25-mm chicken wire (Trustees of the Royal Botanic Gardens, Kew).

Seed viability is affected by the moisture content; dry seeds live longer. Seed dried to equilibrium with ambient conditions of < 75% RH will be dry enough to store without going mouldy, but viability will start to drop within a year. Seeds dried to equilibrium with ambient conditions of < 50% RH will survive several years without loss of viability. If the seed can be dried to < 25 per cent RH, then storage life may typically be several decades, however, this is costly and probably unnecessary for most seed producers.

Drying to < 75% RH is relatively easy. Most crops are dried under similar conditions and if the seed is to be sown relatively quickly, this is generally adequate. If, however, the seed is to be kept for any length of time, a lower moisture content is required. For example, Yellow-rattle is particularly short-lived if not dried, as are some other meadow species.

Drying to < 50% RH can be achieved using ambient conditions: this can occur on a warm sunny afternoon and seed will lose moisture accordingly, but at night the percentage of RH will rise and the seeds will gain moisture again. If seed is bagged up in plastic sacks during the late afternoon when conditions are generally at their driest, the plastic will prevent them picking up the moisture as moisture level rises in the evening. Seeds in plastic sacks will not sweat or suffer from condensation if they are dried first and then kept in a stable temperature.

If ambient conditions are not suitable for drying (i.e. wet rainy days or later in the year when the percentage of RH naturally rises) then some form of assistance may be required. Avoid heat drying wherever possible (normal grain driers); dehumidifiers or commercial drying units in enclosed spaces such as sheds or shipping containers can produce satisfactory results.

Seed cleaning

Seed cleaning is best performed when the harvested material is dry. The use of a drying rack (see earlier) makes this job easier because much of the unwanted material is left behind as the seed dries and falls through the mesh. After a week to ten days of drying, turning daily, the loose unwanted material can be lifted/ shaken with a fork to allow the seed to fall through. Repeat as necessary until all loose seed has been separated; this is best done by transferring material from one tarpaulin to another, and removing the seed left behind before moving the material back again.

This reduces the bulk dramatically and the partially cleaned material can then be sieved through a variety of different mesh sizes until the seed is free from most unwanted material. Mesh sizes starting at 1" and reducing down to ½", ¼" and then finally ⅛" has proved suitable for work done so far. Debris needs to be checked for seed before disposal; if seed is left behind then the previous mesh size should be used. (Yellow-rattle has particularly large seeds.) Rubbing the debris through the sieve with a gloved hand early on in the process will help break up any undispersed seeds in pods, heads or capsules.

Sieves can be made from wire mesh attached to frames for large-scale operations or horticultural size sieves (large) for just a few sacks. This technique will leave some seed-sized debris but for most applications this is satisfactory. Commercial seed cleaning machines can be used but the risk of losing either small seed by aspirating or large seed that will not fit through the mesh on the machines makes these unsuitable for mixed species bulk harvests. If seed is cleaned at the end of the summer or on wet days when the percentage of RH is high, then it will absorb moisture and need to be re-dried. Cleaning is best done immediately after drying and the clean seed spread to dry in the same manner as before.

Seed storage

Once dry (again, this can be checked by using a hygrometer in a small plastic bag of seed), the seed will need to be stored in sealed containers to prevent moisture uptake during the storage period. Containers need to be relatively well sealed and can be plastic sacks, polypropylene drums or kegs, or any suitable container that prevents moisture ingress. These containers will then need to be stored preferably in a cool place out of direct sun and in as stable a temperature as possible. Alternatively some practitioners may wish to use a temperature and RH-controlled environment to maintain drying and temperature stability. In this case choice of container is less critical.

Seed sowing

Seed produced and stored in this way is best sown using broadcast type machines that have an adjustable aperture. Hand sowing for small areas, pedestrian-propelled drop spreaders or rotary spreaders are equally suitable for small- to medium-sized areas and vehicle-towed rotary spreader type machines are best for field-sized plots. The sowing rate is more even when ground wheel-driven options are used. Light raking of the plot post-sowing can assist with an even spread and may prevent seed being blown away in windy conditions. Avoid burying seed too deep as this may affect germination. Autumn sowing is best, as this will provide the cold, moist conditions required by some species to break seed dormancy. Spring sowing of seed material can also risk predation by birds at a time when food for seed-eaters is scarce, in which case some protection may be required.

For further information on post-harvest handling, measuring seed moisture status and more, see our Technical Information Sheets at http://www.kew.org/science-conservation/research-data/resources/millennium-seed-bank-resources. For more information about the Seed Hub and its work visit Kew's website (http://www.kew.org/business-centre/welcome-uk-native-seed).

By John Adams, Ted Chapman and Kate Hardwick (Royal Botanic Gardens, Kew).

6.5 Sowing and the role of soil disturbance

Sowing seeds directly on to grassy vegetation usually fails as there are few available microsites for germination and establishment, depending on the management regime. A closed, compact turf under continuous mowing or grazing will tend to provide too much competition, as will a relatively unmanaged sward in which a thick litter layer has formed, creating a barrier to soil-seed contact. In contrast, intermittent mowing regimes, combined with occasional grazing, should provide more opportunities for seedling recruitment by creating disturbance and reducing competition from surrounding vegetation. A balance between these two attributes is critical to success, whether sowing seed mixtures into bare, cultivated land, or seeding directly into existing grassland.

Vegetation density and structure has a profound influence on the process of seedling recruitment in grass swards. Comparing the microclimate of a fen meadow community with its improved equivalent, a dense, agriculturally improved and productive rush pasture, Isselstein et al. (2002), found a much greater variation in temperature range, lower RH and much less light attenuation in the semi-natural sward. These conditions enhanced seed germination in the fen meadow compared with the rush pasture, but, critically, seedling survival was also better. Hutchings and Booth (1996b) also reported better survival of four chalk grassland perennials after they were sown in an ex-arable pasture which had been either close-mown or entirely cleared of vegetation, compared with the uncut pasture or adjacent chalk grassland. Seedlings were most vulnerable in their first two weeks after germination, when they were most likely to succumb to desiccation. The close-mown pasture provided more favourable conditions for establishment than uncut sites, probably because the sparse vegetation cover moderated temperature fluctuations and retained more moisture, while at the same time providing relatively little competition. Others studies have demonstrated the benefit of increased light penetration after mowing, during the early phase of seedling establishment. In one investigation, mowing at this critical time gave the best chance of survival for eight forb species, recorded 2 years after their introduction into permanent grassland (Hofmann and Isselstein 2004). Kleijn (2003) also suggested that rapid seedling emergence is a key factor, allowing late-successional species to establish in newly established swards and to overcome moderate shading.

An early study by Wathern and Gilbert (1978) demonstrated the importance of creating gaps or microsites for plant recruitment in existing swards, prior to sowing with native forbs. In an experiment set out on a motorway grass verge, they seeded nine native forb species into uncut, pre-cut, herbicide-treated and rotovated strips, oversowing the latter two treatments with a slow-growing grass mixture as a precaution against erosion. Seedling establishment was far superior in the herbicide-treated plots where there was less grass competition, but in both the cut and uncut plots the existing Perennial Rye-grass and fescue sward quickly shaded out the young seedlings. Rotovation was ineffective here because it produced a poor tilth for seedling establishment. In another investigation using combinations of irrigation, soil disturbance and vegetation removal treatments, Isselstein *et al.* (2002) found that soil disturbance had by far the greatest positive effect on the germination of forbs introduced into permanent grassland, probably due to better soil-seed contact and the removal of intervening litter layers.

Some form of artificial disturbance, creating gaps on the soil surface, is therefore critical if introduced or spontaneously colonising species are to germinate and establish themselves effectively. In traditional agricultural management, this function may simply be fulfilled by grass or hay cutting machinery, or poaching by hooves of grazing animals. Often, however, more drastic measures are needed. Several experimental investigations have looked into determining optimum levels of disturbance necessary to achieve good establishment of sown or planted native species. In the UK some of this work was funded by MAFF or Defra to set a framework for agri-environmental schemes such as those in ESAs or Environmental Stewardship projects. Techniques included:

- different intensities and combinations of mowing and grazing (with no soil disturbance);
- overall applications of herbicide, or localised in strips or patches;
- strip or slot seeding, i.e. using precision sowing equipment to sow directly into excised drills, sometimes applying herbicide parallel to the rows (see Section 7.2)
- light harrowing, using chains or disc harrows, but creating little bare ground;
- deeper rotovation, creating up to 50% of bare ground, or;
- turf removal or topsoil stripping, localised in patches or over more extensive areas.

Light harrowing or rotovation may not create sufficient microsites for sown seed when the vegetation already present recovers quickly, or if a weedy seed bank is activated that competes with the desired introduced species. Hoffman and Isselstein (2004) found that harrowing of an improved mesic grassland promoted better emergence of sown seeds than undisturbed control plots, despite the activation of an undesirable seed bank. However, soil disturbance may not always be effective in all circumstances. In another experiment, neither rotovation nor the removal of 10 cm of topsoil from an improved rush pasture (MG10) was successful in establishing a seed mix harvested from a local fen meadow: only 3 of 17 introduced dicotyledonous species survived into the second year (Tallowin *et al.* 2001). The removal of thicker layers of topsoil (to 15–20 cm) significantly improved the survival and cover of introduced, pre-grown transplants after 5 years, compared with less drastic rotovation treatments. Similarly, on improved lowland pastures at several sites in England and Wales, ground preparation treatments causing the greatest disturbance developed the most diverse vegetation from sown seed mixtures (Hopkins *et al.* 1999). The most effective of these was turf stripping to 3 cm, followed by rotovation (to create 50% bare ground[6]) and then slot seeding, whereas light harrowing had little surface penetration and gave poor results. This was similar to the finding by Edwards *et al.* (2007) that power harrowing of an improved chalk pasture to a depth of 5 cm, creating 60% bare ground, gave no better results than uncultivated controls, whereas turf stripping to 10 cm was highly effective in establishing a sown native seed mixture. Another multisite experiment on arable land in five ESAs in lowland England compared natural regeneration, shallow cultivation (using harrows or discs) and deep cultivation to 30–40 cm – the latter reversing the soil profile, thus burying fertile topsoil and the seed bank of undesirable weeds (Pywell *et al.* 2002). Once again the deep cultivation treatment, followed by sowing with species-rich seed mixtures, developed the closest match to reference target vegetation after four growing seasons.

In many respects, establishing sown target species in improved, species-poor grasslands is more of a challenge than on ex-arable sites, because the presence of existing vegetation (or viable vegetative fragments) as well as buried seeds, is a recipe for rapid sward closure. In contrast, cultivation prior to seeding on ex-arable sites means that there is little pre-existing vegetation, and although the weed seed bank is often large, its germination is contemporaneous with the target species and is relatively short-lived, creating a rapid turnover that allows more space for establishment. On large-scale arable restoration sites in the northern Upper Rhine region of Germany, Donath *et al.* (2006; 2007) found that the establishment of species introduced from donor sites was three times greater than in improved grasslands after 4 years, although in the latter case there was a slight improvement when the grasslands were previously rotovated. One grassland restoration experiment, reseeding arable land with a diverse calcareous mixture, found that 'ridge and furrow' ploughing, creating 1.5-m wide × 0.4-m high ridges, significantly improved establishment compared with band spraying with glyphosate to kill approximately 50% of the vegetation, or power harrowing to create approximately 70–80% bare ground (Wagner *et al.* 2014).

An example of the importance of disturbance in establishing sown diverse mixtures of forbs and grasses in productive swards is illustrated in Figure 6.5. At two separate study sites in Devon and Buckinghamshire, species-poor and improved mesotrophic (MG7) grasslands were seeded with diverse mixtures after implementing progressively severe disturbance treatments, comprising: (a) intensive autumn sheep grazing; (b) slot seeding using a tine drill (but without accompanying herbicide (see Section 6.7.1)); (c) power harrowing to 5-cm depth leaving 30–40% bare ground; and (d) turf removal to 10-cm depth using an excavator (Pywell *et al.* 2007a). After 4 years, intensive grazing and

---

[6]   These authors cautioned that more complete and severe treatments, such as deep rotary culti-
       vation, might temporarily increase nutrient mineralisation and fertility, leading to competitive
       exclusion by dominant species able to respond.

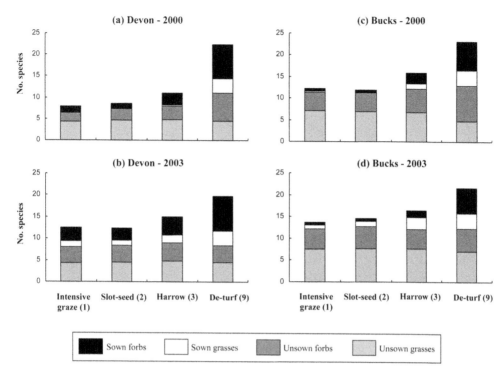

Figure 6.5 Effects of increasing disturbance intensity on plant species richness in improved grasslands at the Devon (a and c) and Buckinghamshire (b and d) study sites, 1 and 4 years after seeding with diverse seed mixtures comprising 4 grasses and 14 forbs (reprinted from Pywell *et al.* 2007a with permission from John Wiley and Sons).

slot seeding were both relatively ineffective in establishing the new species, probably because the gaps created in the sward were too small or too short-lived. On the other hand harrowing and de-turfing were both reliable and effective in establishing a limited number of desirable, generalist species. Turf removal in this case also had little immediate impact on nutrient levels, suggesting that reduced competition, rather than a lowering of fertility, was the key to success. Kiehl and Pfadenhauer (2007) also noted that on ex-arable plots, topsoil removal resulted in a lower standing crop and greater persistence of the target species introduced in green hay applications.

Competition-free gaps can also play a part in diversifying permanent grassland by inserting young, pre-grown transplants into the sward. Transplants offer a means of rapid establishment and promote earlier flowering, especially for those species with limited seed production or unreliable germination. Small gaps created using spot herbicide applications can be effective (Bisgrove and Dixie 1994). In one experiment where two different transplant sizes (2-cm plugs and 9-cm pots) were inserted into gaps of variable size (0-, 15- and 30-cm diameter), the large gaps improved the survival of the smaller transplants, but only in the most productive swards (Davies *et al.* 1999). There was great variation in individual responses of the 15 forb species transplanted, but a clear overall negative correlation was found between transplant survival and increasing levels of P and K, and sward biomass productivity. In practice, very large gaps for transplants could destroy unacceptably large areas of the sward, and may encourage undesirable weeds such as thistle and Common Ragwort. Gaps may also attract the attention of voles and

Rabbits, as well as slugs and snails, especially when occupied by ex-nursery transplants of high nutrient status.

6.6 Sowing practice

6.6.1 Sowing mechanics

Seed can be broadcast, direct drilled or slot seeded, the latter a refinement of direct drilling. So that it runs freely in the seed drill or broadcast spreader, the seed should be dried and cleaned beforehand, and bulk carriers such as barley meal or fine sawdust added to avoid settlement and partitioning between seeds of different shapes and sizes. Direct drills usually operate by cutting a simple groove in the soil surface into which the seeds are deposited, or in grassland by removing narrow strips of turf (which may receive additional rotovation in advance of each seed coulter). The advantage of direct drilling over harrowing and broadcasting is the better contact between the seed and soil; while only minimum cultivation is necessary, the overall surface remains intact. Common drawbacks are that only precision drills can overcome the problem of placing small seeds too deeply, and only certain types of drill can cope with seeding into a dense grass mat.

Specialised slot or strip seeding drills were originally developed to seed clover into upland pastures; the technique has since been adopted to restore or enhance improved and species-poor grasslands. Narrow, parallel strips of turf < 2-cm deep and 20–50 cm apart are opened up, creating slots into which seed is drilled, while the strip boundaries are sometimes simultaneously sprayed with a band of growth retardant or herbicide (Figure 6.6). In theory, these competition-free zones aid seed germination and allow the seedlings to grow to a competitive size before the surrounding sward reinvades the lines. Wells *et al.* (1989) described three early wildflower slot seeding experiments, all on heavy clay soils at separate sites in Cambridgeshire and Bedfordshire, two of which also compared autumn and spring sowing. Seeds were sown approximately 5–15-mm deep into slots about 5-cm wide, with a contact herbicide spray clearing a band about 10-cm wide over the slot area. Survival varied markedly between the forb species sown, some being badly affected by winter waterlogging. In one experiment, slot sowing into 9 different types of amenity grassland had no significant effect on the establishment of the 12 species introduced: after 2 years the average survival was almost 20% and 7 species flowered, with one, Yellow-rattle, spreading rapidly into the adjacent grassland. Another experiment with 14 slot-sown forb species found an average of 10.1% year-end survival after an autumn sowing, compared with 3.7% for the same mixture sown in the following spring. The authors attributed this to the more persistent herbicide bands in the autumn, allowing the emerging seedlings a longer period free of competition. However, in a third experiment no differences were found between autumn and spring sowing: the average establishment of 15 forb species was 3.9% at the end of the first season, but more than half the species subsequently flowered and set seed.

Controlling competition (see Section 6.5) appears just as critical to the success of slot seeding as other forms of seeding and regeneration in established grasslands. For example, Pywell *et al.* (2007b) compared sowing wildflower seeds using either slot seeding treatments (both with and without a herbicide band spray), with harrowing and oversowing, followed by different intensities of aftermath grazing by sheep. Slot seeding without herbicide performed poorly compared with either harrowing and oversowing

Hunters Rotary Strip Seeder: five slots per metre, depth 10–15 mm

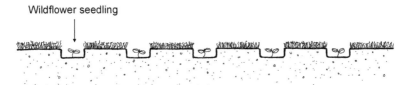

Stanhay precision drill: three slots per metre, depth 5–15 mm

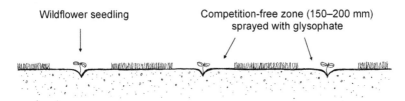

Contraseeder: nine slots per metre, depth 5–15 mm

Figure 6.6 A schematic section through slot-seeded turf taken across the line of trenching (illustrated by Tharada Blakesley).

or slot seeding with herbicide, while sheep grazing intensity had little influence on the outcome, emphasising the importance of controlling weed ingress.

Slot seeding and direct drilling can be considerably cheaper than broadcasting because this avoids the need to prepare a full-scale seedbed. Economies are also improved by the often better germination and establishment rates achieved, and because seeding rates can be considerably reduced (perhaps by a third to a fifth of the broadcast rate in the case of slot seeding, for example, at 1–3 kg/ha). On the other hand specialised slot seeding machinery, usually not available on-farm, is likely to be required for effective sward enhancement. Another drawback is that the lines of developing young seedlings give the sward an unnatural, striped appearance, at least initially. These lines have been described as 'motorways for slugs', and Rabbits and sheep have also been observed to graze preferentially along the enhanced strips. Rolling after seeding can help to control slugs (Natural England 2010a), but slug pellets are likely to be more effective in avoiding considerable damage.

6.6.2   Complete and partial sowing

Seed can be sown by hand in small areas, but for larger sites tractor-mounted machinery will be necessary, using seed or fertiliser broadcasters or seed drills. If the seed is to be broadcast, the experimental evidence is compelling that thorough ground preparation involving repeated cultivation or herbicide applications, is necessary to create a good seedbed (Table 6.6). Most seed suppliers recommend generous sowing rates of 4–5 g/m^2 (40–50 kg/ha), but several studies have found that, given good initial ground preparation, lower rates yield perfectly acceptable results. Stevenson et al. (1995) found that seeding with a calcareous wildflower mix at rates of 0.4, 1.0 and 4.0 g/m^2 on former agricultural land developed similar species compositions after two growing seasons, although the higher rates did eliminate weeds more rapidly. Broadcasting seed in narrow strips or localised patches is another means of reducing seed rates, assuming that the surrounding matrix will colonise from these initials. In an experiment set up in the White Carpathians Protected Landscape Area, Jongepierová et al. (2007) monitored the spread away of species from 2.5-m strips of a regional grass/forb seed mixture, sown centrally, within plots 20-m wide. After 5 years there was some limited spread to the rest of the plots in treatments left to naturally regenerate, but this was much less marked when the plot matrix was sown with a commercial grass mix. After 10 years the cover of sown forbs in adjacent naturally regenerated plots was similar, although less species-rich, to those broadcast with the regional seed mixture. However, the commercial grasses had meantime spread into the unsown plots from their original sown positions (Mitchley et al. 2012). Although in the successful treatments there was some convergence towards the composition of reference ancient grassland sites of the region, at this time they were comparatively less species-rich.

Similarly, Donath et al. (2007) seeded 10-m wide strips with fresh donor plant material, covering 20% of sites, to restore alluvial grassland in the Upper Rhine floodplain of Germany. After 7–8 years, transects were positioned at right angles across the restored strips and up to 10 m into the adjacent species-poor grassland (Burmeier et al. 2011). By this stage over 90% of the species had succeeded in colonising the adjacent area to some extent, while 52% had also entered the soil seed bank, although both the species numbers and their abundance decreased with increasing distance into the hinterland. The most successful colonists tended to be common meadow species, whereas some rarer species remained confined to the areas originally seeded. Although colonisation was slow, the authors suggested that it could be accelerated by orientating the seed-enhanced strips perpendicular to the operating path of agricultural machinery, for example, during mowing, to encourage seed dispersal.

A similar pattern of colonisation was observed in a grassland diversification experiment (Coulson et al. 2001) where slot-sown Oxeye Daisy and Yellow-rattle spread rapidly, but this depended on the management treatments employed and species phenology. Spread was assisted by hay cutting and the direction of cutting, and to some extent the prevailing wind in the case of Yellow-rattle, which advanced more than 6 m in 3 years, compared with less than a metre in an autumn grazing-only treatment. Oxeye Daisy dispersal was much less effective because it was negatively affected by hay cutting before much seed had ripened. These contrasting results illustrate the need to adjust management, and particularly cut dates, to achieve maximum rates of spread of individual species.

6.6.3   Arable reversion and grassland enhancement protocols

Recreating species-rich grassland through arable reversion depends strongly on managing competition with unsown species. If troublesome agricultural weeds are known to be present and abundant in the seed bank, there is strong argument for carrying out complete cultivation, followed by herbicide applications, to create a 'stale' seedbed prior to sowing (Table 6.6). High seeding rates may also help to suppress the weed flora, as may highly diverse species mixes, simply because more sown components add to the 'insurance' that more desirable species will survive to compete with the less desirable ones. On the other hand, if weeds are not considered likely to be a problem, strip sowing may be more economical and convenient. If the objective is to enhance species-poor grasslands, minimising competition is again important, in this case by maintaining a short sward, followed by power harrowing to create bare ground for seeding.

Although autumn sowing is more usual, spring sowing is also feasible. The advantage of autumn sowing is that there is more likelihood of rainfall to encourage rapid germination, while the soil is still warm and while gaps should be present in existing swards following summer grazing. However, this practice does tend to favour fast germinating grasses, whereas spring sowing may favour forbs. Some species such as Cowslip, Pepper-saxifrage and Yellow-rattle also require vernalisation before they will germinate, so they may not receive a long enough period of chilling if sown in late spring. On the other hand, spring sowing can be effective where it avoids winter waterlogging on heavy soils, and also reduces the likelihood of young seedlings undergoing frost heave or predation by birds and invertebrates. The protocol for autumn sowing given in Table 6.6 can be shifted to cultivation and sowing in March and April.

In the first growing season (year 2 after an autumn sowing), many researchers recommend a light mowing or grazing regime to allow the sward to establish, while also spot-treating or topping agricultural weeds such as thistles and docks, and pulling Common Ragwort. Frequent close mowing or overgrazing will control annual and invasive weeds, although this may have a negative impact on some 'desirable' plants and prevent their seeding. For instance, there is a danger of defoliating or overgrazing annuals such as Yellow-rattle with this regime, but after July the sward height can be steadily reduced. From the second growing season onwards (year 3 after an autumn sowing) the long-term management regime can be established either as a pasture or meadow. However, provided these regimes are not too intensive, they may be introduced earlier in the first growing season once the sward has achieved reasonable cover.

Several authorities suggest adding a low proportion of 'nurse' crops with the seed mixture. These are usually fast-growing annuals such as Italian Rye-grass, cereals (e.g. Spring Barley), Flax, Buckwheat and occasionally Cornflower mixtures. The principle is that their rapid canopy development provides shelter for the seedlings of other species while suppressing excessive weed growth while preventing soil erosion on unstable surfaces, such as steep slopes. On the other hand, vigorous nurse crops, especially if sown at the recommended rates of 2–4 g/m², can also prevent establishment of other species if the soils are too fertile (see Mitchley et al. 1996; Manchester et al. 2002; Pywell et al. 2002). In an experiment where a dominant grass, Red Fescue, was sown together with three subordinate forbs Carthusian Pink, Perennial Flax and Yellow Oxeye Daisy, high sowing rates of the grass did suppress the forbs; but in this case the cultivar fescue was less aggressive than a regional source. Rhizomatous fescues also caused no reduction in establishment than the tussock form (Walker et al. 2015). Cutting these crops early in the growing season before the seed is shed can be effective in preventing this. However, Walker et al. (2007) found no detriment, but also little benefit of sowing nurse crops of Italian Rye-grass or Spring Barley to accompany a heathland/acid grassland mix on two ex-arable sites in the Breckland region of East Anglia. Similarly, experiments to restore arable fields flanking the northern Upper Rhine to floodmeadow vegetation showed that

Table 6.6 Management operations in grassland creation or enhancement

Timing	Grassland creation – arable reversion	Grassland enhancement
Year 1		
Late June	Spray off crop residues or perennial weeds with a translocated herbicide, allowing 4–6 weeks for the sward or crop to die off	–
Late July	Plough or rotovate the soil. Cultivate to create a fine seedbed using a power harrow, discs or raking: then roll	Keep the sward short (3–5cm) by close mowing or grazing
August	Allow any weed seeds to germinate, then spray the regrowth with herbicide to create a 'stale' seedbed, avoiding further soil disturbance	Power harrow, disc or rake the grass surface to expose 30–50% of bare ground, or localise cultivation into small patches or strips, removing up to 30% ground cover. Herbicide-treated strips or patches are an alternative to cultivation
September to October	If required, spray again in late September. Surface broadcast the grass/wildflower seed, rolling after sowing to improve soil/seed contact. Direct drilling is another option	Sow wildflower seed thinly on to exposed soil, or insert plug transplants in small areas. Alternatively, slot seed the entire area
October to March (years 1–2)	Keep the sward short (< 10 cm) by regular mowing, removing cuttings if feasible	Keep the grass short (< 10 cm) by mowing or grazing in spring and autumn, removing cuttings if feasible
Year 2[a]		
April to June/July	Control annual weeds by topping, pulling or spot-treating perennial weeds. If Yellow-rattle has been sown, do not close-mow (< 10 cm) before July	
July to September	Mow or lightly graze the sward to 5–10 cm for the remainder of the growing season	
Year 3 onwards		
	Establish the future management regime, i.e. for hay meadow or permanent pasture:	
	Hay meadow: cut for hay in late July or early August, grazing or mowing the aftermath from September to December, ground conditions permitting	*Grazing pasture*: the main grazing period should extend from late summer (allowing plants to flower and set seed beforehand) to autumn/winter, with optional light grazing in spring

[a]Alternatively, proceed immediately to the regime indicated for year 3.

a grass mixture consisting of six species, sown at 5 g/m^2 in addition to seed supplied by hay additions from a donor site, had not significantly affected the recruitment of the target species after 4 years (Donath *et al.* 2007).

If domestic animals are introduced to the restoration site too soon after sowing, young seedlings may be grazed off or trampled. Small mammals, such as Rabbits, Brown Hares, mice and voles may also be a problem locally. However, mollusc grazing can also significantly affect community composition in the early stages (Hanley *et al.* 1996) – slugs are highly selective in their preferences – and therefore control measures, such as dispensing baited pellets containing metaldehyde, are occasionally advocated. When applied in the autumn and spring following sowing, molluscicides can be highly effective: at one experimental site in harrowed or de-turfed treatments, the slug population was more than halved by metaldehyde applications, resulting in increased numbers and abundance of sown forbs compared with untreated plots (Pywell *et al.* 2007a). Similarly, in a subalpine meadow in the Harz Mountains, Germany, Scheidel and Brulheide (2005) found that molluscicide treatment increased the survival of the grassland perennial *Arnica montana* tenfold, when seeded into small gaps. The obvious drawback is that such pesticide use can also be harmful to other species, particularly small mammals and birds, and may also affect the germination of some sown species (Hanley and Fenner 1997).

6.6.4 Transplants

Pot-grown wildflower transplants or seedling plugs, grown either in modular or larger containers, can be used to supplement newly seeded areas or to enhance species-poor grassland, both during the establishment phase or subsequently. An evaluation of plug planting on 13 neutral grassland sites in Oxfordshire, recorded the best establishment in the least intensively managed regimes (late haymaking, aftermath grazing and no fertilisers), whereas fertiliser additions, early and frequent silage cuts or prolonged grazing all reduced the mean transplant survival to < 40% after 2 years (Warman *et al.* 2007). The tolerance of the 12 common grassland forbs tested varied, with more than half responding negatively to the most intensive management treatments and to high background levels of soil N or P. These findings complement other studies showing improved performance of plug plants placed in competition-free gaps.

Plug planting is potentially much more expensive than broadcast seeding. Suitable species are mid- to late-successional species that tend to establish poorly during the first phase of meadow restoration. Typically these are absent, rare or under-represented in seed mixtures, have unreliable germination and tend to be slow growing (Table 6.7). In a study comparing seed sowing with plug planting in Swedish hay meadows, Wallin *et*

Table 6.7 Species with unreliable germination, therefore recommended for plug planting (after Natural England 2009) (© Natural England)

Bugle	Common Rock-rose
Harebell	Horseshoe Vetch
Cuckooflower	Field Scabious
Dropwort	Great Burnet
Meadow Crane's-bill	Meadow Saxifrage
Wood Crane's-bill	Large Thyme

al. (2009) found that plug planting was twice as effective in promoting the establishment of two test species Spotted Cat's-ear and Devil's-bit Scabious after 2 years. Practically, commercially grown four- to five-month old plug plants in 3–5-cm modules can be inserted using a special planting dibber (Figure 6.7), or can be supplied as larger plants in 7.5–9-cm pots. Clumps of 2–4 plants/m², inserted into herbicide or bare soil patches, are often recommended, using two or three species at each station.[7] Small transplants are more vulnerable to frost and drought and watering may be required in dry weather until the plants have rooted: hence autumn planting is usually preferable. Finally, pre-grown wildflower turf containing 20–30 forb species, usually produced in compost on fine plastic netting, is available from a number of specialist suppliers. The limitation here, as with ready-grown transplants, is the high cost compared with conventional seeding.

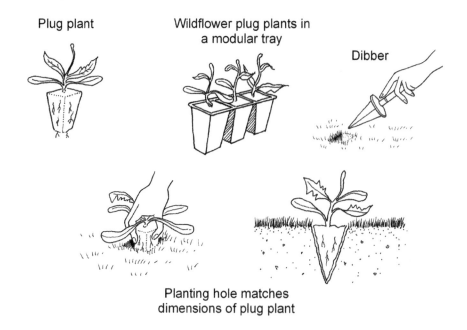

Figure 6.7 Plug planting using modular transplants and a customised dibber (illustrated by Tharada Blakesley).

---

[7] For agri-environment schemes, the maximum density of plug plants supported is unlikely to exceed 1,000/ha.

7. Defining success in grassland restoration

Grassland restoration and creation techniques range widely, from spontaneous succession (with minimum intervention and management), to expensive fertility reduction measures, such as topsoil removal, accompanied by seeding or transfer of plant material from a donor site. The relative effectiveness of these techniques has been the subject of many investigations (e.g. Walker *et al.* 2004b; Hedberg and Kotowski 2010; Kiehl *et al.* 2010; Torok *et al.* 2011), with the general consensus showing that success is often hampered by both a lack of propagules (dispersal limitation), or unsuitable conditions at the restoration site (establishment/recruitment limitation). Ideally, therefore, one would select only those sites on relatively nutrient-poor soils, which in addition retain a remnant seed bank of that community or are positioned close or adjacent to seed sources emanating from it. Careful planning beforehand is needed to identify and prioritise such sites. However, as natural dispersal mechanisms cannot always be relied on – even when donor sites are adjacent and an appropriate management regime is applied – the restoration target may remain incomplete, or take decades to achieve. Realistically, in heavily modified and fragmented agricultural landscapes, it will be necessary to deliberately introduce some of the 'missing' species.

7.1 Plant introductions

Which species, and how many to introduce, will depend on the objectives of restoration, informed by practical management considerations and cost. Low-diversity seed mixes (i.e. fewer than ten species) are relatively inexpensive and often commercially available. These usually contain a nucleus of robust species capable of establishing in a wide range of soils, but ultimately they cannot deliver rarer, late-successional species that are dispersal-limited. The international experiments described by Lepš *et al.* (2007) illustrate how sowing low- or high-diversity seed mixes can significantly alter the successional pathway compared with plots left to colonise naturally. Although these colonising plots had a greater species complement after 8 years, they contained far fewer of the target species whereas high-diversity mixtures not only tended to deliver more of the desired species, but also provided greater insurance, as more species were available to compensate for the failure of others. On large restoration sites, a possible compromise would be to sow an overall matrix of low-diversity seeds, with dispersed patches of high-diversity seed mixtures to reduce the cost (Torok *et al.* 2011). A further refinement would be to include species with known persistence but poor spreading ability into the overall matrix, while the patches would contain species with proven propensity to spread. The former, plus species showing poor initial establishment, can be sown by slot seeding or by plug planting at a later date (Pakeman *et al.* 2002).

Potentially even richer sources of species diversity are the plant materials imported from wild harvesting (Section 6.3.2). Green hay is more efficient than dried hay bales as it contains more seeds, and may even transfer some invertebrates, but it cannot be safely stored for more than 24 h and therefore has a limited transport radius between the donor and receptor sites. Otherwise conventional hay can be used, or the seeds extracted by brush or vacuum harvesters, bearing in mind the evidence that these machines may harvest unevenly from the full sward profile in tall vegetation. Furthermore, all such plant material transfers represent a phenological snapshot of the sward at the time of harvest, and so will lack some of the species present. To obtain a fuller complement, particularly of early- and late-flowering varieties, necessitates staggered harvests, using multiple transfers, unless the seed is extracted, stored and mixed for a single sowing. For somewhat different reasons, several authors have also argued that restoration should be a two-stage process, with the initial seeding or transfers of plant material followed later by introductions of mid- and late-successional species, recognising the importance of plant-soil feedback processes due to long-term changes in fertility and the facilitation of below-ground soil microbial communities.

For species that are difficult to establish or do not germinate freely from seed, a high-cost-option is to propagate plants in a nursery environment before planting out in the field. To moderate the cost, only the most recalcitrant species need be considered

Enrichment planting at Bloomers Valley, Wakehurst Place, undertaken by staff from the Royal Botanic Gardens, Kew (Trustees of the Royal Botanic Gardens, Kew).

With funding provided by the John Ellerman Foundation, the site at Bloomers Valley, which had been managed as a grass field since the Second World War, was harrowed and sown with a wildflower meadow mix of local origin. This method was very effective for species like Common Knapweed, Oxeye Daisy and Yellow-rattle. Subsequently, hundreds of nursery-raised plug plants were planted annually, adding characteristic species such as Harebell (illustrated), Sneezewort and Betony to the sward.

for this method, limiting them to patches within the area as a whole. A final option is to import topsoil or turf from a donor site, but as it involves damaging or destroying existing semi-natural grasslands this is not recommended. Such operations have usually been in compensation for infrastructure developments involving transport, housing and quarrying schemes, but potentially more species can be delivered by this method than any other. Success, determined by how many species of the original complement survive the operation, has been good in some cases, provided there is a good match of the physical conditions of donor and receptor sites, and consistent management post-transfer.

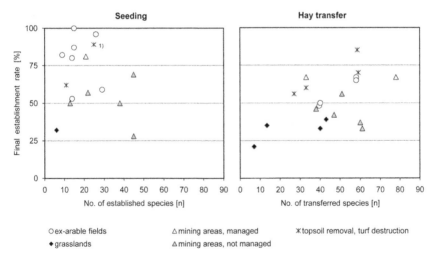

Figure 7.1 Efficiency of species introductions using (a) seeding and (b) green hay transfer recorded from a number of restoration studies. The number of established species is that developing from the original seed mixture, or from species present at the donor site, respectively. The final establishment rate is the percentage of the transferred species successfully established at the end of the observation period (reprinted from Kiehl *et al.* 2010, with permission from Elsevier).

7.2   Preparation for sowing

Creating gaps and microsites for seedling establishment is critical to success. In the majority of restoration experiments, seeding into existing grassland (i.e. diversification) was generally less successful than seeding ex-arable sites (habitat creation). Cultivation that creates substantial soil disturbance appears more effective in preparing a good seedbed and eliminating competition with established vegetation, compared with light harrowing or scarification, particularly where it is employed on grassland diversification sites. A detailed analysis of British and European restoration projects showed that both ex-arable sites and raw soils on restored mining areas were consistently more successful in establishing the sown species (Figure 7.1; Kiehl *et al.* 2010). Topsoil or turf removal was also effective in reducing competition from buried seed banks, while also lowering fertility; however, as a general rule, disturbance appears to be more critical to success than soil fertility per se. Fallowing and mowing are both very long-term options for reducing nutrient levels, although on some soil types the cropping of ex-arable sites for 1 or 2 years, without fertiliser, can be beneficial. Similarly, chemical manipulation or immobilisation with carbon-rich materials can also be effective initially in establishing the target vegetation; however, if possible, selecting restoration sites with low fertility indices in the first place is most likely to give good results.

7.3   Management techniques

A low- to medium-intensity grazing management, accompanied by hay cuts and with nil or minimal fertiliser input appears to offer the best options for encouraging species richness in grasslands. As this requires considerable investment in time and resources, mowing only is often adopted as a relatively economical solution, and is certainly preferable to no management. On uncut sites, competition with tall dominant species, the build-up of litter and ultimately scrub development, invariably results in poorer establishment of any introduced species. Most studies have shown that cutting more than once, for example in July and September, tends to maintain greater species-richness with a higher proportion of forbs present, but very frequent mowing can restrict the species list to shorter-statured, tolerant species. Because cutting is non-selective, it tends to create a uniform sward architecture that is less attractive to invertebrates; the latter also suffer mortality though repeated mechanical damage. Grazing, on the other hand, needs constant inputs in shepherding, veterinary inspections, supplementary feeding and fencing, but tends to produce a greater diversity of sward structure than mowing, together with an uneven disturbance mosaic that provides more niches for invertebrates. If donor sites are available, transporting animals between these and nearby receptor sites appears to have some potential for dispersing propagules from the former to the latter. Action may need to be taken to ensure that propagules of non-target species present on 'improved' sites do not travel in the reverse direction (Mitlacher *et al.* 2002), and that local eutrophication does not result from this dual grazing (Kirkham 2006).

7.4   Long-term vegetation development

Managers with experience of grassland creation and restoration will be well aware that their efforts do not always turn out as predicted. Restorative treatments may trigger several possible trajectories in terms of the eventual species composition, because 'random differences in colonisation can result in alternative mixes of species that are resistant to invasion by others' (Young *et al.* 2001, p. 11). This uncertainty, also known as the 'Alternative Stable States' theory of ecosystem development (Suding *et al.* 2004), allows that vegetation development can follow many different pathways, determined by the availability of species (i.e. those introduced, colonising or already present on-site) and their response to site conditions and soil nutrient levels. On lowland heaths, Mitchell *et al.* (1997) showed that vegetation succession trajectories could tend towards dominance by birch, pine, Bracken, gorse or Rhododendron, respectively, while in grasslands, Kardol *et al.* (2008) found that reducing soil fertility by removing topsoil led to a completely different successional trajectory. Alday and Marrs (2014) used a multivariate approach (principal response curves) to predict the stability of alternative stable states after Bracken control treatments in heather or acid grassland communities, although the technique relies on long-term data records. Given this possibility of multiple stable states, attempts to recreate a precise target community may actually be impossible to achieve. The interaction between plant species, soil microbial communities, soil nutrients and management practices will combine to determine the progress (or otherwise) of the vegetation towards the intended composition, as shown by many investigations (e.g. Pywell *et al.* 1996; 2007a; Smith *et al.* 2008). Extrapolating from the vast tracts of different-aged calcareous grassland on Salisbury Plain in England, Redhead *et al.* (2014) considered that restoration projects were unlikely to achieve complete reassembly of ancient grasslands within a human lifetime, both in terms of community composition and species traits.

Grassland restoration experiments are costly and tend to suffer from a lack of long-term monitoring, so that trends and changes will still be occurring when the work has been concluded; Török *et al.* (2011) calculated a mean experimental duration of less than 4 years. Small experimental plots also rapidly become compromised as the sown species spread between them, leading to a convergence of the vegetation community over time.[1] In cases where monitoring is pursued, declines in species-richness of the introduced communities were observed by Pywell *et al.* (1996) in a 16-year experiment involving sowing 8 different wildflower seed mixtures on a chalky boulder clay soil. Increasing proportions of unsown forbs and grasses developed in the different treatments, followed by a gradual decline both in species richness and sward productivity. In spite of a consistent management regime, large fluctuations in species composition occurred, attributed to a rapid species turnover driven by rainfall and temperature variation between years. The decrease in productivity over time was explained variously as falling soil fertility due to leaching and biomass removal, a decline in N-fixing species and the accumulation of leaf litter. Similarly, Smith *et al.* (2010) noted that short-term changes in wildflower mixes sown on arable field margins did not necessarily predict longer-term shifts in the vegetation community. Over 13 years, the sown species gradually declined because natural colonists, especially rhizomatous, weedy perennials, invaded them, although the vegetation did remain considerably more species-rich than in adjacent, naturally regenerated swards.

---

[1]   Cross-contamination between small experimental plots may give misleading projections of the rate and direction of vegetation development in practical restoration projects undertaken on a larger scale. The problem may be partly overcome with larger experimental plots, buffered by substantial guard rows (Pakeman *et al.* 2002) or using large-scale, multisite investigations (e.g. Pywell *et al.* 2002; 2007a; Prach *et al.* 2013).

Gradually declining species diversity can also be explained by extinctions of subjects that were rare or infrequent in the original seed mix or introduced plant material, and which were then unable to develop viable populations in small, experimental plots. In the long-running meadow management experiment at Colt Park in the Ingleborough NNR, optimum treatments involving late hay cuts, spring and autumn grazing and seed additions took 6 years to approach the target MG3b meadow, with the sown species still being present 18 years later. However, a decline in Yellow-rattle in 2006 coincided with an increase in grass cover scores and fertility values, illustrating the susceptibility of vegetation community to change. Smith (2010) considered that it might take another 20–40 years to achieve a really good example of an upland hay meadow at this site, a time frame similar to that considered by Hirst *et al.* (2005) to be necessary to develop a mesotrophic grassland, and by Pakeman *et al.* (2002) to recreate chalk grassland on an ex-arable site at Royston Heath, Hertfordshire.

Large-scale arable reversion projects set within a landscape still rich in reference grasslands are likely to be more successful than in a region where grasslands are degraded and fragmented. In this context, in the foothills of the White Carpathians, Prach *et al.* (2014; 2015) found that whether sowing simple 'commercial' seed mixtures or diverse, regionally selected seed mixtures, or allowing spontaneous succession, all treatments converged over time to resemble mesic hay meadows. Between 10 and 20 years after sowing or abandonment, surrogate semi-natural grassland formed, irrespective of the method of restoration. Sowing regional mixes was, however, justified in the short-term because they accelerated a progression towards reference grasslands, and because target species from the surroundings spontaneously established more easily in prepared seedbeds in the first 10 years. After this initial phase, Prach *et al.* (2015, p. 187) reported that they could 'reasonably rely on gradual spontaneous colonisation by target species if they occur in the surroundings and site conditions are not adverse.'

Given consistent management, a gradual convergence towards stable communities might be anticipated, but in reality there is always potential for change. A 60-year data series from the Rothamsted Park Grass Experiment showed that several species exhibited either increasing or decreasing trends, or occasional outbreaks (Dodd *et al.* 1995). About half of the 43 species on limed (mesotrophic) plots showed no trends over time, but the minority (6) that did increase tended to flower late, after the June hay crop, and to show more ruderal and outcrossing traits than average, allowing them to spread rapidly, and 10 species showed temporary outbreaks followed by a decline. As the authors remarked, it was impossible to tell whether these observations were part of even longer-term changes (such as global warming), or merely cycles. In practice it may be necessary to carry out occasional introductions of species from donor sites to maintain species richness over time.

7.5 Cost-effectiveness in restoration

Restoration treatments need to be judged from a cost-effective, as well as a nature conservation perspective. Agri-environmental schemes attempt to reflect this balance by offering higher payments for more expensive tiers of treatments and materials, but often lack measurable outputs in terms of their delivery. Various methods have been used to test the success of different restoration treatments. Bullock (1998) used similarity coefficients to compare species composition in pre- and post-translocated communities, while Klimkowska *et al.* (2007) adopted a measure of the 'completeness' or saturation index (SI) of restored meadow communities, based on comparisons with the regional species pool for each particular community. A change in the SI over time, i.e. the difference before and after restoration, gave a relative measure of success. In evaluating prairie restoration success, Martin *et al.* (2005) compared 8–10-year-old restored sites with nearby remnants using diversity measures at different scales. They found that at the within-plot scale, diversity and species numbers were invariably lower on restored sites, but between-plot diversity was higher, probably because the restorations had large, clumped, single-species patches, whereas the species were more highly intermingled on remnant sites.

Other success measures can include: similarity to a target NVC community achieved; numbers of sown species established; proportion of non-native species; species density per unit area; and the proportion of the vegetation accounted for as target species. In their investigation of the feasibility of recreating floodmeadows on arable land, Manchester *et al.* (1999) concluded that although the most complex, 22-species, seed mixture gave the greatest species richness, the best value for money (based on the full range of success criteria outlined earlier) was delivered either with a 9-species mixture, or from seed extracted from hay bales. Natural regeneration and a basic, four-species mixture of common grasses delivered practically no comparable ecological benefit over the three-year time period. The time frame is an important consideration in such evaluations, as it can be argued that even basic mixtures will diversify over time as and when new species arrive; however, without suitable colonising sources, an initial lack of investment will give a very poor return. Even partial restorations that fall short of the target will have a 'new native' value in their own right (Mitchley *et al.* 2012) and can also be seen as a part of our cultural heritage (Prach *et al.* 2015).

The costs of carrying out a given restoration technique vary widely according to grassland type, the scale of the operation, and the relevant country's economy, with such work tending to be much more expensive in Western than Eastern Europe (Klimkowska *et al.* 2007). The costs of various techniques for restoring former croplands in Europe were reviewed by Török *et al.* (2011), ranging from unmanaged, spontaneous succession to the transfer of whole plant communities as intact soil profiles (Figure 7.2). After natural colonisation, where only mowing or grazing maintenance was required, sowing commercially available, low-diversity seed mixes (< 10 species) was usually the most

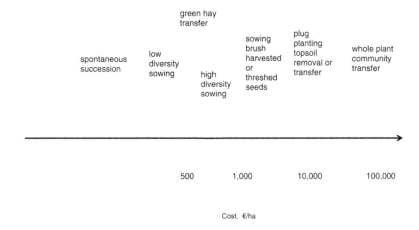

Figure 7.2 Relative direct costs of using different techniques to restore grassland communities on former cropland. Note the notional logarithmic scale.

economical technique, which in turn was about half the cost of high-diversity mixes. Green hay transfer from a donor site could be more or less expensive than sowing, with costs in part determined by the thickness of the hay layer, harvesting machinery and transport, but cheaper than using specialist harvesting or threshing equipment to extract the seed. Raising plant seedlings for transplanting at the restoration site is a high-cost operation, depending on the number of species and the density of planting. Finally, removing a topsoil layer to lower fertility, or transferring it from a donor to a receptor site, was potentially very costly, depending on the thickness, and even more expensive when using machinery capable of stripping and laying whole profiles of intact turves to depths of 50 cm.

7.6 Monitoring success

The objective here is to devise a simple recording methodology for testing whether the sward is following an expected trajectory towards the target vegetation. Monitoring intervals set at 0/1, 2/3, 5 and 10 years should be enough, both to detect and correct any initial problems and to check whether the restoration treatment is on course to satisfy the longer-term objectives. Such judgements are usually based on a number of set attributes, which in turn can be converted into scoring systems that define the progress (or otherwise) of the vegetation over time. These attributes may be positive or negative: for example, a positive development might be an increase in species richness and the number of target species, but negative if there is a progressive decline in the number of sown species or an increase in potentially dominant grasses. Examples of positive and negative attributes that could be applied are shown in Table 7.1.

The 'target species' referred to here would be drawn from selected lists of species particular to the grassland community envisaged for the site, or based on the species introduced through sowing or green hay transfer, or any desirable ones present in the vicinity. 'Presence' of a species can later be quantified in frequency bands, or as mean visual percentage cover.

Table 7.1 Positive and negative attributes to measure vegetation success

Positive	Negative
Presence of target species (sown or colonising)	Presence of pernicious weeds (docks, thistles)
High forb species richness	Low forb species richness
Few competitor species	Many stoloniferous and rhizomatous species (White Clover, Creeping Buttercup) and dominant grasses
Many target species present	Few target species present
Moderately open, swards of low/medium height	Dense, tall, closed swards
Variable vegetation height and cover	Uniform sward height and cover
Sparse litter layers	Dense litter mats
Small patches of bare ground	Extensive areas of bare ground, or none at all

Table 7.2 Practical criteria for assessing success in grassland restoration, based on the Farm Environment Plan Manual (© Natural England 2010)

Assessed category	Competitor species cover[a]	Species/m²	Wildflower, sedge and rush cover	Indicator species[b] abundance thresholds	Typical grass species
Species-rich	<10%	>15	>30%	1–2 frequent, 2–3 occasional	Common grasses, plus habitat-specific species such as Sheep's-fescue, Yellow Oat-grass, Upright Brome, Heath-grass, Quaking-grass, Crested Hair-grass
Semi-improved	<30%	9–15	≥20%	≥4 occasional	Cock's-foot, False Oat-grass, Common Bent, Meadow Foxtail, Crested Dog's-tail, Sweet Vernal-grass, Meadow Fescue, Timothy, Red Fescue, Tufted Hair-grass
Species-poor, improved	>30%	≤8	<10%	<4 occasional	Cock's-foot, Italian Rye-grass, Perennial Rye-grass, Yorkshire-fog, Timothy, Rough Meadow-grass

[a]For example, Perennial Rye-grass, White Clover, Creeping Buttercup, docks, thistles, Bracken, scrub.
[b]Indicator species are those specific to reference habitats, drawn from lists for each semi-natural grassland type.

A *sample-based* methodology is best to make a proper, objective and repeatable assessment. Initially, the site should be demarcated into visually representative areas which are sampled separately. Sampling usually takes the form of a 'W'-shaped, structured walk across the area, stopping at random points to record the vegetation in sample 1 × 1 or 2 × 2 m quadrats. Ten quadrats may suffice in a small, relatively homogeneous area (< 1 ha), but larger or more variable areas will require 20 or more. The overall abundance of a target or 'indicator' species can be expressed as its frequency in survey quadrats – for example, 'frequent' could be presence in > 50% of survey quadrats, 'occasional' 30–50% and 'rare' in < 20% of quadrats. Rules are then applied to determine how closely the grassland undergoing restoration approaches a reference, semi-natural type using the positive or negative attributes and attaching thresholds or scoring criteria in each case. This general approach has been adopted in agri-environmental schemes, such as the HLS scheme operated by Natural England, to assess eligibility of grants for habitat enhancement and management (Natural England 2010). Table 7.2 illustrates how the criteria might apply, with different thresholds set according to a hierarchy of assessment categories, for competitor species, species richness, wildflower cover and frequency of selected indicator species.

The assessed categories in Table 7.2 – species-poor, semi-improved and species-rich – represent a broader continuum that can be further broken down into smaller units for determining changes in sward condition over the monitoring period. Each attribute, positive or negative, can be given an arbitrary score according to whether they meet the threshold criterion; scores are then aggregated to give a measure of 'success' at each time interval. Surveys of grassland creation and restoration sites using these 'indicators of success' to test the effectiveness of agri-environmental schemes have been commissioned by Natural England (Hewins *et al.* 2012; Stevens and Wilson 2012; Hewins 2013; Wilson *et al.* 2013). Bearing in mind that candidate sites were often selected as examples of good practice and were not a random sample, the results nevertheless indicate that such schemes are capable of delivering priority grassland types. In selected parcels of created or restored grasslands, 85% qualified as the former BAP priority habitat, although many lacked the requisite number of indicator species, or had low cover of wildflower species or a high cover of undesirable species such as White Clover. Age since establishment appeared to have relatively little influence on the outcome (many sites had been established for more than 10 years), whereas the original seed sources appeared to be influential. In a smaller random survey of 36 grassland parcels created on arable sites, only about half qualified as BAP priority grassland, of which 17% was considered to be in good condition, with the remainder showing some potential for recovery given appropriate management (Hewins *et al.* 2012).

BOX 7.1 Case Study: Life on the verge – Lincolnshire Wildlife Trust (LWT)

Topic: A Living Landscape Partnership Project co-ordinated by the LWT aimed at surveying and restoring chalk and limestone grassland on roadside verges. The principal aims are to: link grassland nature reserves with species-rich road verges; raise awareness in local communities; work in partnership with local authorities;

and recruit and co-ordinate survey volunteers. Results reported here are for the period 2009–2013.

Location: Lincolnshire and Rutland Limestone Natural Area and the Lincolnshire Wolds, East Midlands.

Area: Evaluation: approximately 2,560 km of road, approximately 900 ha; designation of 146 new Local Wildlife Sites (LWS) along 233 km of road comprising approximately100 ha of roadside grassland.

Designation of site: Roadside verges: some designated as SSSIs and LWSs. Surveys have identified additional candidate stretches for LWS designation. Over 60 Roadside Nature reserves (RNRs) in these areas have already been designated by the local authorities to include all roadside SSSIs and some LWSs by Lincolnshire and Rutland County Councils, which fund their management.

Management and community use: Management: 6 ha of tree and scrub control, approximately 25 ha of hay baling. Twenty new community wildflower meadows and churchyards, over 1 ha in total.

Condition of site: Unfavourable and/or deteriorating in most cases.

Website: http://www.lincstrust.org.uk/

Little Warren Roadside Nature Reserve (© Mark Schofield).

Background

Calcareous soils overlie the Lincolnshire and Rutland Limestone Natural Area, also known as the 'Jurassic Limestone Uplands'. These Uplands begin north of Lincoln as a thin ridge with a steep western scarp slope and extend south towards Stamford, widening into a broad plateau dissected by the river systems of the

Witham, Welland and Glen. They cover the south-west part of Lincolnshire, eastern Rutland and the north-east of Leicestershire. Until the start of the Enclosure Acts in the mid-eighteenth century, this area was characterised by extensive tracts of species-rich, semi-natural grassland maintained by sheep and Rabbit grazing.

Enclosure represented the first major wave of habitat loss, as most went under the plough leaving only remnant grassland in steep valleys, on hillsides and alongside roads and tracks. The last 50 years has seen a further decline, estimated to be of the order of 60 ha, almost 40% of limestone grassland in the project area. Both the size and number of sites have decreased dramatically over time, creating the most fragmented example of this type of habitat in the country. Such rapid losses have inevitably led to local extinctions within Lincolnshire. For example, 77 plant species were lost between 1900 and 1985. The Chalkhill Blue was last recorded breeding on Copper Hill SSSI road verge just south of the town of Ancaster in the 1970s, but is now thought to be extinct in the county.

Neither designation, actual or proposed was able to prevent the destruction of several sites, including Waddingham Common SSSI, lost to drainage and ploughing in 1963; the proposed NNR at Hollywell Mound, ploughed and reseeded; Colsterworth Glebe Quarry, once a potential SSSI but used as a refuse tip. The conversion of 80% of Robert's Field SSSI into a conifer plantation in the early 1960s severely damaged the finest butterfly site in the county, but this at least is an example of successful restoration, following work carried out by the LWT since 1993.

Life on the Verge is one of over 100 Living Landscape projects that are now active throughout the UK, under the management of county Wildlife Trusts. These initiatives aim to reverse the trend of habitat fragmentation and to establish 'wildlife corridors'. Life on the Verge is recruiting, training and co-ordinating volunteers to explore and identify those road verges that can link species-rich limestone grassland on a landscape scale, making its associated species better able to disperse. This provides both escape routes in the face of climate change and a means of recolonising disturbed sites. Furthermore, due to the scarcity of calcareous grassland, roadside verges have emerged in their own right as a feature of the landscape vital to its survival. Wide verges along 'drove roads' created during the eighteenth-century Enclosure Acts can provide significant areas of linear hay meadow across the older county thoroughfares.

Surveying and monitoring (community involvement)

A total of over 240 volunteers have undertaken surveys of just over 1,150 km of roadside verges over 5 years, which has involved an estimated 3,000 person-hours of voluntary effort. Volunteers claimed their own survey units and returned their data via an online database, thus minimising administrative overheads. Approximately a further 615 km of road have been surveyed 'virtually' using Google Street View©, which has helped to eliminate verge-less, suburban and heavily shaded sections. The task of botanical surveying was simplified and supported through field classes and the production of an identification guide, highlighting plant indicators of both good- and poor-quality habitat. In addition, LWT has appointed voluntary wardens to monitor the condition of existing RNRs.

Community surveys (© MatthewRoberts.co.uk).

Training community surveyors (© Lincolnshire Wildlife Trust).

Roadside verge management

Hay cutting is essential to grassland diversity on roadside verges because stock grazing is not possible. The best results for wildflowers are obtained by baling hay after mid-July to allow seed heads to mature and set seed. The removal of hay maintains a low soil fertility that is a vital factor in preventing nettles, thistles and rank grasses from smothering species that are less well able to exploit high nutrient levels that would otherwise accumulate. In this way, the effect of natural grazing is simulated. Reciprocating blade cutters or drum mowers cut stems at their base, producing long-stemmed material. This action both minimises damage to invertebrates (which would be harmed by a flail head) and enables baling, which facilitates collection and transportation of plant material from the site. Where the receiver site is nearby, material can be removed with a brush collector or buck rake and transported at a lower density.

'Visibility cuts' are legally required for a 1.1-m width along carriageway edges, as are splays at junctions and on tight bends. Where safety allows, these can be minimised and timed later in the season. The project purchased and used pedestrian-powered hay cutting and mini-baling equipment, which is designed to reach less accessible terrain. Contractors also used small tractor-trailed equipment to complement this approach, and to maximise coverage and efficiency of operations.

Limestone grassland restoration

Using survey data, the course of the Roman road known as Ermine Street was identified as a potentially vital corridor that is under significant threat from advanced scrub and tree encroachment. Tree and scrub removal work was targeted along 10 km totalling approximately 6 ha over the course of three winters. Stump treatment and subsequent scrub control is now planned to return the corridor to species-rich grassland. Rabbit control will also be necessary where warrens have developed under cover.

Thirty-two kilometres of roadside verge were cut and baled for hay between 2009 and 2010, equivalent to nearly 25 ha. Stretches were targeted on the basis of survey data where vegetation indicated a lack of cutting but a high potential for recovery, for example, verges where positive indicator species were being outcompeted by False Oat-grass. Hay has been supplied to local smallholders for animal feed, or provided as an organic input to local farming systems where it has not been of sufficiently good quality for fodder. Ongoing management operations funded by Lincolnshire County Council along high-quality RNRs have provided green hay for species-rich grassland creation on private land elsewhere in the county.

Limestone grassland creation

In 2010, seed harvested from Heydour Lodge RNR, Robert's Field SSSI and Ancaster Valley SSSI was sown along 4 km of verge to enrich the species diversity of King Street LWS. The stretch targeted for restoration was identified as species-poor but with the potential to act as a link between botanically richer areas on either side. Seed was harvested both mechanically by a pedestrian-powered brush collector and hand-collected by volunteers. Mechanical collection favoured Yellow-rattle, Kidney Vetch, Ribwort Plantain and grasses, whereas Small Scabious, Field Scabious, Great and Common Knapweed and Cowslip could only be collected effectively by hand. The seedbed was prepared by a small tractor using several passes of a small disc harrow followed by a pedestrian-powered belt rake to create 40–50% bare soil which was then manually sown and finally rolled.

More fertile stretches are best stripped of turf 1 m in from the carriageway and field boundaries with turves moved to the hedgerow base. Regrowth can later be sprayed off and then sown at about 5 g/m².

Wildflower seed harvesting (© Dave Vandome).

Hay baling (© Mark Schofield).

Continuity and partnership work

The newly designated roadside LWSs have been mapped on a geographic information system (GIS) for the attention of the local Highways Authority, which now has the opportunity to incorporate sympathetic management into its county-wide cutting schedules. An 'Advised Code of Practice' leaflet has been developed

Delivery of hay to local small holders (© Mark Schofield).

for landowners and work with Lincolnshire County Council is revising guidance documentation for highways maintenance contractors and utilities companies.

A third phase of road verge surveys was scheduled between Lincoln and the Humber during 2015–2016. A follow-on project named 'Lincolnshire's Wildflower Meadow Network' received funding from the Heritage Lottery Fund in the spring of 2014. 'Champion' communities have been identified and assisted in plug planting, seeding and maintaining their public green spaces with wildflowers of local provenance to improve green infrastructure in both rural and urban areas. Since 2009 23 community wildflower meadows totalling over 1 ha in area have so far been created in 20 towns and villages including 11 churchyards, with the help of over 215 community volunteers and wildflower propagators using seed hand-gathered from LWT nature reserves.

Work continues with the East Midlands in Bloom competition to ensure that wildflower-rich road verges and public green spaces are valued in judges' assessments, and local farms and garden centres are helping with the wildflower propagation effort. Partnership working with Caring for God's Acre has been very useful in accessing opportunities to recreate and restore wildflower meadow grassland on favourable soil conditions while reducing management costs. A showcase wildflower meadow is now being created on Lincoln University campus as a project and resource for staff and students.

Lincolnshire County Council has commissioned a feasibility desk study into biomass harvesting from road verges and public green spaces for low carbon energy. It is envisaged that such a programme would offset the operational costs of safety cuts while lowering the carbon footprint and benefiting biodiversity. Rotational forage harvesting, anaerobic digestion, pyrolysis, solid fuel briquetting, biochar, bio-ethanol and compost production are being investigated. In 2015 roadside harvesting equipment from Brittany was demonstrated at Riseholme Campus; initial biomass sampling was undertaken and findings of an initial desk study was published by Peakhill Associates and presented to representatives of local authorities, highways maintenance contractors and biomass businesses. Large-scale harvesting trials and more detailed chemical sampling are planned for 2016. Lincolnshire Wildlife Trust is advising on cutting methods and timescales to ensure vegetation management is beneficial to grassland biodiversity as far as is practicable. It is hoped that by identifying such biomass as a resource, it can be shown that the cutting and clearing of rough grassland and scrub without inputs can make grassland conservation possible on a landscape scale. Harvesting catchments would be designed to include the maximum area of road verges with the highest potential conservation value outside of the most sensitively managed RNRs. Care will be taken not to encourage fertilised cropping of verges. Golf courses, Ministry of Defence bases and marginal arable land may also benefit. Only once fertility and invasive vegetation were under control could green hay from neighbouring RNRs be used to improve species diversity.

Volunteers from The Lincoln Conservation Group enhancing the floral diversity of an important archaeological site at Navenby, adjacent to the Roman road (Ermine Street). The Roman road has been identified as a principal wildlife corridor running from north to south across the limestone area (© Mark Schofield).

Information provided by Mark Schofield (Lincolnshire Wildlife Trust).

References

Ainsworth, A.M., Smith, J.H., Boddy, L., Dentinger, B.T.M., Jordan, M., Parfitt, D., Rogers, H.J. and Skeates, S.J. (2013) *Red List of Fungi for Great Britain: Boletaceae: A pilot conservation assessment based on national database records, fruit body morphology and DNA barcoding.* JNCC Species Status 1. Peterborough: JNCC.

Alday, J.G. and Marrs, R.H. (2014) A simple test for alternative states in ecological restoration: the use of principal response curves. *Applied Vegetation Science* 17 (2): 302–311.

Alexander, K.N.A. (2003) *A review of the invertebrates associated with lowland calcareous grassland.* English Nature research reports, no. 512. Peterborough: English Nature.

Allison, M. and Ausden, M. (2004) Successful use of topsoil and soil amelioration to create heathland vegetation. *Biological Conservation* 120 (2): 221–228.

Ameloot, E., Verheyen, K. and Hermy, M. (2005) Meta-analysis of standing crop reduction by *Rhinanthus* spp. and its effect on vegetation structure. *Folia Geobotanica* 40 (2): 289–310.

Antonsen, H. and Olsson, P.A. (2005) Relative importance of burning, mowing and species translocation in the restoration of a former boreal hayfield: responses of plant diversity and the microbial community *Journal of Applied Ecology* 42 (2): 337–347.

Ash, H.J., Bennett, R. and Scott, R. (1992) *Flowers in the Grass: Creating and managing grasslands with wild flowers.* Peterborough: English Nature.

Asher, J., Warren, M., Fox, R., Harding, P., Jeffcoate, G. and Jeffcoate, S. (2001) *The Millenium Atlas of Butterflies in Britain and Ireland.* Oxford: Oxford University Press.

Auestad, I., Austad, I. and Rydgren, K. (2015) Nature will have its way: local vegetation trumps restoration treatments in semi-natural grassland. *Applied Vegetation Science* 18 (2): 190–196.

Auffret, A.G. (2012) Can seed dispersal by human activity play a useful role for the conservation of European grasslands? *Applied Vegetation Science* 14 (3): 291–303.

Ausden, M., Allinson, M., Bradley, P., Coates, M., Kemp, M.and Phillips, N. (2010) Increasing the resilience of our lowland dry heaths and acid grasslands. *British Wildlife* 22 (2): 101–109.

Ausden, M., Bradbury, R., Brown, A., Eaton, M., Lock, L. and Pearce-Higgins, J. (2015) Climate change and Britain's birdlife: what might we expect? *British Wildlife* 26 (3): 161–174.

Averett, J.M., Klips, R.A., Nave, L.E., Frey, S.D. and Curtis, P.S. (2004) Effects of soil carbon amendment on nitrogen availability and plant growth in an experimental tallgrass prairie restoration. *Restoration Ecology* 12 (4): 568–574.

Babai, D. and Molnár, Z. (2014) Small-scale traditional management of highly species-rich grasslands in the Carpathians. *Agriculture, Ecosystems and Environment* 182: 123–130.

Backshall, J., Manley, J. and Rebane, M. (eds) (2001) *The Upland Management Handbook.* Peterborough: English Nature.

Baer, S.G., Kitchen, D.J., Blair, J.M. and Rice, C.W. (2002) Changes in ecosystem structure and function along a chronosequence of restored grasslands. *Ecological Applications* 12 (6): 1688–1701.

Bakker, E.S. and Olff, H. (2003) Impact of different-sized herbivores on recruitment opportunities for subordinate herbs in grassland. *Journal of Vegetation Science* 14 (4): 465–474.

Bakker, J.P. (1987b) Restoration of species-rich grassland after a period of fertilizer application. In J. van Andel, J.P. Bakker and R.W. Snaydon (eds) *Disturbance in grasslands: Causes, effects, and processes.* Dordrecht: Dr W. Junk Publishers. pp. 185–200.

Bakker, J.P. and Berendse, F. (1999) Constraints in the restoration of ecological diversity in grassland and heathland communities. *Trends in Ecology and Evolution* 14 (2): 63–68.

Bakker, J.P., Elzinga, J.A. and de Vries, Y. (2002) Effects of long-term cutting in a grassland system: perspectives for restoration of plant communities on nutrient-poor soils. *Applied Vegetation Science* 5 (1): 107–120.

Barbaro, L., Dutoit, T. and Cozic, P. (2001) A six-year experimental restoration of biodiversity by shrub-clearing and grazing in calcareous grasslands of the French Prealps. *Biodiversity and Conservation* 10 (1): 119–135.

Bardgett, R.D. and McAlister, E. (1999) The measurement of soil fungal:bacterial biomass ratios as an indicator of ecosystem self-regulation in temperate meadow grasslands. *Biology and Fertility of Soils* 29 (3): 282–290.

Bardgett, R.D., Smith, R.S., Shiel, R.S., Peacock, S., Simkin, J.M., Quirk, H. and Hobbs, P.J. (2006) Parasitic plants indirectly regulate below-ground properties in grassland ecosystems. *Nature* 439 (7079): 969–972.

Bat Conservation Trust. (2014) *The National Bat Monitoring Programme: Annual report 2013.* London: Bat Conservation Trust.

Battersby, J. (ed.) and Tracking Mammals Partnership (TMP). (2005) *UK Mammals: Species, status and population trends. First Report by the Tracking Mammals Partnership.* Peterborough: Joint Nature Conservation Committee on behalf of the Tracking Mammals Partnership.

Beebee, T.J.C. (1994) Amphibian breeding and climate. *Nature* 374: 219–220.

Bekker, R.M., Verweij, G.L., Bakker, J.P. and Fresco, L.F.M. (2000) Soil seed dynamics in hayfield succession. *Journal of Ecology* 88 (4): 594–607.

Bekker, R.M., Verweij, G.L., Smith, R.E.N., Reine, R., Bakker, J.P. and Schneider, S. (1997) Soil seed bank in European grasslands: does land use affect regeneration perspectives? *Journal of Applied Ecology* 34 (5): 1293–1310.

Bennie, J., Hill, M.O., Baxter, R. and Huntley, B. (2006) Influence of slope and aspect on long-term vegetation change in British chalk grasslands. *Journal of Ecology* 94 (2): 355–368.

Bezemer, T.M., Lawson, C.S., Hedlund, K., Edwards, A.R., Brook, A.J., Igual, J.M., Mortimer, S.R. and van der Putten, W.H. (2006) Plant species and functional group effects on abiotic and microbial soil properties and plant-soil feedback responses in two grasslands. *Journal of Ecology* 94 (5): 893–904.

Biesmeijer, J.C., Roberts, S.P.M., Reemer, M., Ohlemuller, R., Edwards, M., Peeters, T., Schaffers, A.P., Potts, S.G., Kleukers, R., Thomas, C.D., Settele, J. and Kunin, W.E. (2006) Parallel declines in pollinators and insect-pollinated plants in Britain and the Netherlands. *Science* 313 (5785): 351–354.

Birks, H.J.B. (1973) *Past and Present Vegetation of the Isle of Skye: A paleoecological study.* Cambridge: Cambridge University Press.

Bisgrove, R. and Dixie, G. (1994) Wild flowers: plugging the gap. *Enact* 2: 18–20.

Blackstock, T.H., Howe, E.A., Stevens, J.P., Burrows, C.R. and Jones, P.S. (2010) *Habitats of Wales: A comprehensive field survey, 1979–1997.* Cardiff: University of Wales Press.

Blakesley, D. and Buckley, P. (2010) *Woodland Creation for Wildlife and People in a Changing Climate: Principles and practice.* Newbury: Pisces Publications.

Bluementhal, D.M., Jordan, N.R. and Russelle, M.P. (2003) Soil carbon addition controls weeds and facilitates prairie restoration. *Ecological Applications* 13 (3): 605–615.

Bobbink, R., Hornung, M. and Roelofs, J.G.M. (1998) The effects of air-borne nitrogen pollutants on species diversity in natural and semi-natural European vegetation. *Journal of Ecology* 86 (5): 717–738.

Bossuyt, B. and Honnay, O. (2008) Can the seed bank be used for ecological restoration? An overview of seed bank characteristics in European communities. *Journal of Vegetation Science* 19 (6): 875–884.

Bourn, N., Thomas, J., Stewart, K. and Clarke, R. (2002) Importance of habitat quality and isolation. Implications for the management of butterflies in fragmented landscapes. *British Wildlife* 13 (6): 398–403.

Brown, A.F. and Grice, P.V. (2005) *Birds in England.* London: T. & A.D. Poyser.

Brown, R. and Robinson, R. (1997) *Bracken Management Handbook. Integrated Bracken Management: A guide to best practice.* Ongar: Rhône-Poulenc Agriculture.

Brown, R.J. (1989) Wild flower seed mixtures: supply and demand in the horticultural industry. In G.P. Buckley (ed.) *Biological Habitat Reconstruction.* London: Belhaven Press. Pp. 201–220.

Buckland, P.C., Buckland, P.I. and Hughes, D. (2005) Palaeoecological evidence for the Vera hypothesis? In: K.H. Hodder, J.M. Bullock. P.C. Buckland and K.J. Kirby (eds) *Large Herbivores in the Wildwood and Modern Naturalistic Grazing Systems.* English Nature research reports, no. 648. Peterborough: English Nature. pp. 62–116.

Bullock, J.M. (1998) Community translocation in Britain: setting objectives and measuring consequences. *Biological Conservation* 84 (3): 199–214.

Bullock, J.M. and Pywell, R.F. (2005) *Rhinanthus*: a tool for restoring diverse grassland? *Folia Geobotanica* 40 (2) 273–288.

Bullock, J.M., Pywell, R.F. and Walker, K.J. (2007) Long-term enhancement of agricultural production by restoration of biodiversity. *Journal of Applied Ecology* 44 (1): 6–12.

Bullock, J.M., Franklin, J., Stevenson, M.J., Silvertown, J., Coulson, S.J., Gregory, S.J. and Tofts, R. (2001). A plant trait analysis of responses to grazing in a long-term experiment. *Journal of Applied Ecology* 38 (2): 253–267.

Bullock, J.M., Jefferson, R.G., Blackstock, T.H., Pakeman, R.J., Emmett, B.A., Pywell, R.J., Grime, J.P. and Silvertown, J. (2011) Semi-natural Grasslands. In *UK National Ecosystem Assessment: Technical report.* Cambridge: UNEP-WCMC. pp. 161–195.

Buri, P., Arlettaz, R. and Humbert, J.-Y. (2013) Delaying mowing and leaving uncut refuges boosts orthopterans in extensively managed meadows: evidence drawn from field-scale experimentation. *Agriculture, Ecosystems and Environment* 181: 22–30.

Burke, M.J.W. and Grime, J.P. (1996) An experimental study of plant community invasibility. *Ecology* 77 (3): 776–790.

Burmeier, S., Eckstein, R.L., Otte, A. and Donath, T.W. (2011) Spatially-restricted plant material application creates colonization initials for flood-meadow restoration. *Biological Conservation* 144 (1): 212–219.

Burnside, N.G., Smith, R.F. and Waite, S. (2003) Recent historical land use change on the South Downs, United Kingdom. *Environmental Conservation* 30 (1): 52–60.

Burton, J.F. and Sparks, T.H. (2002) Flying earlier in the year: the phenological responses of butterflies and moths to climate change. *British Wildlife* 13 (5): 305–311.

Carvell, C., Meek, W.R., Pywell, R.F. and Nowakowski, M. (2004) The response of foraging bumblebees to successional change in newly created arable field margins. *Biological Conservation* 118 (3): 327–339.

Carvell, C., Westrich, P., Meek, W.R., Pywell, R.F, Nowakowski, M. (2006) Assessing the value of annual and perennial forage mixtures for bumblebees by direct observation and pollen analysis. *Apidologie* 37 (3): 326–340.

Catchpole, R. (2011) *A National Climate Change Vulnerability Assessment.* Natural England Technical Information Note TIN095. Peterborough: Natural England.

Centre for Ecology & Hydrology. (2009) *Countryside Survey: England results from 2007.* Swindon: Natural Environment Research Council. p. 119.

Centre for Ecology & Hydrology. (2015) Modular Analysis of Vegetation Information (MAVIS). Lancaster: Lancaster Environment Centre. Accessed at: http://www.ceh.ac.uk/services/modular-analysis-vegetation-information-system-mavis (4 October 2015).

Ceulemans, T., Merckx, R., Hens, M. and Honnay, O. (2013) Plant species loss from European semi-natural grasslands following nutrient enrichment: is it nitrogen or is it phosphorus? *Global Ecology and Biogeography* 22 (1): 73–82.

Chauvat, M., Wolters, V. and Dauber, J. (2007) Response of collembolan communities to land-use change and grassland succession. *Ecography* 30 (2): 183–192.

Chytrý, M., Hejcman, M., Hennekens, S.M. and Schellberg, J. (2009) Changes in vegetation types and Ellenberg indicator values after 65 years of fertilizer application in the Rengen Grassland Experiment, Germany. *Applied Vegetation Science* 12 (2): 167–176.

Clark, J., Darlington, J. and Fairclough, G. (2004) *Using Historic Landscape Characterisation*. London: English Heritage & Lancashire County Council.

Coiffait-Gombault, C., Buisson, E. and Dutoit, T. (2012) Using a two-phase sowing approach in restoration: sowing foundation species to restore, and subordinate species to evaluate restoration success. *Applied Vegetation Science* 15 (2): 277–289.

Collinson, N. and Sparks, T. (2008) Phenology – nature's calendar: an overview of results from the UK phenology network. *Arboricultural Journal* 30 (4): 271–278.

Conrad, M.K. and Tischew, S. (2011) Grassland restoration in practice: do we achieve the targets? A case study from Saxony-Anhalt/Germany. *Ecological Engineering* 37 (8): 1149–1157.

Cosyns, E., Claerbout, S., Lamoot, I. and Hoffmann, M. (2005) Endozoochorous seed dispersal by cattle and horse in a spatially heterogeneous European landscape. *Plant Ecology* 178 (2): 149–162.

Cotton, P.A. (2003) Avian migration phenology and global climate change. *Proceedings of the National Academy of Sciences of the United States of America* 100 (21): 12219–12222.

Coulson, S.J., Bullock, J.M., Stevenson, M.J. and Pywell, R.F. (2001) Colonisation of grassland by sown species: dispersal versus microsite limitation in responses to management. *Journal of Applied Ecology* 38 (1): 204–216.

Couvreur, M., Christiaen, B., Verheyen, K. and Hermy, M. (2004) Large herbivores as mobile links between isolated nature reserves through adhesive seed dispersal. *Applied Vegetation Science* 7 (2): 229–236.

Critchley, C.N.R., Burke, M.J.W. and Stevens, D.P. (2004) Conservation of lowland semi-natural grasslands in the UK: A review of botanical monitoring results from agri-environment schemes. *Biological Conservation* 115: 263–278.

Critchley, C.N.R., Chambers, B.J., Fowbert, J.A., Bhogal, A., Rose, S.C. and Sanderson, R.A. (2002a) Plant species richness, functional type and soil properties of grasslands and allied vegetation in English Environmentally Sensitive Areas. *Grass & Forage Science* 57 (2): 82–92.

Critchley, C.N.R., Chambers, B.J., Fowbert, J.A., Sanderson, R.A., Bhogal, A. and Rose, S.C. (2002b) Association between lowland grassland plant communities and soil properties. *Biological Conservation* 105 (2): 199–215.

Crofts, A. and Jefferson, R.G. (eds) (1999) *The Lowland Grassland Management Handbook*, 2nd edn. Peterborough: English Nature/Wildlife Trusts.

Davies A. and Waite S. (1998) The persistence of calcareous grassland species in the soil seed bank under developing and established scrub. *Plant Ecology* 136 (1): 27–39.

Davies, A., Dunnett, N.P. and Kendle, A. (1999) The importance of transplant size and gap width in the botanical enrichment of species-poor grasslands in Britain. *Restoration Ecology* 7 (3): 271–280.

Davies, Z.G., Wilson, R.J., Coles, S. and Thomas, C.D. (2006) Changing habitat associations of a thermally constrained species, the silver-spotted skipper butterfly, in response to climate warming. *Journal of Animal Ecology* 75 (1): 247–256.

Davis, M.A., Grime, J.P. and Thompson, K. (2000) Fluctuating resources in plant communities: a general theory of invasibility. *Journal of Ecology* 88 (3): 528–534.

Day, J., Symes, N. and Robertson, P. (2003) *The Scrub Management Handbook: Guidance on the management of scrub on nature conservation sites*. Wetherby: English Nature.

Defra. (2001) *Traditional Farming in the Modern Environment*. London: Defra.

Defra. (2005) *Guidance on the Disposal Options for Common Ragwort*. London: Defra.

Defra. (2006a) *Pesticides: Code of practice for using plant protection products*. London: Defra.

Defra. (2006b) *Sustainable Management Strategies for Creeping Thistle*. Research Project BD1449. London: Defra.

Defra. (2004) *Code of Practice on How to Prevent the Spread of Ragwort*. London: Defra.

Defra. (2007) *Guidance Note on the Methods That Can Be Used to Control Harmful Weeds*. London: Defra.

Defra. (2009) *Protecting Our Water, Soil and Air: A code of good agricultural practice for farmers, growers and land managers.* Norwich: The Stationery Office.

Defra. (2010) *Fertilizer Manual (RB209)*, 8th edn. Norwich: The Stationery Office.

Defra. (2011) *Biodiversity 2020: A strategy for England's wildlife and ecosystem services.* London, Defra.

Defra and Natural England. (2008) *Environmental Stewardship: Review of progress.* London: Defra.

Department of Transport. (1993) The wildflower handbook. In *Design Manual for Roads and Bridges, Vol 10, Section 3, Part 1 HA 67/93.* London: The Stationery Office. pp. 1–126.

Diaz, A., Green, I. and Evans, D. (2011) Heathland restoration techniques: ecological consequences for plant-soil and plant-animal interactions. *ISRN Ecology* 1–8. DOI:10.5402/2011/961807.

Dodd, M.E., Silvertown, J., McConway, K., Potts, J. and Crawley, M. (1994a) Application of the British National Vegetation Classification to the communities of the Park Grass Experiment through time. *Folia Geobotanica et Phytotaxonomica* 29 (3): 321–334.

Dodd, M.E., Silvertown, J., McConway, K., Potts, J. and Crawley, M. (1994b) Stability in the plant communities of the Park Grass Experiment: the relationships between species richness, soil pH and biomass variability. *Philosophical Transactions of the Royal Society B: Biological Sciences* 346 (1316): 185–193.

Donath, T.W. and Eckstein, R.L. (2012) Litter effects on seedling establishment interact with seed position and earthworm activity. *Plant Biology (Stuttgart, Germany)* 14 (1): 163–170.

Donath, T.W., Bissels, S. and Otte, A. (2007) Large-scale application of diaspore transfer with plant material in restoration practice: impact of seed and microsite limitation. *Biological Conservation* 138 (1–2): 224–234.

Donath, T.W., Hölzel, N. and Otte, A. (2006) Influence of competition by sown grass, disturbance and litter on recruitment of rare flood-meadow species. *Biological Conservation* 130 (3): 315–323.

Dumont, B., Carrère, P., Ginane, C., Farruggia, A., Lanore, L., Tardif, A., Decuq, F., Darsonville, O. and Louault, F. (2011) Plant–herbivore interactions affect the initial direction of community changes in an ecosystem manipulation experiment. *Basic and Applied Ecology* 12 (3): 187–194.

Dunsford, S.J., Free, A.J. and Davy, A.J. (1998) Acidifying peat as an aid to the reconstruction of lowland heath on arable land: a field experiment. *Journal of Applied Ecology* 35 (5): 660–672.

Duranel, A.J., Acreman, M.C., Stratford, C.J.,Thompson, J.R. and Mould, D.J. (2007) Assessing the hydrological suitability of floodplains for species-rich meadow restoration: a case study of the Thames floodplain, UK. *Hydrology and Earth System Sciences* 11 (1): 170–179.

Eaton, M.A., Aebischer, N.J., Brown, A.F., Hearn, R.D., Lock, L., Musgrove, A.J., Noble, D.G., Stroud, D.J. and Gregory, R.D. (2015) Birds of Conservation Concern 4: the population status of birds in the UK, Channel Islands and Isle of Man. *British Birds* 108: 708–746.

Edgar, P., Foster, J. and Baker, J. (2010) *Reptile Habitat Management Handbook.* Bournemouth: Amphibian and Reptile Conservation.

Edwards, A.R. and Younger, A. (2006) The dispersal of traditionally managed hay meadow plants by farmyard manure application. *Seed Science Research* 16 (2): 137–147.

Edwards, A.R., Mortimer, S.R., Lawson, C.S., Westbury, D.B., Harris, S.J., Woodcock, B.A. and Brown, V.K. (2007) Hay strewing, brush harvesting of seed and soil disturbance as tools for the enhancement of botanical diversity in grasslands. *Biological Conservation* 134 (3): 372–382.

Ellenberg, H. (1988) *Vegetation Ecology of Central Europe*, 4th edn. Cambridge: Cambridge University Press.

English Heritage. (2004) *Farming the Historic Landscape: Caring for archaeological sites in grassland.* London: English Heritage.

English Nature. (2003) *The Herbicide Handbook: Guidance on the use of herbicides on nature conservation sites.* Wetherby: English Nature in association with FACT.

English Nature. (2004) *Reptiles: Guidelines for developers.* Peterborough: English Nature.

English Nature. (2005) *The Importance of Livestock Grazing for Wildlife Conservation.* Peterborough: English Nature.

Eriksson, A. (2001) Arbuscular mycorrhiza in relation to management history, soil nutrients and plant species diversity. *Plant Ecology* 155 (2): 129–137.

Eschen, R. Mortimer, S.R., Lawson, C.S., Edwards, A.R., Brook, A.J., Iqual, J.M., Hedlund, K. and Schaffner, U. (2007) Carbon addition alters vegetation composition on ex-arable fields. *Journal of Applied Ecology* 44 (1): 95–104.

Eschen, R., Müller-Schärer, H. and Schaffner, U. (2006) Soil carbon addition affects plant growth in a species-specific way. *Journal of Applied Ecology* 43 (1): 35–42.

European Environment Agency. (2013) *The European Grassland Butterfly Indicator: 1990–2011.* Copenhagen: European Environment Agency.

Evans, S. (2003) *Waxcap-grasslands: An assessment of English sites.* English Nature research reports, no. 555. Peterborough: English Nature.

Evans, S.E., Henrici, A. and Ing, B. (2006) *Preliminary Assessment: The Red Data List of threatened British fungi.* Manchester: British Mycological Society.

Fagan, K.C., Pywell, R.F., Bullock, J.M. and Marrs, R.H. (2008) Do restored calcareous grasslands on former arable fields resemble ancient targets? The effect of time, methods and environment on outcomes. *Journal of Applied Ecology* 45 (4): 1293–1303.

Field, C.B., Barros, V.R. and Intergovernmental Panel on Climate Change, Working Group II. (2014) *Climate Change 2014: Impacts, adaptation and vulnerability.* Working Group II contribution to the fifth assessment report of the Intergovernmental Panel on Climate Change. New York: Cambridge University Press.

Fischer, S.F., Poschlod, P. and Beinlich, B. (1996) Experimental studies on the dispersal of plants and animals on sheep in calcareous grasslands. *Journal of Applied Ecology* 33 (5): 1206–1222.

Fittter, A.H. and Fitter, R.S. (2002) Rapid changes in flowering time in British plants. *Science* 296 (5573): 1689–1691.

Foster, B.L., Murphy, C.A., Keller, K.R., Aschenbach, T.A., Questad, E.J. and Kindscher, K. (2007) Restoration of prairie community structure and ecosystem function in an abandoned hayfield: a sowing experiment. *Restoration Ecology* 15 (4): 652–661.

Fox, R., Warren, M.S., Brereton, T.M. (2010) *A new Red List of British butterflies.* Species Status. Peterborough: Joint Nature Conservation Committee. pp. 1–32.

Fox, R., Asher, J., Brereton, T., Roy, D. and Warren, M. (2006) *The State of Butterflies in Britain and Ireland.* Newbury: Pisces Publications for Butterfly Conservation and the Centre for Ecology and Hydrology.

Fox, R., Brereton, T.M., Asher, J., Botham, M.S., Middlebrook, I., Roy, D.B. and Warren, M.S. (2011) *The State of the UK's Butterflies 2011.* Wareham: Butterfly Conservation and the Centre for Ecology and Hydrology.

Fox, R., Warren, M.S., Asher, J., Brereton, T.H. and Roy, D.B. (2007) *The State of Britain's Butterflies 2007.* Wareham: Butterfly Conservation and the Centre for Ecology and Hydrology.

Fox, R., Brereton, T.M., Asher, J., August, T.A., Botham, M.S., Bourn, N.A.D., Cruickshanks, K.L., Bulman, C.R., Ellis, S., Harrower, C.A., Middlebrook, I., Noble, D.G., Powney, G.D., Randle, Z., Warren, M.S. and Roy, D.B. (2015) *The State of the UK's Butterflies 2015.* Wareham: Butterfly Conservation and the Centre for Ecology and Hydrology.

Franco, A.M.A., Hill, J.K., Kitschke, C., Collingham, Y.C., Roy, D.B., Fox, R., Huntley, B. and Thomas C.D. (2006) Impacts of climate warming and habitat loss on extinctions at species' low-altitude range boundaries. *Global Change Biology* 12 (8): 1545–1553.

Fuller, R.J. (1982) *Bird Habitats in Britain.* London: T. & A.D. Poyser.

Fuller, R.J. (1996) *Relationships between grazing and birds with particular reference to sheep in the British Uplands.* BTO research report, no. 164. Thetford: British Trust for Ornithology.

Fuller, R.J. (2012) Avian responses to transitional habitats in temperate cultural landscapes: woodland edges and young-growth. In: R.J. Fuller (ed.) *Birds and Habitat: Relationships in changing landscapes*. Cambridge: Cambridge University Press. pp. 125–149.

Fuller, R.M. (1987) The changing extent and conservation interest of lowland grasslands in England and Wales: a review of grassland surveys 1930–1984. *Biological Conservation* 40: 281–300.

Gardiner, T. and Hill, J. (2006) Mortality of Orthoptera caused by mechanised mowing of grassland. *British Journal of Entomology and Natural History* 19 (1): 38–40.

Gent, T. and Gibson, S. (2012) *Herpetofauna Workers' Manual*. Exeter: Pelagic Publishing.

Gibson, C.C. and Watkinson, A.R. (1989) The host range and selectivity of a parasitic plant: *Rhinanthus minor* L. *Oecologia* 78 (3): 401–406.

Gibson, C.W.D. and Brown, V.K. (1991) The nature and rate of development of calcareous grassland in southern Britain. *Biological Conservation* 58 (3): 297–316.

Gibson, C.W.D. and Brown, V.K. (1992) Grazing and vegetation change: deflected or modified succession? *Journal of Applied Ecology* 29 (1): 120–131.

Gilbert, J., Gowing, D. and Wallace, H. (2009) Available soil phosphorus in semi-natural grasslands: assessment methods and community tolerances. *Biological Conservation* 142 (5): 1074–1083.

Gilbert, J.C., Gowing, D.J.G. and Bullock, R.J. (2003) Influence of seed mixture and hydrological regime on the establishment of a diverse grassland sward at a site with high phosphorus availability. *Restoration Ecology* 11 (4): 424–435.

Gilbert, O. (2000) *Lichens*. The New Naturalist Library. London: HarperCollins Publishers.

Gordon, H., Haygarth, P.M. and Bardgett, R.D. (2008) Drying and rewetting effects on soil microbial community composition and nutrient leaching. *Soil Biology and Biochemistry* 40 (2): 302–311.

Gormsen, D., Hedlund, K., Korthals, G.W., Mortimer, S.R., Pizl, V., Smilauerova, M. and Sugg, E. (2004) Management of plant communities on set-aside land and its effects on earthworm communities. *European Journal of Soil Biology* 40 (3–4): 123–128.

Gough, M.W. and Marrs, R.H. (1990) A comparison of soil fertility between semi-natural and agricultural plant communities: implications for the creation of species-rich grassland on abandoned agricultural land. *Biological Conservation* 51 (2): 83–96.

Goulson, D., Hanley, M.E., Darvill, B., Ellis, J.S. and Knight, M.E. (2005) Causes of rarity in bumblebees. *Biological Conservation* 122 (1): 1–8.

Gowing, D.J.G., Lawson, C.S., Youngs, E.G., Barber, K.R., Rodwell, J.S., Prosser, M.V., Wallace, H.L., Mountford, J.O. and Spoor, G. (2002) *The Water Regime Requirements and the Response to Hydrological Change of Grassland Plant Communities*. Final report to the Department for Environment, Food and Rural Affairs. Bedford: Institute of Water and Environment.

Graham, D.J. and Hutchings, M.J. (1988) A field investigation of germination from the seed bank of a chalk grassland ley on former arable land. *Journal of Applied Ecology* 25 (1): 253–263.

Grazing Animals Project. (2001a) *The Breed Profiles Handbook: A guide to the selection of livestock breeds for grazing wildlife sites*. Accessed at: http://www.grazinganimalsproject.org.uk/breed_profiles_handbook.html (7 October 2015).

Grazing Animals Project. (2001b) *A Guide to Animal Welfare in Nature Conservation Grazing*. Available at: http://www.knepp.co.uk/Other_docs/interesting_articles_papers/Livestock/a_guide_to_animal_welfare_in_nature_conservation_grazing[1].pdf (7 October 2015).

Grazing Animals Project. (2007a) *Animal Health Plans*. GAP Information Leaflet 6. Available at: http://www.grazinganimalsproject.org.uk/animal_welfare.html (7 October 2015).

Grazing Animals Project. (2007b) *Grazing Stock On Sites With Public Access*. GAP Information Leaflet. Available at: http://www.grazinganimalsproject.org.uk/stock_management.html (7 October 2015).

Grazing Animals Project. (2007c) *Cattle Handling Facilities*. GAP Information Leaflet. Available at: http://www.grazinganimalsproject.org.uk/stock_management.html (7 October 2015)

Grazing Animals Project. (2007d) *Equine Handling Facilities*. GAP Information Leaflet. Available at: http://www.grazinganimalsproject.org.uk/stock_management.html (7 October 2015).

Grazing Animals Project. (2007e) *Dogs and Grazing*. GAP Information Leaflet. Available at: http://www.grazinganimalsproject.org.uk/animal_welfare.html (7 October 2015).

Griffith, G.W., Bratton, J.H. and Easton, G. (2004) Charismatic megafungi: the conservation of waxcap grasslands. *British Wildlife* 16 (1): 31–43.

Griffith, G.W., Roderick, K., Graham, A. and Causton, D.R. (2012) Sward management influences fruiting of grassland basidiomycete fungi. *Biological Conservation* 145 (1): 234–240.

Grime, J.P. (1979) *Plant Strategies and Vegetation Processes*. Chichester: John Wiley and Sons.

Grime, J.P., Fridley, J.D., Askew, A.P., Thompson, K., Hodgson, J.G. and Bennett, C.R. (2008) Long-term resistance to simulated climate change in an infertile grassland. *Proceedings of the National Academy of Sciences of the United States of America* 105 (29): 1028–1032.

Hanley, M.E. and Fenner, M. (1997) Effects of molluscicides on seedlings of four grassland plant species. *Journal of Applied Ecology* 34 (6): 1479–1483.

Hanley, M.E., Fenner, M. and Edwards, P.J. (1996) The effect of mollusc grazing on seedling recruitment in artificially created grassland gaps. *Oecologica* 106 (2): 240–246.

Harris, S., Morris, P., Wray, S. and Yalden, D.W. (1995) *A Review of British Mammals: Population estimates and conservation status of British mammals other than cetaceans*. Peterborough: Joint Nature Conservation Committee.

Haslgrübler, P., Krautzer, B., Blaschka, A., Graiss W. and Pötsch E. M. (2014) Quality and germination capacity of seed material harvested from an Arrhenatherion meadow. *Grass and Forage Science* 69 (3): 454–461.

Hayes, G.F. and Holl, K.D. (2011) Manipulating disturbance regimes and seeding to restore mesic Mediterranean grasslands. *Applied Vegetation Science* 14 (3): 304–315.

Hayes, M.J. and Sackville Hamilton, N.R. (2001) *The Effect of Sward Management on the Restoration of Species-Rich Grassland: A reassessment of IGER's lowland grassland restoration experiment, Trawsgoed*. Contract Science Report, no. 438. Bangor: Countryside Council for Wales.

Hector, A., Schmid, B., Beierkuhnlein, C., Caldeira, M.C., Diemer, M., Dimitrakopoulos, P.G., Finn, J.A., Freitas, H., Giller, P.S., Good, J., Harris, R., Högberg, P., Huss-Danell, K., Joshi, J., Jumpponen, A., Körner, C., Leadley, P.W., Loreau, M., Minns, A., Mulder, C.P.H., O'Donovan, G., Otway, S.J., Pereira, J.S., Prinz, A., Read, D.J., Scherer-Lorenzen, M., Schulze, E.-D., Siamantziouras, A.-S.D., Spehn, E.M., Terry, A.C., Troumbis, A.Y., Woodward, F.I., Yachi, S. and Lawton, J.H. (1999) Plant diversity and productivity experiments in european grasslands. *Science* 286 (5442): 1123–1127.

Hedberg, P. and Kotowski, W. (2010) New nature by sowing? The current state of species introduction in grassland restoration, and the road ahead. *Journal for Nature Conservation* 18 (4): 304–308.

Hegland, S.J, Nielsen, A., Lázaro, A., Bjerknes, A-L. and Totland, Ø. (2009) How does climate warming affect plant-pollinator interactions. *Ecology Letters* 12 (2): 184–195.

Hejcman, M., Schellberg, J. and Pavlů, V. (2010) Long-term effects of cutting frequency and liming on soil chemical properties, biomass production and plant species composition of *Lolio-Cynosuretum* grassland after the cessation of fertilizer application. *Applied Vegetation Science* 13 (3): 257–269.

Hewins, E.J., Pinches, C. and Cooke, A.I. (2012) Creation of species-rich grassland: evidence for the effectiveness of Environmental Stewardship. *Aspects of Applied Biology* 115, 89–96.

Hewins, E. (2013) *A survey of selected agri-environment grassland creation and restoration sites*. Part 1 – 2010 survey. Natural England Commissioned Report, no. NECR107. Available at: http://publications.naturalengland.org.uk/publication/4538148 (7 October 2015).

Hewins, E.J., Pinches, C., Arnold, J., Lush, M., Robertson, H. and Escott, S. (2005) *The condition *of lowland BAP priority grasslands: Results from a sample survey of non-statutory stands*. English Nature Research Reports, no. 636. Peterborough: English Nature.

Hickling, R., Roy, D.B., Hill, J.K., Fox, R. and Thomas, C.D. (2006) The distributions of a wide range of taxonomic groups are expanding polewards. *Global Change Biology* 12 (): 450–455.

Higgins, S.I., Nathan, R. and Cain, M.L. (2003) Are long-distance dispersal events in plants usually caused by nonstandard means of dispersal? *Ecology* 84 (8): 1945–1956.

Hill, D., Fasham M., Tucker G., Shewry, M. and Shaw, P. (eds) (2005) *Handbook of Biodiversity Methods: Survey, evaluation and monitoring*. Cambridge: Cambirdge University Press.

Hill, M.O. (1996) *TABLEFIT Version 1.0: For identification of vegetation types*. Program Manual. Huntingdon: Institute of Terrestrial Ecology.

Hill, M.O., Evans, D.F. and Bell, S.A. (1992) Long term effects of excluding sheep from hill pastures in North Wales. *Journal of Applied Ecology* 80 (1): 1–13.

Hill, M.O., Mountford, J.O., Roy, D.B. and Bunce, R.G.H. (1999) *Ellenberg's Indicator Values for British Plants*. ECOFACT Volume 2 Technical Annex. Huntingdon: Institute of Terrestrial Ecology.

Hirst, R.A., Pywell, R.F., Marrs, R.H. and Putwain, P.D. (2005) The resilience of calcareous and mesotrophic grasslands following disturbance. *Journal of Applied Ecology* 42 (3): 498–506.

Hodgson, J.G. (1989) Selecting and managing plant materials used in habitat construction. In G. P. Buckley (ed.) *Biological Habitat Reconstruction*. London: Belhaven Press.

Hofmann, M. and Isselstein, J. (2004) Seedling recruitment on agriculturally improved mesic grassland: the influence of disturbance and management schemes. *Applied Vegetation Science* 7 (2): 193–200.

Hölzel, N. and Otte, A. (2003) Restoration of a species-rich flood meadow by topsoil removal and diaspore transfer with plant material. *Applied Vegetation Science* 6 (2): 131–140.

Hölzel, N. and Otte, A. (2004) Assessing soil seed bank persistence in flood-meadows: the search for reliable traits. *Journal of Vegetation Science* 15 (1): 93–100.

Hooper, D. U., Chapin, F.S., Ewel, J.J., Hector, A., Inchausti, P., Lavorel, S., Lawton. J.H., Lodge, D.M., Loreau, M., Naeem, S., Schmid, B., Setälä, H., Symstad, A.J., Vandermeer, J. and Wardle, D.A. (2005) Effect of biodiversity on ecosystem functioning: a consensus of current knowledge. *Ecological Monographs* 75 (1): 3–35.

Hopkins, A., Pywell, R.F., Peel, S., Johnson, R.H. and Bowling, P.J. (1999) Enhancement of botanical diversity of permanent grassland and impact on hay production in Environmentally Sensitive Areas in the UK. *Grass and Forage Science* 54 (2): 163–173.

Hopkins, J. (2003) Some aspects of geology and the British flora. *British Wildlife* 14 (3): 186–194.

Hopkins, J. (2007) British wildlife and climate change 2: adapting to climate change. *British Wildlife* 18(6): 381–387.

Hopkins, J.J, Allison, H.M., Walmsley, C.A., Gaywood, M. and Thurgate, G. (2007). *Conserving Biodiversity in a Changing Climate: Guidance on building capacity to adapt*. Published by Defra on behalf of the UK Biodiversity Partnership. London: Defra.

Humbert, J-Y., Ghazoul, J. and Walter, T. (2009) Meadow harvesting techniques and their impacts on field fauna. *Agriculture, Ecosystems & Environment* 130 (1–2): 1–8.

Humphrey, J.W. and Patterson, G.S. (2000) Effects of late summer grazing on the diversity of riparian pasture vegetation in an upland conifer forest. *Journal of Applied Ecology* 37 (6): 986–996.

Huntley, B., Green, R.E., Collingham, Y.C. and Willis, S.G. (2007) *A Climatic Atlas of European Breeding Birds*. Barcelona: Lynx.

Hutchings, M.J. and Booth, K.D. (1996a) Studies on the feasibility of re-creating chalk grassland vegetation on ex-arable land. I. The potential roles of the seed bank and seed rain. *Journal of Applied Ecology* 33 (5): 1171–1181.

Hutchings, M.J. and Booth, K.D. (1996b) Studies of the feasibility of re-creating chalk grassland vegetation on ex-arable land. II. Germination and early survivorship of seedlings under different management regimes. *Journal of Applied Ecology* 33 (5): 1182–1190.

Hutchings, M.J. and Stewart, A.J.A. (2002). Calcareous grasslands. In: M.R. Perrow and A.J.

Davy (eds) *Handbook of Ecological Restoration*. Volume 2: Restoration in Practice. Cambridge: Cambridge University Press. pp. 419–422.

Hutson, A.M. (1993) *Action Plan for the Conservation of Bats in the United Kingdom*. London: Bat Conservation Trust.

Institute of Environmental Assessment. (1995) *Guidelines for Baseline Ecological Assessment*. Institute of Environmental Assessment. London: E & F Spon.

Isselstein, J., Tallowin, J.R.B. and Smith, R.E.N. (2002) Factors affecting seed germination and seedling establishment of fen-meadow species. *Restoration Ecology* 10 (2): 173–184.

Jacquemyn, H., Brys, R. and Hermy, M. (2003) Short-term effects of different management regimes on the response of calcareous grassland vegetation to increased nitrogen. *Biological Conservation* 111 (2): 137–147.

Janssens, F., Peeters, A., Tallowin, J.R.B., Bakker, J.P., Bekker, R.M., Fillat, F. and Oomes, M.J.M. (1998) Relationship between soil chemical factors and grassland diversity. *Plant and Soil* 202 (1): 69–78.

Jaunatre, R., Buisson, E. and Dutoit, T. (2014) Can ecological engineering restore Mediterranean rangeland afterintensive cultivation? A large-scale experiment in southern France. *Ecological Engineering* 64: 202–212.

Johanidesová, E., Fajmon, K., Jongepierová, I. and Prach, K. (2014) Spontaneous colonization of restored dry grasslands by target species: restoration proceeds beyond sowing regional seed mixtures. *Grass and Forage Science*. DOI: 10.1111/gfs.12144.

Johnston, A.E. and Poulton, P.R. (1977) *Yields on the Exhaustion Land and Changes in the NPK Content of the Soils Due to Cropping and Manuring, 1852–1975*. Rothamsted Experimental Station Report for 1976 Part 2. pp. 53–86.

Joint Nature Conservation Committee. (2004) *Common Standards Monitoring Guidance for Birds*. Peterborough: Joint Nature Conservation Committe.

Jones, A.T. and Hayes, M.J. (1999) Increasing floristic diversity in grassland: the effects of management regime and provenance on species introduction. *Biological Conservation* 87 (3): 381–390.

Jones, G.H., Trueman, I.C. and Millett, P. (1995) The use of hay strewing to create species-rich grasslands. I. General principles and hay strewing versus seed mixes. *Land Contamination & Reclamation* 2: 104–110.

Jongepierová, I., Mitchley, J. and Tzanopoulos, J. (2007) A field experiment to recreate species rich hay meadows using regional seed mixtures. *Biological Conservation* 139 (3–4): 297–305

Kahmen, S., Poschlod, P. and Schreiber, K.-F. (2002) Conservation management of calcareous grasslands. Changes in plant species composition and response of functional traits during 25 years. *Biological Conservation* 104 (3): 319–328.

Kalamees, R. and Zobel, M. (1997) The seed bank in an Estonian calcareous grassland: comparison of different successional stages. *Folia Geobotanica* 32 (1): 1–14.

Kardol, P., Bezemer, T.M. and Van der Putten, W.H. (2009) Soil organism and plant introductions in restoration of species-rich grassland communities. *Restoration Ecology* 17 (2): 258–269.

Kardol, P., Bezemer, T.M., Van der Wal, A. and Van der Putten, W.H. (2005) Successional trajectories of soil nematode and plant communities in a chronosequence of ex-arable lands. *Biological Conservation* 126 (3): 317–327.

Kardol, P., Van der Wal, A., Bezemer, T.M., de Boer, W., Duyts, H., Holtkamp, R. and Van der Putten, W.H. (2008) Restoration of species-rich grasslands on ex-arable land: seed addition outweighs soil fertility reduction. *Biological Conservation* 141 (9): 2208–2217.

Kelemen, A., Török, P., Valkó, O., Deák, B., Miglécz, T., Tóth, K., Ölvedi, T. and Tóthmérész, B. (2014) Sustaining recovered grasslands is not likely without proper management: vegetation changes after cessation of mowing. *Biodiversity and Conservation* 23 (3): 741–751.

Kiehl, K. and Pfadenhauer, J. (2007) Establishment and persistence of target species in newly created calcareous grasslands on former arable fields. *Plant Ecology* 189 (1): 31–48.

Kiehl, K. and Wagner, C. (2006) Effect of hay transfer on long-term establishment of vegetation and grasshoppers on former arable fields. *Restoration Ecology* 14 (1): 157–166.

Kiehl, K., Thormann, A. and Pfadenhauer, J. (2006) Evaluation of initial restoration measures during the restoration of calcareous grasslands on former arable fields. *Restoration Ecology* 14 (1): 148–156.

Kiehl, K., Kirmer, A., Donath, T.W., Rasran, L. and Hölzel, N. (2010) Species introduction in restoration projects: evaluation of different techniques for the establishment of semi-natural grasslands in Central and Northwestern Europe. *Basic and Applied Ecology* 11 (4): 285–299.

Kirby, K.J. (2005) Was the wildwood closed forest or savannah and does it matter for modern conservation: some conclusions. In K.H. Hodder, J.M. Bullock, P.C. Buckland and K.J. Kirby (eds) *Large Herbivores in the Wildwood and in Modern Naturalistic Grazing Systems.* English Nature research reports, no. 648. Peterborough: English Nature. pp. 169–177.

Kirby, P. (2001) *Habitat Management for Invertebrates: A practical handbook.* Sandy: Royal Society for the Protection of Birds.

Kirkham, F.W. (2006) *The Potential Effects of Nutrient Enrichment in Semi-Natural Lowland Grasslands Through Mixed Habitat Grazing or Supplementary Feeding.* Commissioned report, no. 192. Edinburgh: Scottish Natural Heritage.

Kirkham, F.W. and Tallowin, J.R.B. (1995) The influence of cutting date and previous fertiliser treatment on the productivity and botanical composition of species-rich hay meadows on the Somerset Levels. *Grass and Forage Science* 50 (4): 365–377.

Kirkham, F.W., Tallowin, J.R.B., Dunn, R.M., Bhogal, A., Chambers, B.J. and Bardgett, R.D. (2014) Ecologically-sustainable fertility management for the maintenance of species-rich hay meadows: a 12 year fertilizer and lime experiment. *Journal of Applied Ecology* 51 (1): 152–161.

Kirkham, F.W., Tallowin, J.R.B., Sanderson, R.A., Bhogal, A., Chambers, B.J. and Stevens, D.P. (2008) The impact of organic and inorganic fertilizers and lime on the species-richness and plant functional characteristics of hay meadow communities. *Biological Conservation* 141 (5): 1411–1427.

Kirmer, A. and Tischew, S. (2010) The EU-SALVERE Project: producing native seeds using threshing material and species-rich hay from grasslands. Proceedings of SER2010, the Seventh European Conference on Ecological Restoration, 23–27 August 2010, Avignon, France. Washington: Society for Ecological Restoration. pp. 1–4.

Kirmer, A., Baasch, A. and Tischew, S. (2012) Sowing of low and high diversity seed mixtures in ecological restoration of surface mined-land. *Applied Vegetation Science* 15 (2): 198–207.

Kirmer, A., Mann, S., Stolle, M., Tischew, S. and Kiehl, K. (2009) Near-natural restoration methods for high nature value areas. SALVERE – Regional Workshop, Poznań University of Life Sciences, Department of Grassland Sciences, Poland. pp. 21–28.

Kleijn, D. (2003) Can establishment characteristics explain the poor colonisation success of late successional species on ex-arable land? *Restoration Ecology* 11 (2): 131–138.

Kleijn, D. and Sutherland, W.J. (2003) How effective are European agri-environment schemes in conserving and promoting biodiversity? *Journal of Applied Ecology* 40 (6): 947–969.

Klimkowska, A., Van Diggelen, R., Bakker, J.P. and Grootjans, A.P. (2007) Wet meadow restoration in Western Europe: a quantitative assessment of the effectiveness of several techniques. *Biological Conservation* 140 (3–4): 318–328.

Krautzer B., Bartel A., Kirmer A., Tischew S., Feucht B., Wieden M., Haslgrübler P., Rieger E. and Pötsch E.M. (2011) Establishment and use of high nature value farmland. In E.M. Pötsch, B. Krautzer and A Hopkins (eds) *Grassland Farming and Land Management Systems in Mountainous Regions.* EGF 2011 Symposium, Agricultural Research and Education Centre Raumberg-Gumpenstein, 29–31 August 2011, Austria. pp 457–469.

Kruess, A. and Tscharntke, T. (2002) Contrasting responses of plant and insect diversity to variation in grazing intensity. *Biological Conservation* 106 (3): 293–302.

Lawson, C.S., Ford, M.A. and Mitchley, J. (2004) The influence of seed addition and cutting regime on the success of grassland restoration on former arable land. *Applied Vegetation Science* 7 (2): 259–266.

Lawton, J.H., Brotherton, P.N.M., Brown, V.K., Elphick, C., Fitter, A.H., Forshaw, J., Haddow, R.W., Hilborne, S., Leafe, R.N., Mace, G.M., Southgate, M.P., Sutherland, W.J., Tew, T.E., Varley, J. and Wynne, G.R. (2010) *Making Space for Nature: A review of England's wildlife sites and ecological network*. Report to Defra. London: Defra.

Leng, X., Musters, C.J.M. and de Snoo, G.R. (2011) Effects of mowing date on the opportunities of seed dispersal of ditch bank plant species under different management regimes. *Journal for Nature Conservation* 19 (3): 166–174.

Lengyel, S., Varga, K., Kosztyi, B., Lontay. L., Déri, E., Török, P. and Tóthmérész, B. (2012) Grassland restoration to conserve landscape-level biodiversity: a synthesis of early results and experiences from a large-scale project. *Applied Vegetation Science* 15 (2): 264–276.

Lepš, J., Doležal, J., Bezemer, T.M., Brown, V.K., Hedlund, K., Igual Arroyo, M., Jörgensen, H.B., Lawson, C.S., Mortimer, S.R., Peix Geldart, A., Rodríguez Barrueco, C., Santa Regina, I., Šmilauer, P. and van der Putten, W.H. (2007) Long-term effectiveness of sowing high and low diversity seed mixtures to enhance plant community development on ex-arable fields. *Applied Vegetation Science* 10 (1): 97–110.

Lewis, R.J., Pakeman, R.J., Angus, S. and Marrs, R.H. (2014) Using compositional and functional indicators for biodiversity conservation monitoring of semi-natural grasslands in Scotland. *Biological Conservation* 175: 82–93.

Lewis, S.L., Brando, P.M., Phillips, O.L., van der Heijden, G.M.F. and Nepstad, D. (2011) The 2010 Amazon Drought. *Science* 331 (6017): 554.

Loreau, M., Naeem S., Inchausti, P., Bengtsson, J., Grime, J.P., Hector, A., Hooper, D.U., Huston, M.A., Raffaelli, D., Schmid, B., Tilman, D. and Wardle D.A. (2001) Biodiversity and ecosystem functioning: current knowledge and future challenges. *Science* 294(5543): 804–808.

McDonald, A.W. (2001) Succession during the re-creation of a flood-meadow 1985–1999. *Applied Vegetation Science* 4 (2): 167–176.

Maclean, N. (2010) *Silent Summer: The state of wildlife in Britain and Ireland*. Cambridge: Cambridge University Press.

Maddock, A. (ed.) (2008) *UK Biodiversity Action Plan: Priority habitat descriptions*. London: Defra.

Manchester, S.J., McNally, S., Treweek, J.R., Sparks,T.H. and Mountford, J.O. (1999) The cost and practicality of techniques for the reversion of arable land to lowland wet grassland: an experimental study and review. *Journal of Environmental Management* 55 (2): 91–109.

Marren, P. (1999) *Britain's Rare Flowers*. London: T. & A.D. Poyser.

Marrs, R.H. (1985) Techniques for reducing soil fertility for nature conservation purposes: a review in relation to research at Roper's Heath, Suffolk, England. *Biological Conservation* 34 (4): 307–332.

Marrs, R.H. (1993) Soil fertility and nature conservation in Europe: theoretical considerations and practical management solutions. *Advances in Ecological Research* 24: 241–300.

Marrs, R.H. (2002) Manipulating the chemical environment of the soil. In M.R. Perrow and A.J. Davy (eds) *Handbook of Ecological Restoration. Vol 1: Principles of Restoration*. Cambridge: Cambridge University Press. pp. 155–183.

Marrs, R.H., Gough, M.W. and Griffiths, M. (1991) Soil chemistry and leaching losses of nutrients from semi-natural grassland and arable soils on three contrasting parent materials. *Biological Conservation* 57 (3): 257–271.

Marrs, R.H., Snow, C.S.R., Owen, K.M. and Evans, C.E. (1998) Heathland and acid grassland creation on arable soils at Minsmere: identification of potential problems and a test of cropping to impoverish soils. *Biological Conservation* 85 (1–2): 69–82.

Martin, L.M. and Wilsey, B.J. (2006) Assessing grassland restoration success: relative roles of seed additions and native ungulate activities. *Journal of Applied Ecology* 43 (6): 1098–1109.

Martin, L.M., Moloney, K.A. and Wilsey, B.J. (2005) An assessment of grassland restoration success using species diversity components. *Journal of Applied Ecology* 42: 327–336.

Maskell, L.C., Smart, S.M., Bullock, J.M., Thompson, K. and Stevens, C.J. (2010) Nitrogen deposition causes widespread loss of species richness in British habitats. *Global Change Biology* 16 (2): 671–679.Menzel, A., Sparks, T.H., Estrella, N., Koch, E., Aasa, A., Ahas, R., Alm-Kübler, K., Bissolli, P., Braslavská, O., Briede, A., Chmielewski, F.M., Crepinsek, Z., Curnel, Y., Dahl, Å., Defila, C., Donnelly, A., Filella, Y., Jatczak, K., Måge, F., Mestre, A., Nordli, Ø., Peñuelas, J., Pirinen, P., Remišová, V., Scheifinger, H., Striz, M., Susnik, A., Van Vliet, A.J.H., Wielgolaski, F.-E. and Zach, S. (2006) European phenological response to climate change matches the warming pattern. *Global Change Biology* 12: 1969–1976.

Mérő, T.O., Bocz, R., Polyák, L., Horváth, G. and Lengyel, S. (2015) Local habitat management and landscape-scale restoration influence small-mammal communities in grasslands. *Animal Conservation*. DOI: 10.1111/acv.12191.

Ministry of Agriculture, Fisheries and Food (MAFF). (1986) *The Analysis of Agricultural Materials*, 3rd edn. London: HMSO.

Mitchell, F.J.G. (2005). How open were European primeval forests? Hypothesis testing using palaeoecological data. *Journal of Ecology* 93 (1): 168–177.

Mitchell, R.J., Marrs, R.H., Le Duc, M.G. and Auld, M.H.D. (1997) A study of succession on lowland heaths in Dorset, southern England: changes in vegetation and soil chemical properties. *Journal of Applied Ecology* 34 (6): 1426–1444.

Mitchell, R.J., Morecroft, M.D., Acreman, M., Crick, H.Q.P., Frost, M., Harley, M., Maclean, I.D.M., Mountford, O., Piper, J., Pointier, H., Rehfisch, M.M., Ross, L.C., Smithers, R.J., Stott, A., Walmsley, C.A., Watts, O. and Wilson, E. (2007) *England Biodiversity Strategy: Towards adaptation to climate change*. Final Report to Defra for contract CR0327. London: Defra.

Mitchley, J., Buckley G.P. and Helliwell, D.R. (1996) Vegetation establishment on chalk marl spoil: the role of nurse grass species and fertiliser application. *Journal of Vegetation Science* 7: 543–548.

Mitchley, J., Jongepierová, I. and Fajmon, K. (2012) Regional seed mixtures for the recreation of species-rich meadows in the White Carpathian Mountains: results of a 10-yr experiment. *Applied Vegetation Science* 15 (2): 253–263.

Mitlacher, K., Poschlod, O., Rosén, E. and Bakker, J.P. (2002) Restoration of wooded meadows – comparative analysis along a chronosequence on Öland (Sweden). *Applied Vegetation Science* 5 (1): 63–73.

Mortimer, S.R., Booth, R.G., Harris, S.J. and Brown, V.K. (2002a) Effects of initial site management on the Coleoptera assemblages colonising newly established chalk grassland on ex-arable land. *Biological Conservation* 104 (3): 301–313.

Mortimer, S.R., Toller, A., Masters, G.J. and Brown, V.K. (2002b) Efficacy of hay spreading to increase the botanical diversity of agriculturally-improved chalk grassland. In J. Frame (ed.) *Conservation Pays?* Reading: British Grassland Society. pp. 141–144.

Mortimer, S.R., Turner, A.J., Brown, V.K., Fuller, R.J., Good, J.E.G., Bell, S.A., Stevens, P.A., Norris, D., Bayfield, N. and Ward, L.K. (2000) *The Nature Conservation Value of scrub in Britain*. JNCC report, no. 308. Peterborough: Joint Nature Conservation Committee.

Mouissie, A.M., Van der Veen, C.E.J., Veen, G.F. and Van Diggelen, R. (2005) Ecological correlates of seed survival after ingestion by fallow deer. *Functional Ecology* 19 (2): 284–290.

Mountford, J.O., Lakhani, K.H. and Holland, R.J. (1996) Reversion of vegetation following the cessation of fertiliser application. *Journal of Vegetation Science* 7(2): 219–228.

Mudrák, O., Mládek, J., Blažek, P., Lepš, J., Doležal, J., Nekvapilová, E. and Těšsitel, J. (2014) Establishment of hemiparasitic *Rhinanthus* spp. in grassland restoration: lessons learned from sowing experiments. *Applied Vegetation Science* 17(2): 274–287.

Murphy, J.M., Sexton, D., Jenkins, G., Boorman, P., Booth, B., Brown, K., Clark, R., Collins, M., Harris, G., Kendon, L. and Met Office Hadley Centre. (2009) *UK Climate Projections Science Report: Climate change projections*. Exeter: Met Office Hadley Centre.

Natural England. (2008a) *Bracken Management: Ecological, archaeological and landscape issues and priorities*. Natural England Technical Information Note TIN047. Peterborough: Natural England.

Natural England. (2008b) *Bracken Management and Control*. Natural England Technical Information Note TIN048. Peterborough: Natural England.

Natural England. (2008c) *Soil Sampling for Habitat Recreation and Restoration*. Natural England Technical Information Note TIN035. Peterborough: Natural England.

Natural England. (2009) *Sward Enhancement: Diversifying grassland using pot-grown wildflowers or seedling plugs*. Natural England Technical Information Note TIN065. Peterborough: Natural England.

Natural England. (2010a) *Higher Level Stewardship: Farm Environment Plan (FEP) Manual*, 3rd edn. Peterborough: Natural England.

Natural England. (2010b) *Grazing our common*. Commons Factsheet No. 11. Peterborough: Natural England.

Natural Environment Research Council. (1977) *Amenity grassland – the needs for research*. Natural Environment Research Council publication series 'C' No.19. London: Natural Environment Research Council.

Norton, L.R., Murphy, J., Reynolds, B., Marks, S. and Mackey, E.C. (2009) *Countryside Survey: Scotland results from 2007*. Swindon: NERC/Centre for Ecology & Hydrology.

Oates, M. and Bullock, D. (1997) Browsers and grazers. *Enact* 5 (4): 15–18.

Öckinger, E. and Smith, H.G. (2007) Semi-natural grasslands as population sources for pollinating insects in agricultural landscapes. *Journal of Applied Ecology* 44 (1): 50–59.

Oomes, M.J.M. (1992) Yield and species density of grasslands during restoration management. *Journal of Vegetation Science* 3 (2): 271–274.

O'Reilly, J.O. (2010) The state of upland hay meadows in the North Pennines. *British Wildlife* 21 (3): 184–192.

Öster, M., Ask, K., Cousins, S.A.O. and Eriksson, O. (2009) Dispersal and establishment limitation reduces the potential for successful restoration of semi-natural grassland communities on former arable fields. *Journal of Applied Ecology* 46 (6): 1266–1274.

Owen, K.M. and Marrs, R.H. (2001) The use of mixtures of sulphur and bracken litter to reduce pH of former arable soils and control ruderal species. *Restoration Ecology* 9 (4): 397–409.

Owen, K.M., Marrs, R.H., Snow, C.S.R. and Evans, C.E. (1999) Soil acidification – the use of sulphur and acidic plant materials to acidify arable soils for the recreation of heathland and acid grassland at Minsmere, UK. *Biological Conservation* 87 (1): 105–121.

Pakeman, R.J. (2001) Plant migration rates and seed dispersal mechanisms. *Journal of Biogeography* 28 (6): 795–800.

Pakeman, R.J. and Small, J.L. (2009) Potential and realised contribution of endozoochory to seedling establishment. *Basic and Applied Ecology* 10 (7): 656–661.

Pakeman, R.J., Pywell, R.F. and Wells, T.C.E. (2002) Species spread and persistence: implications for experimental design and habitat re-creation. *Applied vegetation Science* 5 (1): 75–86.

Parr, T.W. and Way, J.M. (1988) Management of roadside vegetation: the long-term effects of cutting. *Journal of Applied Ecology* 25 (3): 1073–1087.

Patzelt, A., Wild, U. and Pfadenhauer, J. (2001) Restoration of wet fen meadows by topsoil removal: vegetation development and germination biology of fen species. *Restoration Ecology* 9 (2): 127–136.

Pavlů, V., Schellberg, J. and Hejcman, M. (2011) Cutting frequency vs. N application: effect of a 20-year management in *Lolio-Cynosuretum* grassland. *Grass and Forage Science* 66 (4): 501–515.

Perry, C. and Gamble, D. (2012) Hay Time: Analysis of survey data 2006–2011. Clapham: Yorkshire Dales Millennium Trust.

Peterken, G. (2009) Woodland origins of meadows. *British Wildlife* 20 (3): 161–170.

Peterken, G.F. (2013) *Meadows*. Dorset: British Wildlife Publishing.

Pigott, C.D. (1958) Biological Flora of the British Isles: *Polemonium caeruleum* L. *Journal of Ecology* 46: 507–525.

Pigott, M.E. and Pigott, C.D. (1959) Stratigraphy and pollen analysis of Malham Tarn and Tarn Moss. *Field Studies* 1: 84–101.

Pinches, C.E., Gowing, D.J.G., Stevens, C.J., Fagan, K. and Brotherton, P.N.M. (2013) *Upland Hay Meadows: What management regimes maintain the diversity of meadow flora and populations of breeding birds?* Natural England Evidence Review, no. 005. Peterborough: Natural England.

Plantlife International. (2008) *Saving the forgotten kingdom. A strategy for the conservation of the UK's fungi: 2008–2015.* Salisbury: Plantlife International.

Porter, K. (1994) Seed harvesting – a hay meadow dilemma? *Enact* 2 (1): 4–5.

Poschlod, P., Kiefer, S., Tränkle, U., Fischer, S. and Bonn, S. (1998) Plant species richness in calcareous grasslands as affected by dispersability in space and time. *Applied Vegetation Science* 1 (1): 75–91.

Pöyry, J., Lindgren, S., Salminen, J. and Kuussaari, M. (2005) Responses of butterfly and moth species to restored cattle grazing in semi-natural grasslands. *Biological Conservation* 122 (3): 465–478.

Prach, K. and Hobbs, R.J. (2008) Spontaneous succession versus technical reclamation in the restoration of disturbed sites. *Restoration Ecology* 16 (3): 363–366.

Prach, K., Fajmon, K., Jongepierová, I. and Řehounková, K. (2015) Landscape context in colonization of restored dry grasslands by target species. *Applied Vegetation Science* 18 (2): 181–189.

Prach, K., Jongepierová, I. and Řehounková, K. (2013) Large-scale restoration of dry grasslands on ex-arable land using a regional seed mixture: establishment of target species. *Restoration Ecology* 21 (1): 33–39.

Prach, K., Jongepierov, I., Řehounková, K. and Fajmon, K. (2014) Restoration of grasslands on ex-arable land using regional and commercial seed mixtures and spontaneous succession: successional trajectories and changes in species richness. *Agriculture, Ecosystems & Environment* 182: 131–136.

Pykälä, J. (2003) Effects of restoration with cattle grazing on plant species composition and richness of semi-natural grasslands. *Biodiversity and Conservation* 12 (11): 2211–2226.

Pywell, R.F., Wells, T.C.E. and Sparks, T.H. (1996) Long-term dynamics of reconstructed species-rich grassland communities. *Aspects of Applied Biology* 44: 369–376.

Pywell, R.F., Bullock, J.M., Hopkins, A., Walker, K.J., Sparks, T.H., Burke, M.J.W. and Peel, S. (2002) Restoration of species-rich grassland on arable land: assessing the limiting processes using a multi-site experiment. *Journal of Applied Ecology* 39 (2): 294–309.

Pywell, R.F., Bullock, J.M., Roy, D.B., Warman, L., Walker, K.J. and Rothery, P. (2003) Plant traits as predictors of performance in ecological restoration. *Journal of Applied Ecology* 40 (1): 65–77.

Pywell, R.F., Bullock, J.M., Tallowin, J.R.B. and Masters, G. (2004a) *Practical Techniques to Increase the Biodiversity of Agriculturally Improved Grasslands.* Research report for Defra project BD1425. London: Defra.

Pywell, R.F., Bullock, J.M., Tallowin, J.B., Walker, K.J., Warman, E.A. and Masters, G. (2007a) Enhancing diversity of species-poor grasslands: an experimental assessment of multiple constraints. *Journal of Applied Ecology* 44 (1): 81–94.

Pywell, R.F., Bullock, J.M., Walker, K.J., Coulson, S.J., Gregory, S.J. and Stevenson, M.J. (2004b) Facilitating grassland diversification using the hemiparasitic plant *Rhinanthus minor*. *Journal of Applied Ecology* 41 (5): 880–887.

Pywell, R.F., Hayes, M.J., Tallowin, J.B., Walker, K.J., Meek, W.R., Carvell, C., Warman, L.A. and Bullock, J.M. (2010) Minimizing environmental impacts of grassland weed management: can *Cirsium arvense* be controlled without herbicides? *Grass and Forage Science* 65 (2): 159–174.

Pywell, R.F., Warman, E.A., Walker, K.J. and Bullock, J.M. (2007b) Enhancing grassland diversity using slot-seeding and harrowing. *Aspects of Applied Biology* 82: 191–198.

Rackham, O. (1990) *Trees and Woodland in the British Landscape: The complete history of Britain's trees, woods & hedgerows.* London: J.M. Dent and Sons.

Redhead, J.W., Sheail, J., Bullock, J.M., Ferreruela, A., Walker, K.J. and Pywell, R.F. (2014) The natural regeneration of calcareous grassland at a landscape scale: 150 years of plant community re-assembly on Salisbury Plain, UK. *Applied Vegetation Science* 17 (3): 408–418.

Riley, J.D., Craft, I.W., Rimmer, D.L. and Smith, R.S. (2004) Restoration of Magnesian limestone grassland: optimising the time for seed collection by vacuum harvesting. *Restoration Ecology* 12 (3): 311–317.

Rodwell, J.S. (ed.) (1991a) *British Plant Communities,* Vol 1 *(Woodlands and Scrub).* Cambridge: Cambridge University Press.

Rodwell, J.S. (ed.) (1991b) *British Plant Communities,* Vol 2 *(Heaths and Mires).* Cambridge: Cambridge University Press.

Rodwell, J.S. (ed.) (1992) *British Plant Communities,* Vol 3 *(Grasslands and Montane Communities).* Cambridge: Cambridge University Press.

Rodwell, J.S. (ed.) (1995) *British Plant Communities,* Vol 4 *(Aquatic Communities, Swamps and Tall-herb Fens.* Cambridge: Cambridge University Press.

Rodwell, J.S. (ed.) (2000) *British Plant Communities,* Vol 5 *(Maritime and Weed Communities and Vegetation of Open Habitats.* Cambridge: Cambridge University Press.

Rodwell, J.S. (2006) *National Vegetation Classification: Users' handbook.* Peterborough: JNCC.

Rodwell, J.S., Dring, J.C., Averis, A.B.G., Proctor, M.C.F., Malloch, A.J.C., Schaminée, J.N.J. and Dargie, T.C.D. (2000) *Review of Coverage of the National Vegetation Classification.* JNCC Report, no. 302. Peterborough: Joint Nature Conservation Committee.

Rodwell, J.S., Morgan, V., Jefferson, R.G. and Moss, D. (2007) *The European Context of British Lowland Grasslands.* JNCC Report, no. 394. Peterborough: Joint Nature Conservation Committee.

Rotheroe, M. (2001) *A Preliminary Survey of Selected Semi-Natural Grasslands in Carmarthenshire.* Contract Science Report, no. 340. Bangor: Countryside Council for Wales.

Roy, D.B. and Sparks, T.H. (2000) Phenology of British butterflies and climate change. *Global Change Biology* 6 (4): 407–416.

Royal Society for the Protection of Birds (RSPB). (2006) *Lowland dry acid grassland management on RSPB reserves.* Sandy: RSPB.

Ruprecht, E. (2006) Successfully recovered grassland: a promising example from Romanian old-fields. *Restoration Ecology* 14 (3): 473–480.

Ruprecht, E., Enyedi, M.Z., Eckstein, R.L. and Donath, T.W. (2010) Restorative removal of plant litter and vegetation 40 years after abandonment enhances re-emergence of steppe grassland vegetation. *Biological Conservation* 143 (2): 449–456.

Sanderson, N.A. (1998) *A Review of the Extent, Conservation Interest and Management of Lowland Acid Grassland in England.* English Nature Research Reports, no. 259. Peterborough: English Nature.

Scheidel, U. and Brulheide, H. (2005) Effects of slug herbivory on the seedling establishment of two montane Asteraceae species. *Flora* 200 (4): 309–320.

Schmiede, R., Donath, T.W. and Otte, A. (2009) Seed bank development after the restoration of alluvial grassland via transfer of seed-containing plant material. *Biological Conservation* 142 (2): 404–413.

Scott, R. and Ash, H. (1990) *The Conservation of Urban Grassland.* St. Helens: The Groundwork Trust.

Scottish Government. (2008) *Scottish Government Guidance on How to Prevent the Spread of Ragwort.* Edinburgh: Scottish Government.

Scotton, M., Dal Buono, C. and Timoni, A. (2011) *Semi-Natural Grassland as a Source of Native Seed: Comparison of different harvesting methods.* SER Europe Workshop, 18–20 May 2011, Bernburg, Germany. Bernburg: SALVERE.

Scotton, M., Kirmer, A. and Krautzer, B. (eds) (2012) *Practical Handbook for Seed Harvest and Ecological Restoration of Species-Rich Grasslands*. Padova: Cooperativa Libraria Editrice Universitá di Padova.

Scotton, M., Piccinin, L., Dainese, M. and Sancin, F. (2009) Seed harvesting for ecological restoration: efficiency of haymaking and seed-stripping on different grassland types in the eastern Italian Alps. *Ecological Restoration* 27 (1): 66–75.

SEARS. (2008) *Bracken Control: A guide to best practice*. Natural Scotland.

Settele, J., Kudrna, O., Harpke, A., Kühn, I., van Swaay, C., Verovnik, R., Warren, M., Wiemers, M., Hanspach, J., Hickler, T., Kühn, E., van Halder, I., Veling, K., Vliegenthart, A., Wynhoff, I. and Schweiger, O. (2008) *Climatic Risk Atlas of European Butterflies*. Biorisk 1 (Special Issue). Sofia: Pensoft.

Shiel, R.S. and Batten, J.C. (1988) Redistribution of nitrogen and phosphorus on Palace Leas meadow hay plots as a result of aftermath grazing. *Grass and Forage Science* 43 (2): 105–110.

Silvertown, J., Poulton, P., Johnston, E., Edwards, G., Heard, M. and Biss, P.M. (2006) The Park Grass Experiment 1856–2006: its contribution to ecology. *Journal of Ecology* 94 (4): 801–814.

Silvertown, J., Tallowin, J., Stevens, C., Power, S.A., Morgan, V., Emmett, B., Hester, A., Grime, P.J., Morecroft, M., Buxton, R., Poulton, P., Jinks, R. and Bardgett, R. (2010) Environmental myopia: a diagnosis and a remedy. *Trends in Ecology and Evolution* 25 (10): 556–561.

Simpson, N.A. and Jefferson, R.G. 1996. *Use of Farmyard Manure on Semi-Natural (Meadow) Grassland*. English Nature Research Report, no. 150. Peterborough: English Nature.

Smart, S.M., Allen, D., Murphy, J., Carey, P.D., Emmett, B.A., Reynolds, B., Simpson, I.C., Evans, R.A., Skates, J., Scott, W.A., Maskell, L.C., Norton, L.R., Rossall, M.J. and Wood C. (2009) *Countryside Survey: Wales Results from 2007*. CEH Project Number: C03259. Swindon: NERC/Centre for Ecology & Hydrology. p. 94.

Smith, H., Feber, R.E., Morecroft, M.D., Taylor, M.E. and Macdonald, D.W. (2010) Short-term successional change does not predict long-term conservation value of managed arable field margins. *Biological Conservation* 143 (3): 813–822.

Smith, J., Potts, S.G., Woodcock, B.A. and Eggleton, P. (2008) Can arable field margins be managed to enhance their biodiversity, conservation and functional value for soil macrofauna? *Journal of Applied Ecology* 45 (1): 269–278.

Smith, R.S. (2005) *Ecological Mechanisms Affecting the Restoration of Diversity in Agriculturally Improved Meadow Grassland*. Defra Project BD1439. London: Defra.

Smith, R.S. (2010) Understanding grassland systems: the scientific evidence base for conservation management of plant diversity. In D. Gamble and T. St. Pierre (eds) *Hay time in the Yorkshire Dales: The natural, cultural and land management history of hay meadows*. Clapham: Yorkshire Dales Millennium Trust. pp. 145–177.

Smith, R.S. and Jones, L. (1991) The phenology of mesotrophic grassland in the Pennine Dales, northern England: historic hay cutting dates, vegetation variation and plant species phenologies. *Journal of Applied Ecology* 28 (1): 42–59.

Smith, R.S., Buckingham, H., Bullard, M.J., Shiel, R.S. and Younger, A. (1996) The conservation management of mesotrophic (meadow) grassland in northern England. I. Effects of grazing, hay cut date and fertilizer on the vegetation of a traditionally managed sward. *Grass and Forage Science* 51 (3): 278–291.

Smith, R.S., Shiel, R.S., Bardgett, R.D., Millward, D., Corkhill, P., Evans, P., Quirk, H., Hobbs, P.J. and Kometa, S.T. (2008) Long term change in vegetation and soil microbial communities during the phased restoration of meadow grassland. *Journal of Applied Ecology* 45 (2): 670–679.

Smith, R.S., Shiel, R.S., Bardgett, R.D., Millward, D., Corkhill, P., Rolph, G., Hobbs, P.J. and Peacock, S. (2003) Soil microbial community, fertility, vegetation and diversity as targets in the restoration management of meadow grassland. *Journal of Applied Ecology* 40 (1): 51–64.

Smith, R.S., Sheil, R.S., Millward, D. and Corkhill, P. (2000) The interactive effects of management on the productivity and plant community structure of an upland meadow: an 8-year field trial. *Journal of Applied Ecology* 37 (6): 1029–1043.

Smith, R.S., Shiel, R.S., Millward, D., Corkhill, P. and Sanderson, R.A. (2002) Soil seed banks and the effects of meadow management on vegetation change in a 10-year meadow field trial. *Journal of Applied Ecology* 39 (2): 279–293.

Smithers, R.J., O'Hanley, J.R., Harrison, P.A. and Berry, P.M. (2007) Appendix 2: Limits to climate space modelling. In C.A. Walmsley, R.J. Smithers, P.M. Berry, M. Harley, M.J. Stevenson and R. Catchpole (eds) *MONARCH (Modelling Natural PM Resource Responses to Climate Change): A synthesis for biodiversity conservation*. Oxford: UKCIP.

Snow, C.S.R., Marrs, R.H. and Merrick, L. (1997) Trends in soil chemistry and and floristics associated with the establishment of a low-input meadow system on an arable clay soil in Essex. *Biological Conservation* 79 (1): 35–41.

Sojneková, M. and Chytrý, M. (2015) From arable land to species-rich semi-natural grasslands: Succession in abandoned fields in a dry region of central Europe. *Ecological Engineering* 77: 373–381.

Soons, M.B. and Heil, G.W. (2002) Reduced colonization capacity in fragmented populations of wind-dispersed grassland forbs. *Journal of Ecology* 90 (6):1033–1043.

Spooner, B.M. and Roberts, P. (2005) *Fungi*. The New Naturalist Library. London: Collins.

Stace, C. (2010) *New Flora of the British Isles*, 3rd edn. Cambridge: Cambridge University Press.

Stampfli, A. and Zeiter, M. (2004) Plant regeneration directs changes in grassland composition after extreme drought: a 13-year study in southern Switzerland. *Journal of Ecology* 92 (4): 568–576.

Stebbings, R.E. (1988) *Conservation of European Bats*. London: Christopher Helm.

Steffan-Dewenter, I. and Tscharntke, T. (2002) Insect communities and biotic interactions on fragmented calcareous grasslands: a mini review. *Biological Conservation* 104 (3): 275–284.

Stein, C., Auge, H., Fischer, M., Weisser, W.W. and Prati, D. (2008) Dispersal and seed limitation affect diversity and productivity of montane grasslands. *Oikos* 117 (10): 1469–1478.

Stevens, C.J., Dise, N.B., Mountford, J.O. and Gowing, D.J. (2004) Impact of nitrogen deposition on the species richness of grasslands. *Science* 303 (5665): 1876–1879.

Stevens, D.P., Smith, S.L.N., Blackstock, T.H., Bosanquet, S.D.S. and Jones, P.S. (2010) *Grasslands of Wales: A survey of lowland species-rich grasslands, 1987–2004*. Cardiff: University of Wales Press.

Stevens, P. and Wilson, P. (2012) Species-rich grassland re-creation projects. A route to success? *Aspects of Applied Biology* 115: 53–60.

Stevenson, M.J., Bullock, J.M. and Ward, L.K. (1995) Re-creating semi-natural communities: effect of sowing rate on establishment of calcareous grassland. *Restoration Ecology* 3 (4): 279–289.

Stevenson, M.J., Ward, L.K. and Pywell, R.F. (1997) Re-creating semi-natural communities: vacuum harvesting and hand collection of seed on calcareous grassland. *Restoration Ecology* 5 (1): 66–76.

Stewart, G.B. and Pullin, A.S. (2008) The relative importance of grazing stock type and grazing intensity for conservation of mesotrophic 'old meadow' pasture. *Journal for Nature Conservation* 16 (3): 175–185.

Strykstra, R.J., Verweij, G.L and Bakker, J.P. (1997) Seed dispersal by mowing machinery in a Dutch brook valley system. *Acta Botanica Neerlandica* 46 (4): 387–401.

Suding, K.N., Gross, K.L. and Houseman, G.R. (2004) Alternative states and positive feedbacks in restoration ecology. *TRENDS in Ecology and Evolution* 19 (1), 46–53.

Tallowin, J.R.B. and Jefferson, R.G. (1999) Hay production from lowland semi-natural grasslands: a review of implications for ruminant livestock systems. *Grass and Forage Science* 54 (2): 99–115.

Tallowin, J.R.B. and Smith, R.E.N. (2001) Restoration of a *Cirsio-Molinietum* fen meadow on an agriculturally improved pasture. *Restoration Ecology* 9 (2): 167–178.

Tallowin, J.R.B., Smith, R.E.N., Pywell, R.F., Goodyear, J. and Martyn, T. (2002) Use of fertiliser nitrogen and potassium to reduce soil phosphorus availability. In J. Frame (ed.) *Conservation Pays? Reconciling Environmental Benefits with Profitable Grassland Systems*. Proceedings of the Joint British Grassland Society/British Ecological Society Conference, University of Lancaster, 15–17 April 2002, UK. Reading: British Grassland Society. pp. 163–166.

Taylor, S., Knight, M. and Harfoot, A. (2014) *National Biodiversity Climate Change Vulnerability Model*. Natural England research report NER054. Peterborough: Natural England.

Thomas, C.D. (2009) A speculative history of open-country species in Britain and northern Europe. *British Wildlife* 20 (5): 21–25.

Thomas, C.D., Thomas, J.A. and Warren, M.S. (1992) Distributions of occupied and vacant butterfly habitats in fragmented landscapes. *Oecologia* 92 (4): 563–567.

Thomas, J.A., Bourn, N.A.D., Clarke, R.T., Stewart, K.E., Simcox, D.J., Pearman, G.S., Curtis, R. And Goodger, B. (2001) The quality and isolation of habitat patches both determine where butterflies persist in fragmented landscapes. *Proceedings of the Royal Society. B: Biological Sciences* 268 (1478): 1791–1796.

Tilman, D. (1997) Community invasibility, recruitment limitation, and grassland biodiversity. *Ecology* 78 (1): 81–92.

Tilman, D., Wedin, D. and Knops, J. (1996) Productivity and sustainability influenced by biodiversity in grassland ecosystems. *Nature* 379 (6567): 718–720.

Török, K., Szili-Kovács, T., Halassy, M., Tóth, T., Hayek, Z., Paschke, M.W. and Wardell, L.J. (2000) Immobilization of soil nitrogen as a possible method for the restoration of sandy grassland. *Applied Vegetation Science* 3 (1): 7–14.

Török, P., Deák, B., Vida, E., Valkó, O., Lengyel, S. and Tóthmérész, B. (2010) Restoring grassland biodiversity: sowing low-diversity seed mixtures can lead to rapid favourable changes. *Biological Conservation* 143 (3): 806–812.

Török, P., Vida, E., Deák, B., Lengyel, S. and Tóthmérész, B. (2011) Grassland restoration on former croplands in Europe: an assessment of applicability of techniques and costs. *Biodiversity and Conservation* 20 (11): 2311–2332.

Tracking Mammals Partnership. (2009) *UK Mammals: Update 2009*. Peterborough: Joint Nature Conservation Committee/Tracking Mammals Partnership.

Trueman, I.C. and Millett, P. (2003) Creating wildflower meadows by strewing green hay. *British Wildlife* 15: 37–44.

Turnbull, L.A., Crawley, M.J. and Rees, M. (2000) Are plant populations seed-limited? A review of sowing experiments. *Oikos* 88 (2): 225–238

UK Biodiversity Action Plan (UK BAP). (2006) UK BAP Targets Review (2006). Available at: http://tna.europarchive.org/20110303145238/http://www.ukbap.org.uk/BAPGroupPage.aspx?id=98 (Habitat Action Plan Targets).

UK Biodiversity Group. (1998a) *Tranche 2 Action Plans: Vertebrates and vascular plants (Vol 1)*. Peterborough: English Nature.

UK Biodiversity Group. (1998b) *Tranche 2 Action Plans: Terrestrial and freshwater habitats (Vol 2)*. Peterborough: English Nature.

University of Newcastle upon Tyne. (2013) *Influence of Spring Grazing Regime on the Floristic Diversity and Restorative Potential of Upland Hay Meadows*. Defra EVID4 Evidence Project Final Report (Rev. 06/11). Newcastle upon Tyne: University of Newcastle upon Tyne.

van der Heijden, M.G.A., Klironomos, J.N., Ursic, M., Moutoglis, P., Streitwolf-Engel, R., Boller, T., Wiemken, A. and Sanders, I.R. (1998) Mycorrhizal fungal diversity determines plant biodiversity, ecosystem variability and productivity. *Nature* 396 (6706): 69–72.

Van der Putten, W.H., Mortimer, S.R., Hedlund, K., van Dijk, C., Brown, V.K., Lepä, J., Rodriguez-Barrueco, C., Roy, J, Diaz Len, T.A., Gormsen, D., Korthals, G.W., Lavorel, S., Santa Regina, I and Smilauer, P. (2000) Plant species diversity as a driver of early succession in abandoned fields: a multi-site approach. *Oecologia* 124 (1): 91–99.

Van Dorp, D., van den Hoek, W.P.M. and Daleboudt, C. (1996) Seed dispersal capacity of six perennial grassland species measured in a wind tunnel at varying wind speed and height. *Canadian Journal of Botany* 74 (12): 1956–1963.

Vera, F.W.M. (2000) *Grazing Ecology and Forest History*. New York, NY: CABI Publishing.

Verhagen, R., Klooker, J., Bakker, J.P. and Van Diggelen, R. (2001) Restoration success of low-production plant communities on former agricultural soils after top-soil removal. *Applied Vegetation Science* 4 (1): 75–82.

Verkaar, H.J., Schenkeveld, A.J. and van de Klashorst, M.P. (1983) The ecology of short-lived forbs in chalk grassland: dispersal of seeds. *New Phytologist* 95 (2): 335–344.

Vickery, J.A., Tallowin, J.R., Feber, R.E., Asteraki, E.J., Atkinson, P.W., Fuller, R.J. and Brown, V.K. (2001) The management of lowland neutral grasslands in Britain: effects of agricultural practices on birds and their food resources. *Journal of Applied Ecology* 38 (3): 647–664.

von Blanckenhagen, B. and Poschlod, P. (2005) Restoration of calcareous grasslands: the role of the soil seed bank and seed dispersal for recolonisation processes. *Biotechnology, Agronomy and Society and Environment* 9 (2): 143–149.

Wagner, C. (2004) Passive dispersal of *Metrioptera bicolor* (Phillipi 1830) (Orthopteroidea: Ensifera: Tettigoniidae) by transfer of hay. *Journal of Insect Conservation* 8 (4): 287–296.

Wagner, M., Bullock, J.M, Meek, W.R., Walker, K.J., Stevens, C.J., Heard, M.S. and Pywell, R.F. (2014) How do pre-sowing disturbance and post-establishment management affect restoration progress in ex-arable calcareous grassland? In A. Hopkins, R.P. Collins, M.D. Fraser, V.R. King, D.C. Lloyd, J.M. Moorby and P.R.H. Robson (eds) *EGF at 50: The Future of European Grasslands*. Proceedings of the 25th General Meeting of the European Grassland Federation, Aberystwyth University, 7–11 September 2004, UK. pp 254–256.

Walker, E.A., Hermann, J.-M. and Kollman, J. (2015) Grassland restoration by seeding: seed source and growth form matter more than density. *Applied Vegetation Science* 18 (3): 368–378.

Walker, K.J., Hodder, K.H., Bullock, J.M. and Pywell, R.F. (2004a) *A Review of the Potential Effects of Seed Sowing for Habitat Re-creation on the Conservation of Intraspecific Biodiversity.* Defra Contract BD1447. Monks Wood: Centre for Ecology & Hydrology.

Walker, K.J., Stevens, P.A., Stevens, D.P., Mountford, J.O., Manchester, S.J. and Pywell, R.F. (2004b) The restoration and re-creation of species-rich lowland grassland on land formerly managed for intensive agriculture in the UK. *Biological Conservation* 119 (1): 1–18.

Walker, K.J., Warman, E.A., Bhogal, A., Cross, R.B., Pywell, R.F., Meek, B.R., Chambers, B.J. and Pakeman, R. (2007) Recreation of lowland heathland on ex-arable land: assessing the limiting processes on two sites with contrasting soil fertility and pH. *Journal of Applied Ecology* 44 (3): 573–582.

Wallin, L., Svensson, B.M. and Lönn, M. (2009) Artificial dispersal as a restoration tool in meadows: sowing or planting? *Restoration Ecology* 17 (2): 270–279.

Walmsley, C.A., Smithers, R.J., Berry, P.M., Harley, M., Stevenson, M.J. and Catchpole, R. (eds) (2007) *Monarch (Modelling Natural Resources to Climate Change): a synthesis for biodiversity conservation.* Oxford: UKCIP.

Walsh, G., Peel, S. and Jefferson, R.G. (2011) *The Use of Lime on Semi-Natural Grassland in Agri-Environment Schemes.* Natural England Technical Information Note TIN045. Sheffield: Natural England.

Waring, P. (1990) Observations on invertebrates collected up during wild flower seed harvesting in a hay meadow with particular reference to the butterflies and moths. *British Journal of Entomology and Natural History* 3: 143–152.

Warman, E.A., Pywell, R.F., Walker, K.J. and Bullock, J.M. (2007) Plug plants survival under different grassland management regimes. *Aspects of Applied Biology* 82: 109–116.

Warren, J., Christal, A. and Wilson, F. (2002) Effects of sowing and management on vegetation succession during habitat grassland restoration. *Agriculture, Ecosystems & Environment* 93 (1–3): 393–402.

Warren, J.M. (2000) The role of white clover in the loss of diversity in grassland habitat restoration. *Restoration Ecology* 8 (3): 318–323.

Wathern, P. and Gilbert, O.L. (1978) Artificial diversification of grassland with native herbs. *Journal of Environmental Management* 7: 29–42.

Wells, T., Bell, S. and Frost, A. (1981) *Creating Attractive Grasslands Using Native Plant Species.* London: Nature Conservancy Council.

Wells, T.C.E. (1987) The establishment of floral grasslands. *Acta Horticulturae* 195: 59–70.

Wells, T.C.E., Cox, R. and Frost, A. (1989) *Focus on Nature Conservation no. 21. The establishment and management of wildflower meadows.* Peterborough: Nature Conservancy Council.

Wells, T.C.E., Frost, A. and Bell, S.A. (1986) *Focus on Nature Conservation no. 15. Wild flower grasslands from crop-grown seed and hay-bales.* Peterborough: Nature Conservancy Council.

Wells, T.C.E., Sheail, J., Ball, D.F. and Ward, L.K. (1976) Ecological studies on the Porton Ranges: relationships between vegetation, soil and land-use history. *Journal of Ecology* 64 (2): 589–626.

Welsh Government. (2011) *Code of Practice to Prevent and Control the Spread of Ragwort.* Aberystwyth: Welsh Government.

Westbury, D.B. and Dunnett, N.P. (2008) The promotion of grassland forb abundance: a chemical or biological solution? *Basic and Applied Ecology* 9 (6): 653–662.

Westbury, D.B., Davies, A., Woodcock, B.A., Dunnett, N.P. and Fraser, L. (2006) Seeds of change: the value of using *Rhinanthus minor* in grassland restoration. *Journal of Vegetation Science* 17 (4): 435–446.

Wildlife Trusts. (2007) *Living Landscapes. A call to restore the UK's battered ecosystems, for wildlife and people.* Newark: The Wildlife Trusts.

Willems, J.H. (2001) Problems, approaches, and results in restoration of Dutch calcareous grassland during the last 30 years. *Restoration Ecology* 9 (2): 147–154.

Williams, J.M. (ed.) (2006) *Common Standards Monitoring for Designated Sites: summary: first six year report.* Peterborough: Joint Nature Conservation Committee.

Wilson, J.D., Evans, A.D. and Grice, P.V. (2009) *Bird Conservation and Agriculture.* Cambridge: Cambridge University Press.

Wilson, P., Wheeler, B., Reed, M and Strange, A. 2013. *A Survey of Selected Agrienvironment Grassland and Heathland Creation and Restoration Sites: Part 2.* Natural England Commissioned Reports, no. 107. London: Natural England.

Wilson, S.D., Bakker, J.D., Christian, J.M., Li, X.D., Ambrose, L.G. and Waddington, J. (2004) Semiarid old-field restoration: is neighbor control needed? *Ecological Applications* 14 (2): 476–484.

Winsa, M., Bommarco, R., Lindborg, R., Marini, L., Őckinger, E. and Schwabe-Kratochwil, A. (2015) Recovery of plant diversity in restored semi-natural pastures depends on adjacent land use. *Applied Vegetation Science* 18 (3): 413–422.

Winspear, R. and Davies, G. (2005) *A Management Guide to Birds of Lowland Farmland.* Sandy: Royal Society for the Protection of Birds.

Woodcock, B.A., Edwards, A.R., Lawson, C.S., Westbury, D.B., Brook, A.J., Harris, S.J., Masters, G., Booth, R., Brown, V.K. and Mortimer, S.R. (2010) The restoration of phytophagous beetles in species-rich chalk grasslands. *Restoration Ecology* 18 (5): 638–644.

Woodcock, B.A., McDonald, A.W. and Pywell, R.F. (2011) Can long-term floodplain meadow recreation replicate species composition and functional characteristics of target grasslands? *Journal of Applied Ecology* 48 (5): 1070–1078.

Woodcock, B.A., Potts, S.G., Pilgrim, E., Ramsay, A.J., Tscheulin, T., Parkinson, A., Smith, R.E.N., Gundrey, A.L., Brown, V.K. and Tallowin, J.R. (2007) The potential of grass field margin management for enhancing beetle diversity in intensive livestock farms. *Journal of Applied Ecology* 44 (1): 60–69.

Woodcock, B.A., Westbury, D.B., Brook, A.J., Lawson, C.S., Edwards, A.R., Harris, S.J., Heard, M.S., Brown, V.K. and Mortimer, S.R. (2012) Effects of seed addition on beetle assemblages during the re-creation of species-rich lowland hay meadows. *Insect Conservation and Diversity* 5 (1): 19–26.

Woods, R.G., Stringer, R.N., Evans, D.A. and Chater, A.O. (2015) *Rust Fungus Red Data List and Census Catalogue for Wales.* Aberystwyth: A.O. Chater.

Yalden, D.W. (1999) *The History of British Mammals (Poyser Natural History).* London: T. & A.D. Poyser.

Young, T.P., Chase, J.M. and Huddleston, R.T. (2001) Community succession and assembly:

comparing, contrasting and combining paradigms in the context of ecological restoration. *Ecological Restoration* 19 (1): 5–18.

Zulka, K.P., Abensperg-Traun, M., Milasowszky, N., Bieringer, G., Gereben-Krenn, B-A., Holzinger, W., Hölzler, G., Rabitsch, W., Reischütz, A, Querner, P., Sauberer, N., Schmitzberger, I., Willner, W., Wrbka, T. and Zechmeister, H. (2014) Species richness in dry grassland patches of eastern Austria: a multi-taxon study on the role of local, landscape and habitat quality variables. *Agriculture, Ecosystems and Environment* 182 : 25–36.

Species index

Page numbers in **bold** indicate tables and in *italic* indicate figures, photographs and captions.

Adder (*Vipera berus*) 54, **55**, 67
Adder's-tongue (*Ophioglossum vulgatum*) 20
Admiral, Red (*Vanessa atalanta*) 125
Anemone, Wood (*Anemone nemorosa*) 5, 6
Apple, Crab (*Malus sylvestris*) 31
Argus
    Brown (*Aricia agestis*) **38**, 124, 160
    Northern Brown (*Aricia artaxerxes*) **42**, 125
    Scotch (*Erebia aethiops*) 125
Arnica (*Arnica montana*) 213
Ash (*Fraxinus excelsior*) 28, 31, 127
Auroch (*Bos primigenius*) 3, 57
Avens
    Mountain (*Dryas octopetala*) 3, 4, 20, **21**, 22
    Water (*Geum rivale*) 6

Badger (*Meles meles*) 57, **60**, 113
Balsam, Himalayan (*Impatiens glandulifera*) 94
Baneberry (*Actaea spicata*) 28
Barbastelle (*Barbastella barbastellus*) 62, **62**
Barley, Spring (*Hordeum distichon*) 211
Bartsia, Red (*Odontites vernus*) 181
Bat
    Alcathoe (*Myotis alcathoe*) **61**
    Brandt's (*Myotis brandtii*) **61**
    Daubenton's (*Myotis daubentonii*) **61**
    Greater Horseshoe (*Rhinolophus ferrumequinum*) **61**
    Grey Long-eared (*Plecotus austriacus*) 62
    Leisler's (*Nyctalus leisleri*) **61**
    Lesser Horseshoe (*Rhinolophus hipposideros*) **61**
    Natterer's (*Myotis nattereri*) **61**
    Noctule (*Nyctalus noctula*) **61**

Serotine (*Eptesicus serotinus*) **61**
    Whiskered (*Myotis mystacinus*) **61**
Bear, Brown (*Ursus arctos*) 57
Bedstraw
    Heath (*Galium saxatile*) **11**, 13, **15**, 22, 31, 152, 153
    Lady's (*Galium verum*) 8, 10, 17, 27, 190
    Limestone (*Galium sterneri*) 9, 11, **11**, 20, **21**, 22, **29**, 31
Bee
    Banded Mining (*Andrena gravida*) **37**
    Shrill Carder (*Bombus sylvarum*) **37**
    Wall Mason (*Osmia parietina*) **37**
Beetle
    Hazel Pot (*Cryptocephalus coryli*) 11, **37**
    Violet Ground (*Carabus violaceus*) 144
Bellflower, Clustered (*Campanula glomerata*) 135–6, 160, *160*, 178
Bent
    Bristle (*Agrostis curtisii*) 13, **15**
    Brown (*Agrostis vinealis*) 168
    Common (*Agrostis capillaris*) 5, 11, **11**, 13, 14, **15**, 17, **18**, **21**, 22, 23, 24, 26, 31, **33**, 152, 153, 167, *168*, **187**
    Creeping (*Agrostis stolonifera*) 27
    Highland (*Agrostis castellana*) 184
Betony (*Stachys officinalis*) 19, 23, **24**, 90, 91, *185*, 197, 217
Bindweed, Sea (*Calystegia soldanella*) 7
Birches (*Betula*) 3, 31
Bird's-foot-trefoil, Common (*Lotus corniculatus*) 10, 17, **21**, **24**, **38**, 178, *195*, *196*
Bistort, Alpine (*Persicaria vivipara*) 23, **24**
Bitter-vetch (*Lathyrus linifolius*) 23, **24**
Black-grass (*Alopecurus myosuroides*) 134
Blackthorn (*Prunus spinosa*) 30, 31, 106

Bladder-fern, Brittle (*Cystopteris fragilis*) **29**
Blue
  Adonis (*Polyommatus bellargus*) **37**, 39, **40**, *148*
  Chalkhill (*Polyommatus coridon*) 39, **40**, 160, 226
  Common (*Polyommatus icarus*) **38**
  Large (*Maculinea arion*) **40**, 68
  Long-tailed (*Lampides boeticus*) 126, *126*
  Silver-studded (*Plebeius argus*) 4, *41*, **42**
  Small (*Cupido minimus*) **40**, 43, 136, *148*, 160
Bluebell (*Hyacinthoides non-scripta*) 5, 6, *6*
Box (*Buxus sempervirens*) 30
Bracken (*Pteridium aquilinum*) 14, 17, 76, 94, 100–2, *101*, 113
Bramble (*Rubus fruticosus agg.*) 30, 31, 104, *109*
Brimstone (*Gonepteryx rhamni*) 38, **38**
Brome
  Barren (*Anisantha sterilis*) 134
  False (*Brachypodium sylvaticum*) **29**
  Upright (*Bromopsis erecta*) 10, **10**, 159, 189
Broom (*Cytisus scoparius*) 31
Broomrapes (*Orobanche*) 182
  Bedstraw (*Orobanche caryophyllacea*) *8*
  Common (*Orobanche minor*) 136
Brown
  Meadow (*Maniola jurtina*) **38**
  Wall (*Lasiommata megera*) 38–9, **38**
Buckler-fern, Rigid (*Dryopteris submontana*) 28
Buckthorn (*Rhamnus cathartica*) **38**
  Alder (*Frangula alnus*) **38**
Buckwheat (*Fagopyrum esculentum*) 211
Bugle (*Ajuga reptans*) **213**
Bumblebees 143
  Buff-tailed (*Bombus terrestris*) *144*
  Large Garden (*Bombus ruderatus*) **37**
Bunting
  Cirl (*Emberiza cirlus*) **49**, 126
  Corn (*Miliaria calandra*) 66
  Reed (*Emberiza schoeniclus*) 46, **49**

Burnet
  Great (*Sanguisorba officinalis*) 3, 17, **18**, 23, **24**, 89, 141, *150*, 157, **213**
  Salad (*Sanguisorba minor*) 2, 10, **10**, *135*
Burnet Moth, Six-spot (*Zygaena filipendulae*) 160
Burnet-saxifrage (*Pimpinella saxifraga*) **24**, **33**
Bush Crickets 126
  Metrioptera spp. 192
  Wart-biter (*Decticus verrucivorus*) 11, **37**, 43
Buttercup
  Bulbous (*Ranunculus bulbosus*) 17
  Creeping (*Ranunculus repens*) 5, 27, *174*
  Meadow (*Ranunculus acris*) 27, 170, 195, *196*
Buzzard, Common (*Buteo buteo*) 47

Campion
  Bladder (*Silene vulgaris*) 26
  Moss (*Silene acaulis*) **21**, 22
  Sea (*Silene uniflora*) 26
Capercaillie (*Tetrao urogallus*) 66
Catchfly, Spanish (*Silene otites*) 14
Cat's-ear (*Hypochaeris radicata*) 17, 195
  Spotted (*Hypochaeris maculata*) 214
Centaury, Common (*Centaurium erythraea*) 7
Chough (*Pyrrhocorax pyrrhocorax*) 47
Clover
  Red (*Trifolium pratense*) 17, 27
  Sulphur (*Trifolium ochroleucon*) 17
  White (*Trifolium repens*) 1, **11**, 13, 14, **15**, 17, 32, **33**, 153, 174, 178, 224
Cock's-foot (*Dactylis glomerata*) 23, **24**, 27, **33**, 170, 184, 189
Comma (*Polygonia c-album*) 124, *124*
Copper, Small (*Lycaena phlaeas*) **38**
Cornflower (*Centaurea cyanus*) 211
Cornicularia aculeata 14, **15**, 63
Couch
  Common (*Elytrigia repens*) 134, 152, 153
  Sand (*Elytrigia juncea*) 7, *7*
Cowslip (*Primula veris*) 17, 136, 144, 160, 178, 228

Crake, Corn (*Crex Crex*) 28, 46, **49**
Crane's-bill
  Bloody (*Geranium sanguineum*) 28
  Meadow (*Geranium pratense*) **213**
  Wood (*Geranium sylvaticum*) 6, 23–4,
    24, *25*, 89, 91, 120, *150*, **213**
Cricket, Field (*Gryllus campestris*) 14, **37**
Ctenicera pectinicornis **37**
Cuckooflower (*Cardamine pratensis*) **213**
Curlew, Eurasian (*Numenius arquata*) 47,
  50, *50*, **51**, **53**, 93
Currant, Downy (*Ribes spicatum*) 28

Daisy (*Bellis perennis*) 27
  Oxeye (*Leucanthemum vulgare*) 17, **24**,
    136, *150*, 195, *196*, 210, 217
  Yellow Oxeye (*Buphthalmum
    salicifolium*) 211
Damselfly, Willow Emerald (*Chalcolestes
    viridis*) 127
Dandelion (*Taraxacum sect. Ruderalia*) 52,
  153
Dead-nettle, White (*Lamium album*) 127
Deer
  Fallow (*Dama dama*) **60**
  Red (*Cervus elaphus*) 57, 58, **60**
  Roe (*Capreolus capreolus*) 57, **60**
  Sika (*Cervus nippon*) 58
Docks (*Rumex*) 103–4
  Broad-leaved (*Rumex obtusifolius*) 94,
    103
  Curled (*Rumex crispus*) 94, 103
Dog-violet
  Common (*Viola riviniana*) **21**
  Heath (*Viola canina*) 13
Dog's-tail, Crested (*Cynosurus
    cristatus*) 3, 9, 13, 17, **18**, 23, **24**, **29**,
    32, **33**, 64, 89, 111, 120, 121, 153, 178
Dogwood (*Cornus sanguinea*) 30, 106
Dormouse (*Muscardinus avellanarius*) 58
Dove
  Stock (*Columba oenas*) 47, **49**
  Turtle (*Streptopelia turtur*) 47, **49**, 51
Dropwort (*Filipendula vulgaris*) 160, *160*,
  213
Duke of Burgundy (*Hamearis lucina*) 39,
  39, **40**
Dunnock (*Prunella modularis*) 46, **49**

Egret
  Great White (*Casmerodius albus*) 126
  Little (*Egretta garzetta*) 126
Elder (*Sambucus nigra*) 30
Elms (*Ulmus*) 30
Eyebrights (*Euphrasia*) 27, 181

Fern, Limestone (*Gymnocarpium
    robertianum*) **29**
Fescue
  Chewing's (*Festuca rubra ssp.
    commutata*) **187**
  Meadow (*Schedonorus pratensis*) **33**
  Red (*Festuca rubra*) 8, **11**, **15**, 17, **18**, **21**,
    23, **24**, 27, **29**, **33**, 153, 178, *196*,
    211
  Tall (*Schedonorus arundinaceus*) 184
Fieldfare (*Turdus pilaris*) 48, **49**, 52
Flax (*Linum usitatissimum*) 211
  Fairy (*Linum catharticum*) **10**, 178
  Perennial (*Linum perenne*) 211
Fleabane, Irish (*Inula salicina*) 136
Fox, Red (*Vulpes vulpes*) 57, 58, **59**
Foxtail, Meadow (*Alopecurus pratensis*) 3,
  17, **18**, **33**, 89, 153, 157, 170
Fritillary butterflies
  Dark Green (*Argynnis aglaja*) 39, **41**,
    136
  Glanville (*Melitaea cinxia*) **41**
  Heath (*Mellicta athalia*) 39, **42**
  Marsh (*Euphydryas aurinia*) 39, **41**, 79,
    126, 160, 161, 162
  Pearl-bordered (*Clossiana
    euphrosyne*) **42**
  Small Pearl-bordered (*Clossiana
    selene*) 39, **42**, 43
Fritillary (*Fritillaria meleagris*) 17, *119*
Frog, Common (*Rana temporaria*) 55, **55**

Gatekeeper (*Pyronia tithonus*) **38**
Gentian
  Early (*Gentianella anglica*) 68
  Field (*Gentianella campestris*) 29
Globeflower (*Trollius europaeus*) 6, 24, **24**,
  25
Goose
  Barnacle (*Branta leucopsis*) 48
  Bean (*Anser fabalis*) 48

Brent (*Branta bernicla*) 48
   Greylag (*Anser anser*) 48
   Pink-footed (*Anser brachyrhynchus*) 48
Gorse (*Ulex europaeus*) 30, 31
   Western (*Ulex gallii*) *168*
Grasshopper, Meadow (*Chorthippus parallelus*) *144*, 192
Grayling (*Hipparchia semele*) 4, 5, **42**
Greenweed, Dyer's (*Genista tinctoria*) 19, 90
Grouse, Black (*Tetrao tetrix*) 50, **51**, **53**

Hair-grass
   Crested (*Koeleria macrantha*) 10, **21**, **29**, 160
   Grey (*Corynephorus canescens*) 14
   Tufted (*Deschampsia cespitosa*) 130
   Wavy (*Deschampsia flexuosa*) 13, 14, **15**
Hairstreak, Green (*Callophrys rubi*) 39, **40**
Hare, Brown (*Lepus capensis*) **60**, 213
Harebell (*Campanula rotundifolia*) **10**, 13, 16, **21**, 27, 160, 190, **213**, *217*
Harrier
   Hen (*Circus cyaneus*) 48, 50, **51**
   Montagu's (*Circus pygargus*) 46
Hawkbits 153
   Autumn (*Scorzoneroides autumnalis*) 17
   Rough (*Leontodon hispidus*) 17, *196*
Hawthorn (*Crataegus monogyna*) 30, 31, 72, *108*, 127
Hazel (*Corylus avellana*) 30, 31
Heath, Small (*Coenonympha pamphilus*) 38–9, **38**, *39*
Heath-grass (*Danthonia decumbens*) **15**, 17
Heather (*Calluna vulgaris*) 13, 152, *168*
   Bell (*Erica cinerea*) 152, *168*
Hedgehog (*Erinaceus europaeus*) 57, **59**
Helleborine
   Dark-red (*Epipactis atrorubens*) 28
   Young's (*Epipactis helleborine* var. *youngiana*) 26
Herb-Paris (*Paris quadrifolia*) 6
Herb-Robert (*Geranium robertianum*) **29**
Heron, Purple (*Ardea purpurea*) 126
Hogweed (*Heracleum sphondylium*) 24, **24**, 162
   Giant (*Heracleum mantegazzianum*) 94
Holly, Sea (*Eryngium maritimum*) *8*

Hoverfly, Phantom (*Doros profuges*) 11, **37**, 43

Ichneuman stramentor 144
Ink-cap, Parasol (*Coprinus plicatilis*) 64
Iris, Yellow (*Iris pseudacorus*) 32
Ivy (*Hedera helix*) 30

Jackdaw (*Corvus monedula*) 47, 51
Jacob's-ladder (*Polemonium caeruleum*) 3
Juniper (*Juniperus communis*) 30, 31, 106

Kestrel (*Falco tinnunculus*) 47, *47*, **48**
Knapweed
   Common (*Centaurea nigra*) 3, 9, 17, **18**, 23, **24**, **29**, 32, **33**, 36, 89, 91, 111, 120, 136, *144*, 157, 178, *185*, 195, *196*, *217*, 228
   Greater (*Centaurea scabiosa*) 160, 228
Knotweed, Japanese (*Fallopia japonica*) 94

Lady's-mantles (*Alchemilla*) 23–4
   Alpine (*Alchemilla alpina*) **21**, 22
   Smooth (*Alchemilla glabra*) **24**
Lapwing, Northern (*Vanellus vanellus*) 28, **48**, 50, *50*, 51, 52, **53**, 66, 93, 132
Lime (*Tilia* × *europaea*) 127
Linnet (*Carduelis cannabina*) 46, **49**, 50, 52, **53**, 132
Lizard
   Sand (*Lacerta agilis*) 14, 54
   Viviparous (*Lacerta vivipara*) 54, *54*, **55**, 67
Lyme-grass (*Leymus arenarius*) 7
Lynx (*Lynx lynx*) 57

Magpie (*Pica pica*) 47
Marigold, Marsh (*Caltha palustris*) 17, **18**, 23, **24**, 89
Marjoram (*Origanum vulgare*) 36
Marram (*Ammophila arenaria*) 7, *7*
Marsh-bedstraw, Common (*Galium palustre*) 2
Marsh-orchid, Northern (*Dactylorhiza purpurella*) 27
Marten, Pine (*Martes martes*) 57, 58, *58*, **60**
Mat-grass (*Nardus stricta*) 2, 22, 153

Meadow-grass, Rough (*Poa trivialis*) 18, 23, **24**, 141, 170
Meadowsweet (*Filipendula ulmaria*) 3, **18**, **24**, 32, **33**, 91, 141
Mercury, Dog's (*Mercurialis perennis*) 6, 28
Merlin (*Falco columbarius*) 48, 50, **51**
Milkwort, Chalk (*Polygala calcarea*) 160
Mole (*Talpa europaea*) 57–8, **59**
Moor-grass
    Blue (*Sesleria caerulea*) 9, 11, **11**, 20, **21**, 22, 23, **29**, 31
    Purple (*Molinia caerulea*) 2, *2*
Moth, Black-veined (*Siona lineata*) *36*
Mouse
    Harvest (*Micromys minutus*) 58, **59**
    Wood (*Apodemus sylvaticus*) **59**
Mouse-ear
    Alpine (*Cerastium alpinum*) **21**
    Common (*Cerastium fontanum*) **15**
Mouse-ear-hawkweed (*Pilosella officinarum*) **11**
    Shetland (*Pilosella flagellaris ssp. bicapitata*) 26, 27
Mushroom, Field (*Agaricus campestris*) 63
Myrmica sabuleti **40**

Nanna brevifrons **37**
Nettle, Common (*Urtica dioica*) **33**
Newt
    Great Crested (*Triturus cristatus*) 55, 56, **56**, 67, 68
    Palmate (*Lissotriton helveticus*) 55, 56, **56**
    Smooth (*Lissotriton vulgaris*) 55, 56, **56**
Nightingale, Common (*Luscinia megarhynchos*) 46, *46*, **49**
Nightjar (*Caprimulgus europaeus*) 46

Oaks (*Quercus*) 127
Oat-grass
    Downy (*Avenula pubescens*) 10, **11**, 160
    False (*Arrhenatherum elatius*) 3, 17, **29**, 32, *32*, **33**, 120, 130, 157, 162, **187**
    Meadow (*Avenula pratense*) 10, **10**, 11, 13
    Yellow (*Trisetum flavescens*) 32, **33**, 140, 178

Omiamima mollina **37**
Onion, Wild (*Allium vineale*) 8
Orchid
    Early-purple (*Orchis mascula*) 5
    Fragrant (*Gymnadenia conopsea*) 12
    Greater Butterfly (*Platanthera chlorantha*) 12
    Green-winged (*Anacamptis morio*) 17, 20, 144
    Lady's-slipper (*Cypripedium calceolus*) 68
    Lizard (*Himantoglossum hircinum*) 8
    Military (*Orchis militaris*) 11
    Monkey (*Orchis simia*) 10, 12
    Musk (*Herminium monorchis*) 12
    Pyramidal (*Anacamptis pyramidalis*) 136
    Small-white (*Pseudorchis albida*) **24**
Owl, Short-eared (*Asio flammeus*) 50, **51**
Oystercatcher (*Haematopus ostralegus*) 50

Painted Lady (*Vanessa cardui*) 38, **38**
Pansy, Mountain (*Viola lutea*) 23, **24**
Parsnip, Wild (*Pastinaca sativa ssp. sylvestris*) 32, **33**
Partridge, Grey (*Perdix perdix*) **49**, 50, 66, 132
Pasqueflower (*Pulsatilla vulgaris*) 11
Pearlwort, Alpine (*Sagina saginoides*) **21**
Penny-cress, Alpine (*Noccaea caerulescens*) 26
Peregrine (*Falco peregrinus*) 48
Pigeon, Wood (*Columba palumbus*) 51
Pignut (*Conopodium majus*) 6, 23, **24**
Pines (*Pinus*) 31
Pink
    Carthusian (*Dianthus carthusianorum*) 211
    Deptford (*Dianthus armeria*) 14
Pipistrelles (*Pipistrellus*) 62
    Common (*Pipistrellus pipistrellus*) **61**
    Nathusius' (*Pipistrellus nathusii*) **61**
    Soprano (*Pipistrellus pygmaeus*) **61**
Pipits (*Anthus*) 47
    Meadow (*Anthus pratensis*) **49**, 50, **53**, 79
    Tree (*Anthus trivialis*) 31, 47

Plantain, Ribwort (*Plantago lanceolata*) **10**, 17, **41**, 157, 178, 195, *196*, 228

Plover, Golden (*Pluvialis apricaria*) 48, 50, 52, **53**

Polecat (*Mustela putorius*) 57, 58, **60**

Primrose (*Primula vulgaris*) *150*
    Bird's-eye (*Primula farinosa*) 23

Puffball, Meadow (*Vascellum pratense*) 63

Quail (*Coturnix coturnix*) 46

Quaking-grass (*Briza media*) 10, **10**, 135, 160, 178

Rabbit, European (*Oryctolagus cuniculus*) 13, **15**, **60**, 83, 113, 150, 160, 161, 208, 213

Ragged-Robin (*Silene flos-cuculi*) *24*, *184*

Ragwort, Common (*Senecio jacobaea*) 75, 82, 94, 95, 97–100, 207

Rampion, Round-headed (*Phyteuma orbiculare*) *14*

Redshank (*Tringa totanus*) 93

Redwing (*Turdus iliacus*) 48, **49**, 52

Restharrow (*Ononis repens*) *8*

Ringlet, Mountain (*Erebia epiphron*) **42**, 125

Robberfly, Hornet (*Asilus crabroniformis*) **37**

Rock-cress, Northern (*Arabidopsis petraea*) 27

Rock-rose
    Common (*Helianthemum nummularium*) 2, 10, **10**, **11**, **21**, **29**, 160, **213**
    Hoary (*Helianthemum oelandicum*) **11**

Rocket, Sea (*Cakile maritima*) *7*

Rook (*Corvus frugilegus*) 48

Rowan (*Sorbus aucuparia*) 28, 31

Ruff (*Philomachus pugnax*) 66

Rushes (*Juncus*) 72, 94
    Jointed (*Juncus articulatus*) 2
    Soft (*Juncus effusus*) 2
    Toad (*Juncus bufonius*) 139, 141

Rye-grass
    Italian (*Lolium multiflorum*) 211
    Perennial (*Lolium perenne*) 1, 18, **24**, 32, **33**, 64, 89, 94, 120, 121, 122, 153, 184

Sandwort
    Arctic (*Arenaria norvegica ssp norvegica*) 26
    Sea (*Honckenya peploides*) *7*
    Spring (*Minuartia verna*) **21**, 26

Saw-wort (*Serratula tinctoria*) 160

Saxifrage
    Meadow (*Saxifraga granulata*) **213**
    Pepper (*Silaum silaus*) 17

Scabious
    Devil's-bit (*Succisa pratensis*) **41**, 160, 161, 214
    Field (*Knautia arvensis*) **213**, 228
    Small (*Scabiosa columbaria*) **11**, *135*, 160, 228

Sedge
    Glaucous (*Carex flacca*) 10, 20, **21**, 22
    Sand (*Carex arenaria*) 7, *8*, 13, 14, **15**, 63

Seed-eater, Brush-thighed (*Harpalus froelichii*) **37**, 44

Selfheal (*Prunella vulgaris*) 27, 157, 178

Sheep's-fescue (*Festuca ovina*) 1, 10, **10**, 11, **11**, 13, 14, **15**, 20, **21**, 22, **24**, 26, **29**, 31, 135, 160, *168*, 178
    Fine-leaved (*Festuca filiformis*) *168*

Shrew
    Common (*Sorex araneus*) **59**
    Pygmy (*Sorex minutus*) **59**

Skipper
    Dingy (*Erynnis tages*) **40**
    Essex (*Thymelicus lineola*) **38**, 124
    Grizzled (*Pyrgus malvae*) **40**
    Large (*Ochlodes venatus*) **38**
    Lulworth (*Thymelicus acteon*) **40**
    Silver-spotted (*Hesperia comma*) **40**, 43, 125, *125*
    Small (*Thymelicus sylvestris*) **38**

Skylark (*Alauda arvensis*) 46, *48*, **49**, 50, 51, 52, **53**, 66, 79, 91, 93

Slow Worm (*Anguis fragilis*) 54, **55**, 67

Small-reed, Wood (*Calamagrostis epigejos*) 72, 130

Snail
    Geyer's Whorl (*Vertigo geyeri*) **37**
    Roman (*Helix pomatia*) 34
    Round-mouthed Whorl (*Vertigo genesii*) **37**

Snake
  Grass (*Natrix natrix*) 54, **55**, 67
  Smooth (*Coronella austriaca*) 14, 54, **55**
Sneezewort (*Achillea ptarmica*) 217
Snipe (*Gallinago gallinago*) 93
Soft-brome (*Bromus hordeaceus*) 174
Solomon's-seal (*Polygonatum multiflorum*) 28
Sorrels (*Rumex*) 52
  Common (*Rumex acetosa*) *185*, 199
  Sheep's (*Rumex acetosella*) 13, 14, **15**, 167, *168*
Sparrow
  House (*Passer domesticus*) 50
  Tree (*Passer montanus*) **49**, 51, 66
Sparrowhawk (*Accipiter nisus*) 48
Speedwell, Thyme-leaved (*Veronica serpyllifolia*) 153
Spider
  Purseweb (*Atypus affinis*) *144*
  Wasp (*Argiope bruennichi*) 136
Spleenwort
  Forked (*Asplenium septentrionale*) 26
  Green (*Asplenium viride*) 28, **29**
  Maidenhair (*Asplenium trichomanes*) **29**
Spoonbill (*Platalea leucorodia*) 126
Spotted-orchid, Common (*Dactylorhiza fuchsii*) *12*, 27, *148*
Spurge
  Leafy (*Euphorbia virgata*) 136
  Portland (*Euphorbia portlandica*) 7
  Sea (*Euphorbia paralias*) 7
Squinancywort (*Asperula cynanchica*) **11**
Squirrel, Red (*Sciurus vulgaris*) 58
Starling, Common (*Sturnus vulgaris*) 48, **49**, 52, **53**
Stoat (*Mustela erminea*) **60**
Stone-curlew (*Burhinus oedicnemus*) 46, **48**, **53**, 79, 132–3, *132*
Stonechat (*Saxicola torquatus*) 31
Stonecrop, Biting (*Sedum acre*) **29**
Swallow, Barn (*Hirundo rustica*) 47, *48*, 52
Swan
  Bewick's (*Cygnus columbianus*) 48
  Whooper (*Cygnus cygnus*) 48
Swift (*Apus apus*) 47, 52

Thistles 94, 95, 103, 134, 207

Carline (*Carlina vulgaris*) 1, **10**
Creeping (*Cirsium arvense*) 94, 103, 152, 162
Meadow (*Cirsium dissectum*) 2
Melancholy (*Cirsium heterophyllum*) 6, 7, 24
Queen Anne's (*Cirsim canum*) 136
Spear (*Cirsium vulgare*) 94, 103
Thrift (*Armeria maritima*) 26
Thrushes (*Turdus*) 47, 48
  Mistle (*Turdus viscivorus*) 47, **49**
  Song (*Turdus philomelos*) **49**, 52, **53**
Thyme
  Large (*Thymus pulegioides*) **213**
  Wild (*Thymus polytrichus*) 7, 10, 11, **11**, 20, **21**, 22, **29**, 31, **40**
Timothy (*Phleum pratense*) 184
Toad
  Common (*Bufo bufo*) 55, **55**, 56, *56*
  Natterjack (*Bufo calamita*) 55–6, **55**, 68
Toadflax, Ivy-leaved (*Cymbalaria muralis*) 127
Toadstool, Fairy-ring (*Marasmius oreades*) 64
Tor-grass (*Brachypodium rupestre*) **10**, 13, *36*, 130, 154, 159, 162
Tormentil (*Potentilla erecta*) 2, 5, **11**, 13, **21**
Twayblade, Common (*Neottia ovata*) 5, *12*
Twite (*Carduelis flavirostris*) 50, **51**

Vernal-grass, Sweet (*Anthoxanthum odoratum*) **11**, **18**, 23–4, **24**, 89, 120, 170, *196*
Vetch
  Horseshoe (*Hippocrepis comosa*) **11**, **213**
  Kidney (*Anthyllis vulneraria*) 27, **40**, 135, 228
Vetchling, Meadow (*Lathyrus pratensis*) 115, 140
Viper's-bugloss (*Echium vulgare*) 7, *8*, 136
Vole
  Bank (*Clethrionomys glareolus*) **60**
  Field (*Microtus agrestis*) 57–8, **59**

Wagtails (*Motacilla*) 47, 52
  Yellow (*Motacilla flava*) 47, 48, **49**, 50, 51, 79, 128, 132
Wall-rue (*Asplenium ruta-muraria*) **29**

Warbler
    Cetti's (*Cettia cetti*) 126
    Dartford (*Sylvia undata*) 31
    Garden (*Sylvia borin*) 47
    Grasshopper (*Locustella naevia*) 46, **49**
    Willow (*Phylloscopus trochilus*) 46, **49**
Waxcaps (*Hygrocybe*) 63–4, *64*
    Pink (*Hygrocybe calyptriformis*) 63
    Vermillion (*Hygrocybe miniata*) 63, *64*
Wayfaring-tree (*Viburnum lantana*) 30
Weasel (*Mustela nivalis*) **59**
Weevil-wasp, Four-banded (*Cerceris quadricincta*) **37**
Wheatear (*Oenanthe oenanthe*) 46, 50, 79
Whimbrel (*Numenius phaeopus*) **51**
Whinchat (*Saxicola rubetra*) 47, **51**, **53**, 93
White, Marbled (*Melanargia galathea*) **38**, *144*
Whitebeams (*Sorbus*) 30
    Common (*Sorbus aria*) 30
Whitethroat, Lesser (*Sylvia curruca*) 47
Wigeon (*Anas penelope*) 48
Wildcat (*Felis silvestris*) 57, 58, **60**, 68
Willows (*Salix*) 3
    Creeping (*Salix repens*) 7
    Downy (*Salix lapponum*) 31
    Net-leaved (*Salix reticulata*) **21**

Wolf (*Canis lupus*) 57
Wood, Speckled (*Pararge aegeria*) 124
Wood-rushes (*Luzula*) 199
    Field (*Luzula campestris*) **11**, **15**
Woodlark (*Lullula arborea*) 46
Woundwort, Hedge (*Stachys sylvatica*) 157

Yarrow (*Achillea millefolium*) **15**, 17, 157, 190
Yellow, Clouded (*Colias croceus*) 125
Yellow-rattle (*Rhinanthus minor*) 17, 93, *135*, 170, 178, 180–1, *180*, 195, *196*, 210, *217*, 228
    Greater (*Rhinanthus angustifolius*) 17
Yellowhammer (*Emberiza citrinella*) 46, **49**, 51, **53**
Yew (*Taxus baccata*) 30
Yorkshire-fog (*Holcus lanatus*) 5, 14, **15**, 17, **18**, 23, **24**, 27, 30, **33**, 153, *174*

Subject index

Page numbers in **bold** indicate tables and in *italic* indicate figures, photographs and captions.

abandoned sites
  and invertebrates 157, 158
  natural succession 14, 17, 131, 136, 138
  restoration of 131, 138, 139–40, *139*,
    156
  scrub encroachment 14, 17, 30
acid grasslands 1–2, *2*
  dry lowland communities *2*, 13–15,
    15, *16*
  fungi 64
  grazing 14, **15**
  invertebrates 36, **37**, 44
  recreation 167
  scrub communities 31
  soil pH 2, 13, 120
  sward diversity 120, *121*
acid soils 2, 111
acidification 167
aerial photographs 104, 112, 114
afforestation 44, 115, 139, *139*
aftermath grazing 25, 62–3, 90, 91–3, *92*,
  156, 157
agri-environment schemes 82, 98, 109,
  147, 149–50, 154, 205, 220, 224
agricultural improvement 13, 17–18, 23,
  24, 44, 89, 115, **116**
alternative stable states 219
AMF *see* arbuscular mycorrhizal fungi
  (AMF)
amphibians 14, 55–7, **55–6**, *56*, 67–8,
  125–6, 128
Animal Health Plans 82
animal welfare 82
AONBs *see* Areas of Outstanding Natural
  Beauty (AONBs)
arable cropping, to reduce fertility 164–5,
  218
arable reversion sites 210–13, **212**
  case study 132–6, *133*, *134*, *135*

cutting 152–3, 181
long-term vegetation
  development 220
natural colonisation 137
seed mixes 179–80
species establishment 206
arbuscular mycorrhizal fungi (AMF) 63,
  169–70
archaeological sites 112–14, *112*, *113*, *114*
  Seven Barrows, Berkshire Downs 158–
    62, *159*, *160*, *161*, *162*
Areas of Outstanding Natural Beauty
  (AONBs) 85, 147
Ash–Rowan–Dog's Mercury community
  (W9) 28
Askrigg Bottoms, Wensleydale *25*, *153*,
  180
assessments
  of archaeological sites 113–14
  of botanical enhancement
    potential *174*
  monitoring success 222–4, **222**, **223**
  of scrub 104–5
  soil analysis 111, 117–18, **117**, 172–3
  see also risk assessments; surveys
Asulam 101, 103
atmospheric nitrogen deposition 14, **116**,
  118, 130, 156
atmospheric sulphur deposition 130
Austria 143
autumn grazing 80, **80**, 91, 94, 162
autumn sowing 211, **212**
avermectin 44, 45

bacteria, soil 169–70
BAP *see* Biodiversity Action Plan (BAP),
  UK
bare ground
  and grazing 69, 70, 74, 75, 101, 150

and hay cutting 90, 93
preparation for sowing 205–8, *207*, 218
and weeds 95, 101
barrows *see* Seven Barrows, Berkshire
Downs
bats 58–62, **61–2**, *62*, 68
Beech Farm Estate, East Sussex 193–7,
193, 194, 196, 197
bees **37**, 44, 126, 128, 143, *144*
beetles 11, **37**, 43, 142, 143, *144*, 158, 169,
171, 192
Belgium 141
Bentley Station Meadow, Hampshire *101*
Bheinn Shuardail, Skye *29*
Biodiversity 2020 148
Biodiversity Action Plan (BAP), UK 1, 14,
20–2, 28, 129, 147, 149, 175
Biodiversity Action Reporting
System 148
biodiversity surveys 64–8
Biological Records Centres 65
biomass harvesting 230–1
birds
and agricultural intensification 143–4
and climate change 127–8
conservation 50–1
and fertiliser application 92–3, 109
and grazing 52, 73, 78, 79, **79**, 92
and hay cutting 91, 128, 191
lowland 46–8, *46, 47*, **48–9**, *48*, **53**
and seed harvesting 191
and silage 52
and site management 51–2, **53**
surveys of 66
and sward structure 52, **53**
upland 50, *50*, **51**, **53**
and weed control 95
Birds of Conservation Concern
(BoCC) 46, 47, **48–9**, 50, **51**, **53**
Blackthorn–Bramble scrub community
(W22) 30
Blue Moor-grass–Limestone Bedstraw
community (CG9) 9, 11, **11**, 20, **21**,
22, **29**, 31
Hoary Rock-rose–Squinancywort
subcommunity (CG9a) **11**
Blue Moor-grass–Small Scabious
community (CG8) **11**

BoCC *see* Birds of Conservation Concern
(BoCC)
botanical enhancement potential *174*
Box Hill, Surrey *31*
Bramble–Yorkshire-fog scrub community
(W24) 30
Breckland 10, **11**, 14, 15, 46, 63, 211
Breed Profiles Handbook 72–3, 74, 76
Bristle Bent community (U3) **15**
British Trust for Ornithology (BTO) 50,
66
broadcasting of seed 134, *134*, 208,
209–10
browsing 74, 76, 106–7, **107**, *108*
brush harvesting 86–7, *87*, 199, *200*
advantages and limitations 186–7,
188, **190**
and invertebrates 191
roadside verge case study 228, *229*
seed quantity and composition 186–7,
187, 188–90, **189**, 195–6, *196*
brushcutters 105, 106, 162
bryophytes 26, **29**, 63, 151, 152
BTO *see* British Trust for Ornithology
(BTO)
bumblebees **37**, 143, *144*
bund construction 145
burial mounds, Bronze Age *114*, 158–62,
159, 160, 161, 162
burning 31, 95, 154
burrowing animals 113, 150, 160
see also rabbits
bush crickets 11, **37**, 43, 126, 192
butterflies 37–43, 136, *148*
and climate change 124–5, *124, 125,
126*, **126**, 127
conservation management 41–3
habitat specialists 39–41, *39*, **40–1**, *41*,
42, *43*
natural colonisation 142
Seven Barrows case study 160, 161,
162
surveys of 67
and sward structure 158
wider countryside species 37–9, **38**, *39*
Buxton Climate experiment *129*

calaminarian grasslands 26, *26, 27*

calcareous grasslands 1, 2, *2*
  arable reversion case study 132–6, *133,*
    134, 135
  climate change impacts 128–9, *129*
  fungi 64
  grazing **10**, 13, 22–3, 161–2
  invertebrates 36, **37**, 43–4, 45
  lichens 63
  lowland communities *2*, 10–13, **10–11**,
    12, 14
  management case study 158–62, *159,*
    160, 161, 162
  roadside verge case study 224–31, *225,*
    227, 229, 230, 231
  scrub communities 30, *31*
  semi-improved 32, **33**
  soil pH 1, 10, 20, 120
  steep slopes 145
  sward diversity 120, *121*, 155–6, *155*
  upland communities *2*, 20–3, **21**, *22*, *23*
calcareous sands 27
calcareous soils 1, 20, 111
carbon-nitrogen (C:N) ratio 169
carbon treatments 169, 171, 218
Carboniferous limestone 10, **10**, **11**, 20,
  21, 28
Caring for God's Acre 230
case studies
  arable reversion to chalk
    grassland 132–6, *133, 134, 135*
  roadside verge grassland
    restoration 224–31, *225, 227, 229,*
    230, 231
  UK Native Seed Hub 198–204, *198,*
    199, 200, 201, 202
  upland hay meadow restoration 84–9,
    84, 86, 87, 88
  whole-crop method of hay
    transfer 193–7, *193, 194, 196, 197*
Castle Hill National Nature Reserve, East
  Sussex 138
Castle Meadows, Oxfordshire 145
cattle 69–70, **71**, 72–3, **72**, *73*, *102*,
  107, *108, 109*, 141, 150, 151, 152,
  156, 161
cemeteries 64
  see also Seven Barrows, Berkshire
    Downs

CG1 (Sheep's-fescue–Carline Thistle)
  community 1, **10**
CG2 (Sheep's-fescue–Meadow Oat-grass)
  community **10**, 13, *14*
CG3 (Upright Brome) community **10**, *14*
CG4 (Tor-grass) community **10**, *12*
CG5 (Upright Brome–Tor-grass)
  community **10**
CG6 (Downy Oat-grass) community **11**
CG7 (Sheep's-fescue–Mouse-ear-
  hawkweed–Wild Thyme)
  community **11**
CG8 (Blue Moor-grass–Small Scabious)
  community **11**
CG9 (Blue Moor-grass–Limestone
  Bedstraw) community 9, 11, **11**, 20,
  21, *22*, **29**, 31
  CG9a (Hoary Rock-rose–
    Squinancywort)
    subcommunity **11**
CG10 (Sheep's-fescue–Common Bent–
  Wild Thyme) community 11, **11**,
  20, **21**, *22*, 31
  CG10a (White Clover– Field Wood-
    rush) subcommunity **11**
CG11 (Sheep's-fescue–Common
  Bent–Alpine Lady's-mantle)
  community **21**, 22
CG12 (Sheep's-fescue–Alpine Lady's-
  mantle–Moss Campion)
  community **21**, 22
CG13 (Mountain Avens–Glaucous Sedge)
  community 20, **21**, 22
CG14 (Mountain Avens–Moss Campion)
  community **21**, 22
chaff, seed-rich 186
chalk 10, 11, 13
chalk grasslands *see* calcareous
  grasslands
chalk heaths 2, 111
chemicals
  ground preparation for sowing 205,
    206, 207
  molluscicides 209, 213
  reducing grass competition 181
  and slot seeding 208–9, *209*
  soil fertility manipulation 167–9, *168,*
    218

stump treatment 105, 106, 160
weed control 96–7, 99, *100*, 101, 103,
    134
    see also fertiliser application; lime
    application
churchyards 17, 64, 189
climate change 3, 13, 14–15, **116**, 123–9,
    148
    and grassland communities 128–9, *129*
    and phenology 127–8
    projections for Britain 123, 126–7, 129
    and species ranges 124–7, *124*, *125*,
    126, *127*
Climatic Risk Atlas of European
    Butterflies 125
clopyralid 103
Codes of Practice 96, 97, 98
collembolans 171
colonisation *see* natural colonisation
Colsterworth Glebe Quarry,
    Lincolnshire 226
Colt Park, Yorkshire Dales 170, 220
combine-harvesting 186, 187, **188**, 189,
    189, **190**
Common Standards Monitoring Guidance for
    Birds 66
community surveys 226, *227*
completeness index 220–1
conservation
    birds 50–1
    butterflies 41–3
    semi-natural dry grassland
        habitats 147–8
conservation designations 98, 147, 175–6
continuous grazing 80–1, **80**
Control of Pesticides Regulations 1986 96
Control of Substances Hazardous to
    Health (COSHH) assessments 96,
    97
Copper Hill SSSI, Lincolnshire 226
cost-effectiveness 220–2
costs
    grazing 81, 82
    green hay transfers 85–6, 222
    of restoration 220–2, *221*
Countryside and Rights of Way Act
    2000 82
Countryside Stewardship Scheme 149

County Wildlife Sites 175
Cow Green, Teesdale 22
Crested Dog's-tail–Common Knapweed
    community (MG5) 3, 9, 17, **18**, *19*,
    23, **24**, **29**, 89, 111, 120, 178
    Heath-grass subcommunity
        (MG5c) 17
    Lady's Bedstraw subcommunity
        (MG5b) 17
    Meadow Vetchling subcommunity
        (MG5a) 115
Crested Dog's-tail–Marsh Marigold
    community (MG8) 17, **18**, 23, **24**, 89
Cricklade North Meadow National
    Nature Reserve, Wiltshire *145*
crushing of Bracken 102
cutting
    scrub management 105–6
    weed control 99, *99*, *101*, 102, 103
    see also hay cutting
Czech Republic 136–7, 210, 220

daily stock checks 82–3
Dalradian limestone 28
damselflies 126, *127*
decomposers 142
deep cultivation 166
deer 3, 23, 57, 58, **60**, 141
Defra guidance
    code of good agricultural practice 95,
        96
    Common Ragwort 98, 99, *99*, 100, *100*
    conservation 147–8
    fertilisers 111
    livestock welfare 82
    weed control 96, 97, *99*, *100*, 103
Department for Environment, Food &
    Rural Affairs *see* Defra guidance
desk studies 114
direct drilling 208, 209
dispersal, propagule 130, 140–2
diversity *see* species diversity
dogs, and livestock 74, 83
domestic animals *see* livestock
donkeys **76**
Downy Oat-grass community (CG6) **11**
dragonflies 126
drainage 2–3, 145

drought 13, 128, 145
dry grasslands *see* semi-natural dry
    grassland communities
dry scrub communities 30–1, *31*
drying seed 200–3, *200, 201, 202*
dung-feeding invertebrates 44, 45, 58, 69
Durness limestone **21**, 28
dwarf-herb communities **21**, 22, 31
dwarf shrub heaths 13

earthworks 112, *112*
earthworms 52, 171
ecological traits 176–8, 185
Ellenberg indicator values 145–6
enclosure 226
Environment Agency 96
Environmental Stewardship 98, 147, 149,
    205
Environmentally Sensitive Areas
    (ESAs) 120–1, *121*, 149, 205
EPS *see* European Protected Species (EPS)
Estonia 140
EU Habitats and Species Directive 11–13,
    14, 17, 24, 26, 28, 30, 62
European Grassland Butterfly
    Indicator 41
European Protected Species (EPS) 54, 56,
    67, 68
eutrophication **116**, 118, 122
    atmospheric nutrient deposition 14,
        116, 118, 130, 156
    grazing 164
    hay meadows 25
    supplementary feeding 70, 75, 78, 164
    see also fertiliser application
extreme weather events 123

fallowing 122, 163–4, 218
False Oat-grass community (MG1) 32, **33**,
    120, 157
    Common Knapweed subcommunity
        (MG1e) 32, **33**
    Common Nettle subcommunity
        (MG1b) **33**
    Meadowsweet subcommunity
        (MG1c) 32, **33**
    Red Fescue subcommunity (MG1a) **33**

Wild Parsnip subcommunity
    (MG1d) 32, **33**
Farm Environment Plan (FEP) Manual 65,
    66, *174*
farmyard manure 52, 92–3, 109–10, *118*,
    141
F:B ratios 169–70
fen meadow communities 2
fencing 83, 113
fertiliser application 94, 107–11, 115, 130
    farmyard manure 52, 92–3, 109–10,
        118, 141
    and fungi 64
    and ground-nesting birds 92–3, 109
    long-term field trials 118, *118, 119,*
        120, 137–8
    removal of residual phosphorus 164–5
    and weeds 95
    see also lime application
field trials 118, *118, 119, 120*, 137–8
fire risks 128
flood meadows 17, **18**, **33**
flooding 145
Flora Locale 149
flowering dates, early 127
fluroxypyr 103
Food and Environment Protection Act
    1985 96
forests 3, 4–5
    see also afforestation
France 140, 151–2, 166
frost heaving 15
fungi 62–4
    macrofungi 63–4, *64*
    mycorrhizal 63, 142, 169–70, 171

GAP *see* Grazing Animals Project (GAP)
geological information 172
Germany 136, 137, 139, 151, 152, 153, 163,
    166, 210, 213
    Rengen Grassland Experiment 118,
        120
glacial conditions 3
Glastir 98, 147
gleyed horizons 173
glyphosate 96, 101, 105, 134, 206
goats **107**, *108, 109*

Gorse–Bramble scrub community (W23) 30, 31
graminicides 181
grasshoppers 126, 142, 143, *144*, 192
grassland enhancement **212**
grazing 69–83
  acid grasslands 14, **15**
  aftermath 25, 62–3, 90, 91–3, *92*, 156, 157
  agreements 82
  animal management 81–3, *82*
  animal welfare 82
  on archaeological sites 113, 161–2
  and birds 52, 73, 78, 79, **79**, 92
  calcareous grasslands **10**, 13, 22–3, 161–2
  cattle 69–70, **71**, 72–3, **72**, 73, *102*, **107**, *108, 109*, 141, 150, 151, 152, 156, 161
  continuous 80–1, **80**
  duration **77**, 78, *78*, 79–81
  and fungi 62–3
  ground preparation for sowing 205, 206–7, *207*
  and herbicide use 97, 101
  horses and ponies **71**, 75–6, **76**, *76*, 141, 161
  and invertebrates 44, 45, 69, 70, 72, 73, 75, 78, **79–80**, 80, 151
  limestone pavements 28
  livestock selection 70–1, **71**
  overgrazing 17, 23, 28, 44, 45, 70, 74, 75, 95, 101, **116**, 122
  poisoning of livestock 76, 82, 97–8, 101, 102
  Rabbits 83
  replacing with cutting 93–4
  scrub management 31, 74, 76, 105, 106–7, **107**, *108*
  seasonal 27, 79–80, **79–80**, *81*, 91–2, 93, 94, 151, 156, 162
  sheep **71**, 73–4, **74**, *75*, **107**, *109*, 141, 150, 151, 152, 156
  and soil fertility 164
  and species diversity 150–2, *151*, 155–7, *155*, 218
  stocking densities 77–8, **77**, *78*, 130, 151–2

and sward structure 78, 80–1, 93, 150–2
  timing 79–81, **79–80**, *81*, 91–2, 162
  undergrazing 14, 17, 23, 26, 28, 44, 70, **116**, 122
  weed control 95, 99, 102, *102*, 103
Grazing Animals Project (GAP) 70–1, 73, 82
Grazon 90 103
Great Orme 4, *5*, *29*
green hay transfers 85–6, *86*, 89, 149
  advantages and limitations 187, **188**, **190**
  costs 85–6, 222
  and invertebrates 192
  seed quantity and composition 185, 188–9, **189**, 190, 195–6, *196*
  and species diversity 216, *217*
  thickness 187
  whole-crop method 185, 193–7, *193*, *194*, *196*, *197*
Green Spleenwort–Brittle Bladder-fern community (OV40) **29**
ground-nesting birds 46, **48–9**, 50, **51**
  and agricultural intensification 143–4
  and fertiliser application 92–3, 109
  and grazing 52, 73, 78, 79, **79**, 92
  and hay cutting 91, 128, 191
  and seed harvesting 191
  and silage 52
  surveys of 66
  and sward structure 52, **53**
  and weed control 95
ground preparation treatments 205–8, *207*, 218
ground water levels 45, 145, *145*
Groundwork Trust 167, 169

habitat destruction 115, **116**
habitat deterioration **116**, 122
habitat fragmentation 13, 70, 115, **116**, 126, 226
  and invertebrates 142–3, *144*
habitat surveys 65
habitats of high invertebrate biodiversity value 66–7
hand-held chemical sprayers 97
hand seed collection 88, 190, 199, *199*, 228

hand sowing 134, *134*
hand tools 105–6
hand weeding 95, 98–9, *99*, 103
Handbook for Phase 1 habitat survey 65
handling facilities for livestock 83
harrowing 156, 194, 195–7, *196*, 205–6,
   207, *207*
Hawthorn–Ivy scrub community
   (W21) 30
hay bales 149, 186, 216, 221
hay concentrate 86, *87*
hay cutting 25, 89–94
   and fungi 62–3
   and ground-nesting birds 91, 128, 191
   and invertebrates 44, 45, 93–4, 191–2,
      192
   roadside verges 227–8
   and soil fertility 91, 164, 218
   and species diversity 152–4, 155–7,
      155, 218
   and sward structure 152
   timing 90–1, *90*, 92, *92*, **116**, 152–4, 181
   and Yellow-rattle 181
   see also hay transfers
hay meadows 2, 6–7, *6*, 9
   aftermath grazing 25, 62–3, 90, 91–3,
      92, 156, 157
   birds 50, 52
   climate change impacts 128
   fertiliser application 108, 109–10
   invertebrates 36, **37**, 44–5
   lime application 110–11
   lowland communities 17–18, **18**, *19*, *20*
   restoration case study 84–9, *84*, *86*, *87*,
      88
   upland communities 23–5, **24**, *25*
   waterlogging 145, *145*
   see also hay cutting
hay quality 89, 94
   ragwort contamination 98
Hay Time Project, Yorkshire Dales 25,
   84–9, *84*, *86*, *87*, *88*
hay transfers 85–6, *86*, 89, 141–2, 149,
   185–91
   advantages and limitations 186–7,
      188, **190**
   adverse impacts of 191–2, *192*
   costs 85–6, 222

and invertebrates 142, 191–2
seed quantity and composition 185–7,
   187, 188–9, **189**, 190–1, 195–6, *196*
and species diversity 216, *217*
thickness 187
timing 190–1
whole-crop method 185, 193–7, *193*,
   194, *196*, *197*
Health and Safety at Work Act 1974 96
heathland recreation 166–7
heavy metals 26
hemiparasitic plants 170, 180–2
Herbicide Handbook, The 96
herbicides
   ground preparation for sowing 205,
      206, 207
   reducing grass competition 181
   and slot seeding 208–9, *209*
   stump treatment 105, 106, 160
   weed control 96–7, 99, *100*, 101, 103,
      134
Herpetofauna Workers Manual 67–8
Historic England 112, 114
Historic Environment Records
   (HERs) 112, 113
historic sites *see* archaeological sites
Hoe Road Memorial Meadow,
   Hampshire 19
horses and ponies **71**, 75–6, **76**, *76*, **107**,
   141, 161
Hortobágy National Park, Hungary 143,
   152, 179
humus types 173
Hungary 136, 143, 152, 179
hybridisation 58
hydrological regimes 145–6, *145*

immobilisation of nutrients 169, 218
improved grasslands 1, 64, 120, 122, **223**
   see also semi-improved grasslands
'in-by' land 23
indicator values, Ellenberg 145–6
inorganic fertilisers 107–8, 110
   see also fertiliser application
insects *see* invertebrates
interglacial periods 3
invertebrates 14, 34–7, **35**, **37**
   and burning 154

dispersal opportunities 142
and grazing 44, 45, 69, 70, 72, 73, 75,
78, **79–80**, 80, 151
and habitat fragmentation 142–3, *144*
and hay cutting 44, 45, 93–4, 191–2,
192
and hay transfers 142, 191–2
machair 28
and restoration management 157–8
and seed harvesting 142, 191–2, *192*
and site management 43–5, 52
soil communities 171
surveys of 66–7
and sward structure 34, **35**, 36–7, 44,
45, 142, 143, 157–8
on weed species 94–5
see also butterflies
Iron Age hill forts *112, 114*

Jarlshof, Shetland *113*
Joint Nature Conservation Committee
(JNCC) 13, 14, 22–3, 26, 28, 50, 122
Juniper heath community (W19) 31
Jurassic limestone **10**, 11

Keen of Hamar, Shetland 26, *26, 27*
knapsack spraying 97, 101

Lammas meadows 91–2
land abandonment *see* abandoned sites
land conversion 115, **116**
Land Information System 172
Landlife 149
landscaping works 144–5
Late Glacial Maximum period 3, 6
Lathkill Dale, Peak District National
Park 22
leaching 94, 110, 111, 122, 163, 164
leaf litter *see* litter removal
ledge communities **21**, 22, 31
Leyburn Old Glebe Nature Reserve,
Yorkshire *6*
lichens **15**, 26, 63, 186
LiDAR 114
Life on the Verge project 224–31, *225,
227, 229, 230, 231*
lime application 110–11, 115, *118, 119,*
167

limestone 1, 4, 10, **10**, 11, **11**, 20, **21**, 28,
33, 186
Limestone Fern–False Oat-grass
community (OV38) **29**
limestone pavements 4, 28, **29**, *29*
Lincolnshire 14
Lincolnshire Wildlife Trust (LWT) 224–
31, *225, 227, 229, 230, 231*
litter removal 90, 156, 181
livestock 13, **15**, 22–3, 69–70
cattle 69–70, **71**, 72–3, **72**, *73*, *102*, **107**,
108, 109, 141, 150, 151, 152, 156,
161
daily checks 82–3
grazing characteristics **71**
horses and ponies **71**, 75–6, **76**, *76*, **107**,
141, 161
management 81–3, *82*
ownership 81
poisoning of 76, 82, 97–8, 101, 102
propagule dispersal 141
scrub management 31, 74, 76, 105,
106–7, **107**, *108*
selection of 70–1, **71**
sheep **71**, 73–4, **74**, *75*, **107**, *109*, 141,
150, 151, 152, 156
stocking densities 77–8, **77**, *78*, 130,
151–2
welfare 82
see also grazing
livestock units (LSUs) 77
Living Landscape Partnership Projects *see*
Life on the Verge project
Local Nature Reserves (LNRs) 147, 175
lowland calcareous grassland
communities 2, 10–13, **10–11**, *12, 14*
lowland dry acid grassland
communities 2, 13–15, **15**, *16*
Lowland Grassland Survey of Wales 115
lowland meadow communities 17–18,
18, *19, 20*
lowland scrub communities 30–1, *31*
LSUs *see* livestock units (LSUs)
LWT *see* Lincolnshire Wildlife Trust
(LWT)

M23 (Soft Rush/Jointed Rush–Common
Marsh-bedstraw) community 2

M24 (Purple Moor-grass–Meadow
     Thistle) community 2
machair 27–8, *27*
machine sowing 133–4, *134*
macrofungi 63–4, *64*
MAGIC map 172, 175
Magnesian limestone **11**, 186
magnesium (Mg) **117**, 120, *121*
Maidenhair Spleenwort–Wall Rue
     community (OV39) **29**
Making Space for Nature 147–8
mammals 57–8, **59–60**, 143, 207–8, 213
  bats 58–62, **61–2**, *62*, 68
  deer 3, 23, 57, 58, **60**, 141
  rabbits 13, **15**, **60**, 83, 113, 150, 160,
     161, 208, 213
  surveys of 67
  see also livestock
management *see* grazing; hay cutting;
     restoration management; semi-
     natural dry grassland management;
     soil fertility reduction techniques;
     weed control
manure 52, 92–3, 109–10, *118*, 141
Marram mobile dune community (SD6) *7*
Mat-grass–Heath Bedstraw community
     (U5) 22, 153
Meadow Foxtail–Great Burnet
     community (MG4) 3, 17, **18**, 89,
     157
meadows *see* hay meadows
measures of success 220–1, 222–4, **222**,
     223
mechanical control
  scrub 105–6
  weeds 95, 98–9, *99*, 101–2, *101*
mesotrophic grasslands 1, 2
  invertebrates 36, **37**
  lime application 111
  lowland meadow communities 17–18,
     18, *19*, 20
  scrub communities 30
  sward diversity 120–2, *121*, 151
  upland meadow communities 23–5,
     24, *25*
  waterlogging 145, *145*
  see also hay meadows
metaldehyde 213

MG1 (False Oat-grass) community 32, **33**,
     120, 157
MG1a (Red Fescue) subcommunity **33**
MG1b (Common Nettle)
     subcommunity **33**
MG1c (Meadowsweet)
     subcommunity 32, **33**
MG1d (Wild Parsnip)
     subcommunity 32, **33**
MG1e (Common Knapweed)
     subcommunity 32, **33**
MG3 (Sweet Vernal-grass–Wood Crane's-
     bill) community 23–4, **24**, 89, 120
MG4 (Meadow Foxtail–Great Burnet)
     community 3, 17, **18**, 89, 157
MG5 (Crested Dog's-tail–Common
     Knapweed) community 3, 9, 17, **18**,
     19, 23, **24**, **29**, 89, 111, 120, 178
MG5a (Meadow Vetchling)
     subcommunity 115
MG5b (Lady's Bedstraw)
     subcommunity 17
MG5c (Heath-grass)
     subcommunity 17
MG6 (Perennial Rye-grass–Crested
     Dog's-tail) community **24**, 32, **33**,
     64, 120, 121, 153
MG6b (Yellow Oat-grass)
     subcommunity 32, **33**
MG7 (Perennial Rye-grass leys)
     community 32, **33**, 120, 121
MG7c (Perennial Rye-grass–Meadow
     Foxtail–Meadow Fescue)
     subcommunity **33**
MG8 (Crested Dog's-tail–Marsh
     Marigold) community 17, **18**, 23,
     24, 89
mice 58, **59**, 143, 213
microfungi 63
microorganisms *see* soil microbial
     communities
Millennium Seed Bank 149
  see also UK Native Seed Hub
mine workings 26
Minsmere, Suffolk 165, 167, *168*
moder humus 173
molluscicides 209, 213
molluscs 34, **37**, 208, 209, 213

montane heath communities **21**, 22, 31
mor humus 173
mosses 90, *168*, 186
moths *36*, 43, 126, 127, 158, *160*
Mountain Avens–Glaucous Sedge
    community (CG13) 20, **21**, 22
Mountain Avens–Moss Campion
    community (CG14) **21**, 22
mowing *see* hay cutting
Muker Meadows, Swaledale *84*
mull humus 173
multiple stable states 219
mycorrhizal fungi 63, 142, 169–70, 171

National Biodiversity Climate Change
    Vulnerability Model 129
National Biodiversity Network (NBN)
    Gateway 65, 67, 176
National Nature Reserves (NNRs) 147, 175
National Parks 147
    see also Yorkshire Dales
National Trust 114
National Vegetation Classification
    (NVC) 3, 9
    surveys 65–6, 173, 176
National Wildflower Centre 149
native ponies 76, **76**, *76*
natural colonisation 136–44
    dispersal opportunities 130, 140–2
    missing trophic levels 142–4, *144*
    seed banks 130–1, 138–40, *139*, 166,
        218
Natural England 65, 66, 98, 115, 129, 175,
    224
Natural Resources Wales (NRW) 98, 175
NBN *see* National Biodiversity Network
    (NBN) Gateway
nematodes 171
Neolithic 4
Netherlands 140, 166
neutral grasslands *see* hay meadows;
    mesotrophic grasslands
neutral lipid fatty acids (NLFAs) 169–70
neutral soils 111
nitrogen deposition 14, **116**, 118, 130, 156
nitrogen (N) 107, 118, 163, 164, 169
NNRs *see* National Nature Reserves
    (NNRs)

Norway 153, 154
NRW *see* Natural Resources Wales
    (NRW)
nurse crops 211
nursery production 184–5, *185*, 216–17
nutrient enrichment *see* eutrophication
nutrient stripping *see* soil fertility
    reduction techniques
nutrients *see* soil fertility
NVC *see* National Vegetation
    Classification (NVC)

Old Winchester Hill, Hampshire *114*
oligotrophic soils 2, **11**, **15**
on-site threshing 186, 187, **188**, 189, **189**,
    190
Oolitic limestone 10, **10**
OV37 (Sheep's-fescue–Spring Sandwort)
    community 26
OV38 (Limestone Fern–False Oat-grass)
    community **29**
OV39 (Maidenhair Spleenwort–Wall Rue)
    community **29**
OV40 (Green Spleenwort–Brittle Bladder-
    fern) community **29**
overgrazing 17, 23, 28, 44, 45, 70, 74, 75,
    95, 101, **116**, 122

Palace Leas Meadow Hay Trial,
    Northumberland 118
Park Gate *81*
Park Grass Experiment, Rothamsted 118,
    118, *119*, 137–8, 220
Parsonage Down 13
pastures 9
    birds 52
    fertiliser application 109, 130
    fungi 62, 63
    invertebrates **37**, 44–5, 58
    lowland dry acid 2, 13–15, **15**, *16*
    mesotrophic 17–18, **18**
    semi-improved 32, **33**
Perennial Rye-grass–Crested Dog's-tail
    community (MG6) **24**, 32, **33**, 64,
    120, 121, 153
    Yellow Oat-grass subcommunity
        (MG6b) 32, **33**

Perennial Rye-grass leys (MG7) 32, **33**, 120, 121
    Perennial Rye-grass–Meadow Foxtail–Meadow Fescue subcommunity (MG7c) **33**
periodic flooding 145
Pesticide Safety Directorate (PSD) 96
pesticides 96, 209, 213
Phase 1 habitat surveys 65
phenology 127–8
phospholipid lipid fatty acids (PLFAs) 169–70
phosphorus (P) 107, **117**, 118, 120–2, *121*, 130, 163–5, 167
phytophagous beetles 142, 192
planning permission 113
plant introductions 215–17, *216*, *217*
plant material selection 176–82
    complex or simple mixtures 178–80, 215–16
    grass:herb ratios 185, 190–1
    hemiparasitic plants 180–2
    using ecological traits 176–8, 185
plant material sources 149, 182–4, **183**
    nursery production 184–5, *185*, 216–17
    see also hay transfers; seed harvesting
Plant Protection Products Regulations 2012 96
plantations 44, 115, 139, *139*
planting
    plugs and pot-grown plants 149, 207–8, 213–14, **213**, *214*
    shrubs 106, 113
    trees 113
    see also seed sowing
plot threshers 187, **188**, 189, **189**
ploughing 101–2, 122, 138, 166
plug planting 149, 207–8, 213–14, **213**, *214*
poaching of ground 70, 72, 74, 75, 78, 95, 113
poisoning of livestock 76, 82, 97–8, 101, 102
pollinators 128, 140, 143
    see also bees; butterflies
ponies *71*, 75–6, **76**, *76*, **107**, 141, 161
Porton Down 13, 132
pot-grown plants 149, 207–8, 213–14, **213**, *214*

potassium (K) 107, **117**, 118, 120, *121*, 130, 163, 164
power harrowing 194, 195–7, *196*, 206
primary succession 7, *7*, *8*
propagule dispersal 130, 140–2
PSD *see* Pesticide Safety Directorate (PSD)
public rights of way 82
pulling weeds 98–9, *99*
Purple Moor-grass–Meadow Thistle community (M24) 2

quarrying 45

Rabbit control 161
Rabbits 13, **15**, **60**, 83, 113, 150, 160, 161, 208, 213
Rare Breeds Survival Trust (RBST) 70, 82
Red Fescue–Lady's Bedstraw fixed dune community (SD8) *8*, 27
    Daisy–Meadow Buttercup subcommunity (SD8d) 27
    Selfheal subcommunity (SD8e) 27
remote sensing 114
Rengen Grassland Experiment, Germany 118, *120*
reptiles 14, 54, *54*, **55**, 67–8, 125–6
restoration management 131, 150–8
    grazing 150–2, *151*, 155–7, *155*
    hay cutting 152–4, 155–7, *155*
    and invertebrates 157–8
    long-term vegetation development 218–20
    plant introductions 215–17, *216*, *217*
    roadside verges 227–8
    and species diversity 150–4, *151*, 155–7, *155*, 218
    see also soil fertility reduction techniques
risk assessments
    chemical use 96, 97, 98, 99
    livestock 83
Roadside Nature Reserves (RNRs) 176, 225
roadside verges 17, 23, 149, 189
    grassland restoration case study 224–31, *225*, *227*, *229*, *230*, *231*
Robert's Field SSSI, Lincolnshire 226
rodents 57–8, **59**, **60**, 143, 207, 213

Romania 136, 156
Rothamsted Exhaustion Land
  Experiment 165
Rothamsted Park Grass Experiment 118,
  118, 119, 137–8, 220
rotovation 205–6
round barrow cemeteries *see* Seven
  Barrows, Berkshire Downs
Roundup 96, 101
Royal Botanic Gardens Kew *82, 185, 216,*
  217
  UK Native Seed Hub 149, 198–204,
    198, 199, 200, 201, 202
Royal Society for the Protection of Birds
  (RSPB) 50, 66, 132–6, *133, 134, 135*
Rural Stewardship Scheme 98
rush-pasture communities *2, 2*

SACs *see* Special Areas of Conservation
  (SACs)
Salisbury Plain 13, 46, 79, 132, 133, 137,
  219
SALVERE project 188, 189
Sand Couch embryo dune community
  (SD4) *7*
sand dunes *7, 7, 8*
Sand Sedge community (SD10b) 14, **15**
Sand Sedge–*Cornicularia aculeata*
  community (SD11b) 14, **15**, 63
Sandwich Bay, Kent *8*
saturation index (SI) 220–1
scarification 156, 181, 186
Scotland Rural Development
  Programme 147
Scottish Natural Heritage 98, 175
Scottish Wildlife Trust 175
scrub
  and birds 46–7, *46*, 52
  communities 30–1, *31*
  enhancement programmes 106
  and invertebrates 45
  management 30, 31, 74, 76, 104–7, **107**,
    108, 113, 160
SD2 (Sea Rocket–Sea Sandwort
  strandline) community *7*
SD4 (Sand Couch embryo dune)
  community *7*
SD6 (Marram mobile dune) community *7*

SD8 (Red Fescue–Lady's Bedstraw fixed
  dune) community *8*, 27
  SD8d (Daisy–Meadow Buttercup)
    subcommunity 27
  SD8e (Selfheal) subcommunity 27
SD10b (Sand Sedge) community 14, **15**
SD11b (Sand Sedge–*Cornicularia aculeata*)
  community 14, **15**, 63
Sea Rocket–Sea Sandwort strandline
  community (SD2) *7*
seasonal events 127–8
seasonal grazing 27, 79–80, **79–80**, *81*,
  91–2, 93, 94, 151, 156, 162
secondary succession 9
seed banks 130–1, 138–40, *139*, 166, 218
seed cleaning 203
seed dispersal 130, 140–2
seed drying 200–3, *200, 201, 202*
seed harvesting 85–8, *87, 88*, 149
  advantages and limitations 186–7,
    188, 190
  adverse impacts of 191–2, *192*
  cleaning seed 203
  drying seed 200–3, *200, 201, 202*
  and invertebrates 142, 191–2, *192*
  roadside verge case study 228–9, *229*
  seed quantity and composition 185–
    91, **187, 189**
  and species diversity 216, *217*
  storage of seed 203–4
  timing 190–1
  UK Native Seed Hub 149, 198–204,
    198, 199, 200, 201, 202
  see also hay transfers
seed mixes 133, 149, 182–4, **183**
  complex or simple 178–80, 215–16
  germination and survival
    differences 177–8
  grass:herb ratios 185, 190–1
  hemiparasitic plants 180–2
  nursery production 184–5, *185*
  see also seed harvesting
seed production 93
seed sowing 133–4, *134*, 156–7, 204–24
  arable reversion sites 210–13, **212**
  complete and partial 209–10
  germination and survival 157, 177–8,
    204–5, 208, 209

seed sowing – *continued*
  grassland enhancement **212**
  ground preparation treatments 205–8,
    207, 218
  methods 208–9, *209*
  nurse crops 211
  in stages 178
  timing 211, **212**
  transplants 149, 207–8, 213–14, **213**,
    214
seed storage 203–4
seed stripping *see* brush harvesting
seed viability 185, 201, 202
seepages and flushes 45
selective chemical spraying 97, 99, *100*,
    101
semi-improved grasslands 1, 14, 32, *32*,
    33, 64, 120, 122, **223**
semi-natural dry grassland
    communities 9–10
  calaminarian grasslands 26, *26*, 27
  climate change impacts 128–9, *129*
  limestone pavements 4, 28, **29**, *29*
  lowland acid *2*, 13–15, **15**, *16*
  lowland calcareous *2*, 10–13, **10–11**,
    12, 14
  lowland meadows 17–18, **18**, *19, 20*
  machair 27–8, *27*
  origins of 3–7
  primary succession 7, *7, 8*
  scrub communities 30–1, *31*
  secondary succession 9
  upland calcareous *2*, 20–3, **21**, *22, 23*
  upland hay meadows 23–5, **24**, *25*
semi-natural dry grassland
    management 9, 69
  and amphibians 57
  and birds 51–2, **53**
  fertiliser application 107–11
  and invertebrates 43–5, 52
  and reptiles 54, 57
  scrub management 30, 31, 74, 76,
    104–7, **107**, *108*, 113, 160
  see also grazing; hay cutting; weed
    control
serpentine exposures 26, *26*, 27
Seven Barrows, Berkshire Downs 158–62,
    159, 160, 161, 162

Seven Sisters, Wye Valley 4
shade-intolerant plants, origins of 3–7
sheep 71, 73–4, **74**, *75*, **107**, *109*, 141, 150,
    151, 152, 156
Sheep's-fescue–Alpine Lady's-mantle–
    Moss Campion community
    (CG12) **21**, 22
Sheep's-fescue–Carline Thistle
    community (CG1) 1, **10**
Sheep's-fescue–Common Bent–Alpine
    Lady's-mantle community
    (CG11) **21**, 22
Sheep's-fescue–Common Bent–Heath
    Bedstraw community (U4) 13, **15**,
    16, 22, 31, 152
  Bitter-vetch–Betony subcommunity
    (U4c) 23, **24**
  Yorkshire-fog–White Clover
    subcommunity (U4b) **15**
Sheep's-fescue–Common Bent–Sheep's
    Sorrel community (U1) 13, 14, **15**,
    16
Sheep's-fescue–Common Bent–Wild
    Thyme community (CG10) 11, **11**,
    20, **21**, 22, 31
  White Clover– Field Wood-rush
    subcommunity (CG10a) **11**
Sheep's-fescue–Meadow Oat-grass
    community (CG2) **10**, 13, *14*
Sheep's-fescue–Mouse-ear-hawkweed–
    Wild Thyme community (CG7) **11**
Sheep's-fescue–Spring Sandwort
    community (OV37) 26
shelter, livestock 83
Shetland 26, *26*, 27, *113*
shrews 58, **59**, 143
shrub planting 106, 113
SI *see* saturation index (SI)
silage 52, 89, 94, 98, 130, 152
Sites of Interest/Importance for Nature
    Conservation (SINCs) 175, 176
Sites of Special Scientific Interest
    (SSSIs) 13, 122, 147, 226
  for fungi 64
  machair 28
  overgrazing 70
  as reference sites 175
  upland hay meadows 24, 25

weed control 98, 100
Skye 6, *29*
slot seeding 205, 206, 208–9, *209*, 211
slug pellets 209, 213
slugs 208, 209, 213
Soft Rush/Jointed Rush–Common Marsh-
bedstraw community (M23) 2
soil analysis 111, 117–18, **117**, 172–3
soil disturbance treatments 205–8, *207*,
218
soil fertility 115–22, **116**
atmospheric nutrient deposition 14,
116, 118, 130, 156
and hay cutting 91, 164, 218
long-term field trials 118, *118*, *119*,
120, 137–8
nutrient indices 117–18, **117**
and species diversity 118, *118*, *119*,
120–2, *121*, 150, 156
and stocking densities **77**
testing 117–18, **117**, 173
see also eutrophication; fertiliser
application
soil fertility reduction techniques 122,
134
arable cropping 164–5, 218
chemical manipulation 167–9, *168*, 218
deep cultivation 166
fallowing 122, 163–4, 218
grazing 164
hay cutting 164, 218
immobilisation of nutrients 169, 218
reinstating soil communities 169–71
soil stripping 165–7, *165*, 218
and Yellow-rattle 181
soil fungal:bacterial (F:B) ratios 169–70
soil information resources 172
soil inoculation 142
soil macrofauna communities 171
soil maps 172
soil microbial communities 108, 118, 130,
142, 169–71, 219
soil pH 1–2
acid grasslands 2, 13, 120
calcareous grasslands 1, 10, 20, 120
chemical manipulation 167–9, *168*
lime application 110–11, 115, *118*, *119*,
167

and soil stripping 166
and species diversity 120, *121*
testing 111, 173
*Soil Sampling for Habitat Recreation and
Restoration* 111
soil seed banks 130–1, 138–40, *139*, 166,
218
soil stripping
seed transfer 182, 206, 207, 217
soil fertility reduction 165–7, *165*, 218
soil types 1–2
South Downs **10**, 13, *14*, 104
Southwick Hill *81*
sowing *see* seed sowing
Special Areas of Conservation
(SACs) 147, 175
Special Protection Areas (SPAs) 147, 175
species diversity
and burning 154
gradual decline in 219–20
and grazing 150–2, *151*, 155–7, *155*,
218
and hay cutting 152–4, 155–7, *155*, 218
plant introductions 215–17, *216*, *217*
and seed harvesting 216
of seed mixes 178–80, 215–16
and soil fertility 118, *118*, *119*, 120–2,
121, 150, 156
and soil pH 120, *121*
success measures **223**
species range changes 124–7, *124*, *125*,
126, *127*
spiders 126, 136, 169
spoil heaps 26
spot treatments 97, 99, *100*, 101, 103
spring grazing **79**, 80, *81*, 91, 93, 94, 151,
156, 162
spring meadows 91
spring sowing 211
SSSIs *see* Sites of Special Scientific Interest
(SSSIs)
statutory designation 98, 147, 175
steppe-tundra vegetation 3
stock checks 82–3
Stock Keep 70, 71
stocking densities 77–8, **77**, *78*, 130, 151–2
stress-tolerant species 89, 145, 151, 163,
177, 178

strip seeding 205, 206, 208–9, *209*, 211
stump removal 105–6
stump treatments 105, 106, 160
success measures 220–1, 222–4, **222, 223**
succession 131
    abandoned sites 14, 17, 131, 136, 138
    alternative stable states 219
    primary 7, *7, 8*
    and restoration management 219
    secondary 9
    see also natural colonisation
sucrose 169
sulphur deposition 130
sulphur (S) 167
summer drought 128, 145
summer grazing **79**, 80, 151
summer meadows 91
supplementary feeding, livestock 70, 77,
        78, 83, 93, 95, 164
surveys
    biodiversity 64–8
    birds 66
    botanical enhancement potential *174*
    invertebrates 66–7
    mammals 67
    NVC 65–6, 173, 176
    Phase 1 habitat 65
    reference sites 175–6, *175*
    reptiles and amphibians 67–8
    restoration sites 88–9, 172–4
    roadside verges 226, *227*
    soil analysis 111, 117–18, **117**, 172–3
    vegetation 65–6, 173, *174*, 176
sward diversity *see* species diversity
sward structure
    and amphibians 56–7
    and birds 52, **53**
    and grazing 78, 80–1, 93, 150–2
    and hay cutting 152
    and invertebrates 34, **35**, 36–7, 44, 45,
        142, 143, 157–8
    and reptiles 54
Sweden 137
Sweet Vernal-grass–Wood Crane's-bill
        community (MG3) 23–4, **24**, 89, 120

threshing, on-site 186, 187, **188**, 189, **189**,
    190

Thrislington Plantation National Nature
        Reserve, Durham 186
timing
    grazing 79–81, **79–80**, *81*, 91–2, 162
    hay cutting 90–1, *90*, 92, *92*, **116**, 152–4,
        181
    seed harvesting 190–1
    seed sowing 211, **212**
topographic modelling work 144–5
topsoil maps 172
topsoil stripping
    seed transfer 182, 206, 207, 217
    soil fertility reduction 165–7, *165*, 218
Tor-grass community (CG4) **10**, *12*
toxic metals 26, 166, 167–9
trampling by livestock 70, 72, 76, **79–80**,
        92, 95, 102, 150
transplants 149, 207–8, 213–14, **213**, *214*
Transylvania 136, 156
tree planting 113
trophic levels 142–4, *144*
tundra vegetation 3
turf inoculation 142
turf stripping 182n, 206, 207, *207*, 217,
        218

U1 (Sheep's-fescue–Common Bent–
        Sheep's Sorrel) community 13, 14,
        15, *16*
U2 (Wavy Hair-grass) community 14, **15**
U3 (Bristle Bent) community **15**
U4 (Sheep's-fescue–Common Bent–Heath
        Bedstraw) community 13, **15**, *16*,
        22, 31, 152
    U4b (Yorkshire-fog–White Clover)
        subcommunity **15**
    U4c (Bitter-vetch–Betony)
        subcommunity 23, **24**
U5 (Mat-grass–Heath Bedstraw)
        community 22, 153
UK BAP *see* Biodiversity Action Plan
        (BAP), UK
UK Habitats Directive Report 13, 14, 17,
        22, 24, 26, 28
UK Native Seed Hub 149, 198–204, *198,
        199, 200, 201, 202*
UK Pesticide Guide 96

undergrazing 14, 17, 23, 26, 28, 44, 70, **116**, 122
United States 153, 157, 169
upland calcareous grassland communities 2, 20–3, **21**, *22*, *23*
upland hay meadows 6–7, 23–5, **24**, *25*
  birds 50
  fertiliser application 110
  invertebrates **37**, 45
  management 25, 89
  restoration case study 84–9, *84, 86, 87, 88*
upland scrub communities 31
Upright Brome community (CG3) **10**, *14*
Upright Brome–Tor-grass community (CG5) **10**

vacuum harvesting 87–8, *88*, 186, 189, **190**
vegetation surveys 65–6, 173, *174*, 176
voles 57–8, **59**, **60**, 143, 207, 213
Voluntary Initiative 96

W9 (Ash–Rowan–Dog's Mercury) community 28
W19 (Juniper heath) community 31
W21 (Hawthorn–Ivy scrub) community 30
W22 (Blackthorn–Bramble scrub) community 30
W23 (Gorse–Bramble scrub) community 30, 31
W24 (Bramble–Yorkshire-fog scrub) community 30
Waddingham Common SSSI, Lincolnshire 226
waders, breeding 28, 47, **48**, 50, *50*, **51**, 52, **53**, 66, 79, 91
Wakehurst Place *82, 185, 216, 217*
  UK Native Seed Hub 149, 198–204, *198, 199, 200, 201, 202*
water levels 45, 145, *145*
water, livestock 83
water meadows 17, **18**, **33**
waterlogging 145, *145*, 173
Wavy Hair-grass community (U2) 14, **15**
weather events, extreme 123
weed control 94–104

Bracken 100–2, *101, 102*, 113
  chemical 96–7, 99, *100*, 101, 103, 134
  Common Ragwort 97–100, *99, 100*
  cutting 99, *99, 101*, 102, 103
  docks 103–4
  grazing 95, 99, 102, *102*, 103
  non-chemical 95
  thistles 103
weed wiping 97, 99, *100*, 101, 103
Weeds Act 1959 94
welfare, animal 82
wet grasslands 2–3, **33**
  lowland 17, **18**
  upland 2, 23, **24**
White Carpathians Protected Landscape Area, Czech Republic 136, 210, 220
whole-crop method of hay transfer 185, 193–7, *193, 194, 196, 197*
Wild Bird Indicator 50–1
wild harvesting *see* seed harvesting
Wildlife and Countryside Act 1981 *34*, **59**, **60**, 66, 67, 68
wildlife corridors *148*
wildlife surveys 65–8
Wildlife Trusts 175–6
  see also Lincolnshire Wildlife Trust (LWT)
wildwood 3, 4–5
Wiltshire Chalk Country Project 132–6, *133, 134, 135*
winter flooding 145
winter grazing 80, **80**, 151
Winterbourne Downs, Wiltshire 132–6, *133, 134, 135*
woodlice 171
Worton Bottoms, Yorkshire Dales *154*
Wye National Nature Reserve, Kent *81*

Yorkshire Dales 6–7, *6*, *25*, 153, *154*, 170, *180*, 220
  Hay Time Project 25, 84–9, *84, 86, 87, 88*
Yorkshire Dales Millennium Trust (YDMT) 84–9
Younger Dryas 3

CPSIA information can be obtained
at www.ICGtesting.com
Printed in the USA
BVOW07*0916120516
446953BV00013B/2/P